Manufacturing
and Business Excellence

Strategies, techniques

and technology

The Manufacturing Practitioner Series

Manufacturing
and Business Excellence
Strategies, techniques
and technology

Ian Warnock

Prentice Hall

London   New York   Toronto   Sydney   Tokyo   Singapore
Madrid   Mexico City   Munich

First published 1996 by
Prentice Hall Europe
Campus 400, Maylands Avenue
Hemel Hempstead
Hertfordshire, HP2 7EZ
A division of
Simon & Schuster International Group

Typeset in 10/12pt Times
by Hands Fotoset, Leicester

Printed and bound in Great Britain by
Hartnolls Limited, Bodmin, Cornwall

Library of Congress Cataloging-in-Publication Data

Available from the publisher.

British Library Cataloguing in Publication Data

A catalogue record for this book is available from
the British Library

ISBN 0-13-292970-8

1 2 3 4 5   00 99 98 97 96

For Ruth, Laura and Geoffrey

Contents

Preface

■ A market-driven revolution

Technology was the driving force behind the first industrial revolution. We are now in the midst of another such revolution. This time technology is part of it; but the main driving force is the market itself. The market has ever-increasing demands and expectations for response, cost and the quality of customer service. This will not stop or slow down, and only those businesses that can adapt and rise to the challenge of achieving excellence will prosper into the next century.

We're different . . . we're unique! Almost every company thinks it has 'unique' functions, procedures and problems. It doesn't. Irrespective of size and type, most businesses have similar problems – *challenges* that can be addressed and solved by the correct application of simple, methodical techniques and a lot of common sense.

Many problems and threats now face an enterprise of the 1990s, both from internal and external directions. Some businesses have sought and achieved improvements in performance and customer service through new methods, systems and technology, but few have achieved the levels or cultures of total quality or the ability to consider themselves a world class competitor. Indeed many improvements have been small or localised, rarely extending across the whole company. Businesses with such a restricted view of improvement will either be reshaped in a piecemeal and ineffective fashion until almost unrecognisable, or else they will die.

One of the first and most important steps a company must take is to identify and create a strategic plan for achieving success, a plan containing the changes and improvements necessary for that success.

In order to compete and survive in the changing global market, every enterprise needs to commit itself fully to overall, continuous improvement through the undertaking of company-wide total quality, world class performance goals. This is often the most difficult and far-reaching mission a business will ever address. The transition will impact on every individual in the company, significantly changing the culture, organisation and operation. Benefits are there for the taking, but this requires unfaltering leadership and integrity from top management, if there is to be a successful attainment of world class manufacturing.

This text considers many problems and pressures of the 1990s including increasing competition and much higher customer expectations of supplier quality, response and flexibility. It looks at the basic principles of this new industrial revolution and charts a course to corporate success; knowledge, commitment and enthusiasm must be obtained early in the path. Companies that do not embrace these ideas are likely to fall by the wayside.

None of this is new, but many companies have yet properly to address their needs for improvement – some have yet to address improvement at all. Despite the increasing numbers of well trained and educated staff, despite awareness of contemporary problems, any response remains slow and divergent. Strategies for moving the business into the twenty-first century are essential, but rarely found outside the sector leaders.

After many years as a management consultant and senior manager, the author continues to be surprised by the number of businesses that lack any truly strategic planning. Some even lack basic formal procedures. What's more, these shortcomings are apparent in many companies with over five hundred employees. And where companies do have effective procedures, their systems are often less than ideal. In short, no function or business is beyond improvement. Recognise that improvement is needed, then act to achieve the improvements. George Plossl has tellingly observed, 'They are too busy drowning to take time to learn to swim.'

■ From the top down

This book approaches the business from the top, emphasising that 80% of problems are not on the factory or office floor, but are in the management of that business. If the corporate decision makers are getting it wrong, then no amount of improvement at the lower levels can save a company. So we need to start at the top, but rapidly move throughout the organisation to review and improve our ways of working. This will search into *every* corner of the company – business re-engineering.

■ Generic business needs: world class and total quality aims apply to all sectors of business and industry

Although this text is primarily concerned with producers, there are similar functions in virtually every type of business where a product or service is provided. The majority of problems, philosophies and improvement methods are applicable across a wide range of business situations. This is particularly true for service industries, where effective use of human resources and controlling costs are normally the most critical factors for profitability. Common functions to be found in any business include

- sales/estimating: offer and sell the product or service; quote and cost out jobs
- design: design or modify the product/service
- planning: plan and control the production or delivery of the product
- production: produce the required product or service
- delivery: pass the final product or service to the customer
- accounts: control and account for sales and purchases; finance

Don't be misled into thinking that production only happens in a manufacturing business. For example, we could be describing the supply of

- engineered items
- artwork
- food
- research services
- medical services

This is a heavily simplified treatment of the activities within companies, however, it does show how similar they can be. Functional units are already being used at different stages of the process. In practice there are many more, and this places ever growing reliance on good communications. And we must carefully consider whether they could operate more effectively as decentralised elements of focused business units or focused cells.

As soon as we have functional units responsible for a specific part of the business cycle, we require agreed operating rules and procedures. Life is straightforward when the company is small and communications are good. But as the company grows, volumes become larger, more people are involved and communication often suffers. *'You can never have too much communication!'*

In each and every type of business, there are customer/product/supplier relationships between the internal functions. For example, sales supplies production (its customer) with defined orders. In turn production supplies distribution with finished products. This is another key factor for improvement.

■ A focus on hybrid and custom producer companies, and on service industries

This book focuses mainly on hybrid, non-standard items produced in small batches and on fully customised or unique products. This is because their many unique problems cannot always be solved by methods or technology that are appropriate in the high volume or bespoke sectors.

There are innumerable texts covering production line/high volume improvements – heavy automation, standard products and fully defined options. Relatively little has been written for the other types of business, which are also often small or medium-sized enterprises (SMEs) with limited resources to call on. Service-based industries, including the supply of information, expertise, skills, labour, maintenance and installation, etc., are also covered in several chapters. The reader will therefore find many parts of the text go into greater amounts of detail on topics where the author considers it necessary. This is to ensure the reader follows the use of a given technique in those areas of application, for example, MRP1 and MRP2 in markets with short product lead time.

■ Computer applications

Computers and the information systems they can provide are used in virtually all modern businesses. They were first introduced into our factories in the late 1950s and used to address specific problems of data processing, such as payroll and inventory control. Their power and range of applications has grown into complete enterprise-wide resource planning (ERP) to provide a vast range of facilities and functions for today's executive, when correctly applied. However, it has been recognised that computer and software power is not enough. Methodology and formalism need to be embraced in order to derive the desired benefits in company performance.

After examining Japanese industry practices and approaches (originally prescribed by American post-war experts) the West rediscovered total quality and just-in-time philosophies. These were initially seen as opposing the high-tech western solutions, and resulted in substantial confusion for a large proportion of manufacturing managers. Some of this confusion remains today.

During the last few years, many perceptive companies and practitioners have realised that both camps should and can be combined to produce the highest levels of performance, responsiveness and flexibility. This can lead to levels of improved customer service and competitiveness that are several orders of magnitude better than previously thought possible. *The West is now using Eastern philosophies, and the East is now using Western technologies.*

■ Hitchhiker's guide

This book attempts to show how and how not to achieve business excellence. It is based on achievable practice, not state-of-the-art theories. It neither catalogues new 'executive' technologies nor paints pictures of 'lights-out' factories. The main focus is on people and the way they work. People can and will use appropriate technology to assist in their business tasks. But to be effective, technology requires to be preceded by improvement programmes and formal procedures across the business. Much of the discussion will be against a manufacturing background, but is equally applicable to service industries and commercial operations. The similarities should be apparent after reading the case studies.

■ Methodologies and technologies

We are surrounded by three-letter acronyms: CIM, JIT, TQM, MRP, etc. Many are established and well proven, others less so. Where these are used in the text they will be explained at the first time of reference, and are also covered in the appendices.

■ Structure and use of the text

The thirteen chapters of the book can be divided into two parts. Part I looks at techniques for improvement and Part II at the technology to apply them. Chapter 1 sets the scene by considering the problems facing modern business. It proposes a solution based on business re-engineering, total quality and world class performance. Chapter 2 delves further into world class performance including the methodologies of total quality and just-in-time, in all business areas. Chapter 3 outlines the development of strategies for each area of a business. Chapter 4 shows that a business review is a key stage in understanding ineffective operation. It explains how to detect inefficiency and gives advice on using external management consultants. Chapter 5 describes several key methods for business analysis and areas of application. Chapter 6 discusses how natural groups can provide the highest level of service in certain types of business. It shows how to apply systems engineering for more effective operation and reorganising. Chapter 7 looks at manufacturing cells and considers how to design them. Concluding Part I, Chapter 8 considers management of change to provide a stimulus for motivation and improvement.

Beginning Part II, Chapter 9 describes enabling technologies CAD/CAM, MRP plus several other computer based systems and discusses their suitability. Chapter 10 is a detailed look at production simulation. Chapter 11 examines trends in computer and communications systems, highlighting their importance to competitiveness. Chapter 12 is devoted to developments in manufacturing resource planning (MRP2). Its particular focus on the finish- and make-to-order sectors, largely ignored by other texts, contains several case studies. Chapter 13 offers guidelines for successful implementation of MRP2, a prerequisite to world class. There are several appendices.

The author hopes that readers will use the text in different ways. Part I is recommended for all first-time readers. Chapters from Part II may be selected according to specific needs or areas of interest. The book can also be used for reference on specific topics. Case studies throughout the text show how certain problems have been approached.

■ Acknowledgements

I have thoroughly enjoyed writing this book, and can only hope that it gives the reader similar enjoyment, some enlightenment and some direction for the future of his or her business. I offer my thanks to the following people for their enthusiasm and commitment to projects I was fortunate to be involved in over the last few years: Ewing Mitchell, John Brassington and the task force; John Scott, John Gethin, Walter McKinly, Jim Hall and Tom Hunn; Ken Cape and the project team; Tom Whyte, Malcolm Bell and the Hussmann ERP/MRP2 team.

Several suppliers also gave valuable assistance with product information. My thanks to the Strathclyde Institute, AT&T Istel, ASK, Minerva/QAD, Microsoft, Intelligent Environments, CACI and Smart Sortware.

Ian G. Warnock

PART 1
Methodology and Techniques for Improvement

From the Top:
A Strategic View

He that will not apply new remedies
must expect new evils; for time is
the greatest innovator.

Francis Bacon, 1621

■ Introduction: rising to the challenge?

This chapter considers many of the problems and pressures facing businesses of the 1990s: flat or reducing markets, increasing competition and much higher customer expectation. None of this is new, but many companies have yet to properly address their needs for improvement. This is despite the increasing numbers of well-trained, professional managers and engineers across industry as a whole. Awareness of impending or current problems may be there, but response remains slow and divergent. Strategies are essential for moving companies into the twenty-first century, but are usually found among only the top few in a given market sector. If senior managers grasp the problems, analyse them and apply the appropriate solutions, your company can rise to this challenge before it's too late.

■ Competition in the marketplace

The days of high volume, high margin products in a fairly uncompetitive market are largely gone. Customer expectations have accelerated significantly in recent years, and will continue to do so. Customers expect

- greater product or service variety
- improved quality at the same or lower cost
- faster response/shorter lead times
- improved overall service from suppliers

In the 1980s business and industry expanded across the globe. Companies responded to increased demand by increasing their capacity to supply products or services. In the recession of the early 1990s the world economy stopped growing, leaving excess capacity in many sectors. Excess capacity leads

3

suppliers to reduce their prices to maintain market share. And reduced prices increase competition.

Most industry sectors are facing severe competition, locally and internationally. As well as traditional competitors, the influence of Japan and the Pacific Rim continues to grow, extending into aerospace, refrigeration and many other markets. Low cost, high quality labour is only a phone call away in the Far East. Technology is improving at a daunting pace. And many businesses have been stranded using outdated methods, outclassed facilities and outmoded performance measures, losing customer after customer to better prepared rivals (Figure 1.1).

Globalisation has come about largely through the information revolution of the last fifteen years, bringing desk-to-desk communication in seconds. Better transportation systems have also played a part and this is leading to the disappearance of local and regional boundaries. Expanding free trade in the European Union, Eastern Europe and North America heralds the worldwide extinction of protectionism.

Global competition has resulted in a major shift in customer demand and expectation in virtually all market sectors. Price, quality, design and function are now being taken for granted by customers who expect ever better response, flexibility, customisation and support from all suppliers. This shift in demand is having a major impact on the way business and industry operate.

In certain sectors, price alone is becoming the customer's overriding concern. Customers on a tight budget are willing to forgo the extras they used to demand. And this price squeeze is likely to continue, because of their own financial situations and urgent need to reduce expenditure.

Quite simply, the way we run our businesses is no longer good enough.

Service*: the supply, installation or maintenance of goods or services carried out by a supplier.*
Level of service*: the standard or quality of service provided to a customer. Level of service may help to win orders against the competition. Six factors are worth considering.*
> *response time*
> *greater product or service variety*
> *competitive or lower cost*
> *shorter lead time*
> *sales support, before and after*
> *willingness or ability to adapt to customer needs.*

With customers continually expecting or demanding more options, faster delivery and improved quality, there is ever increasing pressure on us to reduce our prices, to improve our products and to raise our efficiency. As other companies respond, competition intensifies, requiring us to match or surpass them. Otherwise, loss of market share may weaken our company until its position is critical. To be the best in our field, it is no longer good enough merely to satisfy our customers – now we must *delight* them!

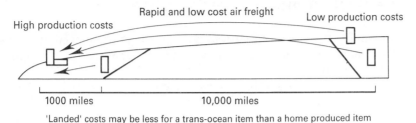

'Landed' costs may be less for a trans-ocean item than a home produced item

Figure 1.1 Continued growth of international competition

As the order-winning factors change with shifts in the market, so our business operations have to change, and change rapidly. Structures created for the old market may not work in the new. They need to be replaced. Perhaps we no longer need large supervisory and middle management structures to transfer information and control. Businesses are now replacing traditional middle management with teams of knowledge workers, who operate and improve the business in every area. This improvement is not confined to manufacturing, or any single area. Many firms have carried out improvement projects in an attempt to respond to these changing market demands.

CAD/CAM, MRP2, CNC, quality assurance (QA) and just-in-time are often short-lived and relatively unsuccessful. Many initiatives are inadequate for the problems facing a business or address only small areas of performance.

■ Initiatives that start then fall by the wayside: quality circles/QA, just-in-time.
■ New computer systems that fail to deliver the promised performance or functionality.
■ High technology plant and equipment that does not provide the expected increases in output: flexible manufacturing cells, MRP, CAD/CAM.
■ Reorganisation that fails to provide any real improvement.

The challenge for each business is to become a highly competitive, total quality enterprise. This involves identifying and implementing appropriate philosophies, organisational changes and technology to provide the levels of flexibility and response required by the market. This text aims to describe the details of world class manufacturing (WCM) philosophy and component methods and techniques that will be used to effect the transformation.

As a prelude, let us further examine the basic aims, threats and opportunities of modern enterprise. And let us see why many companies have not achieved or sustained the levels of improvement necessary to move from 'second class' to become world class performers.

■ The basic aims of an enterprise

If we are aiming to delight our customers, this must be achieved in harmony with other, more basic aims of every business making, and continuing to make money. Our enterprise needs to be efficient and flexible enough to produce goods or services at the right time, quality and cost.

Perpetual profit is the primary aim

Profits maintain the business, allowing it to grow, to fund reinvestment and to satisfy owners or shareholders.

profit = sales value − production costs − overheads

(I include sales and marketing costs in the overheads.) Profit can be improved by addressing several interrelated areas

- selling more of the same item, provided the customer is delighted
- selling different products with higher profit margins
- reducing production costs
- reducing overheads

These should all be addressed where there is a need to improve profitability. 'But we are profitable' I can imagine you saying. 'So we don't have a problem. Our bottom line is healthy and the Profit and Loss account has never been better!'

European companies, survivors of the 1980s, often believe their profits are healthy. But all too rarely has this position been achieved in a planned and clinical fashion. It has usually happened by doing what they have always done, together with some investment in plant, equipment and new systems. For the twenty-first century this approach is far from adequate for the changing and rising expectations of customers. I am surprised and concerned at how many companies are operating near zero profit. But I find it even more worrying that many take no positive steps to improve their profit margins. Even profitable businesses experience competition. Current profitability is no guarantee of future profitability. Such complacency can lead to sudden failure when other companies increase their standards of service.

Long-term improvements foster growth

A business must continually be developed to maintain profitability and competitiveness. But profit alone cannot sustain a business. Avoid sacrificing longer-term business for short-term profitability. Much of this text is about making improvements to foster longer-term success and growth.

Short-term fixes can produce long-term failures

In many cases profit-making can be easily achieved over a short term by cutting staff, reducing expenditure and investment, increasing sales via discounting, and so on. For many businesses, short-term profit is being achieved with little or no regard for longer-term profitability and business success. One-year business plans are the measure; five-year plans are merely given lip service. Most plans are 90% financial and rarely address strategic goals, such as product development or market needs.

Achieving long-term profitability is a much more difficult task, involving careful planning, analysis and execution. Frequently it requires changes in the company's operations and organisational structure.

Internal inertia resists structural change

A year is a very short time for most businesses considering the planning and implementation of substantial change. When facing changing market pressures, a business may often find its operations have considerable inertia; they are very slow to change or reorganise, even with full commitment and funding. The majority of firms need two to three years to substantially change their manufacturing systems, and three to four years to change their internal culture.

To cope with these time-lags, your company needs to be looking *NOW* at what the market and thus its operations will be doing in four years time to plan four years ahead. Otherwise you'll be way behind the market or maybe out of business.

■ Market intelligence: trend analysis

Past, present and future market trends and competitor performances are better indicators of your long-term survivability than annual financial statements. Ongoing analysis of long-term trends is more valuable than making short-term measurements.

■ How do poor and inefficient practices develop?

Before discussing the proposed way forward for each company, it is worthwhile to consider some causes of inefficiency

- lack of vision
- stagnation and low morale
- complacency

7

- change in company market sector
- lack of effective direction
- lack of professional management
- lack of training and personnel development
- lack of investment and/or incorrect investment

According to one theory, these causes arise from prevailing market conditions. This implies that every business is wholly determined by external factors – highly unlikely and not shown in practice. It is the management that directs, drives and motivates any business, almost irrespective of the external market conditions.

■ Problems with conventional systems

Most businesses grow from small operations into larger, more complex and often unwieldy organisations. The original procedures and systems were designed to deal with the original problems, perhaps with smaller volumes and fewer competitors. As a business grows, its procedures and systems may need to change. Some businesses respond by training their staff to deal with such changes. Many do not.

Common fix-it problems

Many companies deal with perceived changes by developing prescriptive 'fix it' and piecemeal add-ons or modifications to established procedures and systems. Others buy in or create company-wide information and control systems, such as manufacturing resource planning (MRP2). The results can be highly complex and difficult to operate, requiring many staff and generating much paperwork. Often these types of system can be found where other problems are evident.

- High numbers of overhead/indirect staff to support detailed requirements for information, planning and material sourcing.
- Undertrained but very specialised staff with fluctuating loads and periods of inactivity.
- Wasted investment in high levels of inventory, warehousing and associated control systems.
- Poor levels of customer satisfaction: missed due dates, shortages and non-conformance.
- Complex production and material control: large batch quantities, overproduction and inflexibility.
- Detailed scheduling and monitoring of each process, operator and machine.

Figure 1.2 shows the circle of events.

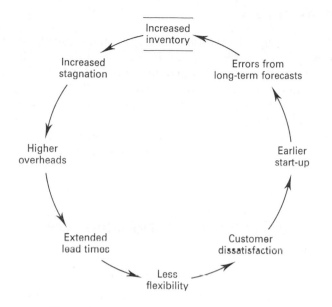

Figure 1.2 Knock-on effects and problems.

Advanced manufacturing technology

There are many advocates for powerful computer-based solutions. But beware, often they begin a dangerous spiral of complexity. *Complex problems need complex solutions, then the more complex problems need even more complex solutions, and so on* (Figure 1.2).

Consider this extract from *Business Week*, 16 March, 1987.

General Motors: what went wrong?
GM believed that if it spent enough on computers and robots, increased efficiency would be assured. It has found out the hard way that new technology only pays off when coupled with changes in the way work is organised. What GM learned is that simply organising work more efficiently and giving workers more say can produce more impressive results than millions of dollars worth of robots.

This is typical of many businesses. Full of expectation, they invest heavily in the latest equipment. But the expected improvements rarely materialise. An army equipped with the most modern weapons is ineffective unless the troops are properly trained. They may have to learn new, markedly different tactics to gain advantage using the new weapons. In this respect warfare and manufacturing are the same.

Money and investment

Managers often see capital investment as a relatively quick solution to a problem: throw money and people at it. Perhaps this is understandable. Managers want to take some action that will have a rapid effect. But signing a cheque or appointing a bright, new employee takes little real commitment. Real commitment requires a sustained programme of appropriate improvements geared towards long-term survival and competitiveness. Real commitment means undertaking the voyage to world class status, delighting customers along the way. A satisfied customer tells three other businesses; a dissatisfied customer tells ten others!

■ A strategic direction: world class, continuous improvement

To respond to these demands each and every business needs to maintain significant development in every area and aspect of its operation. As it grows more and more competitive, it eventually can be compete and win against all other contenders, becoming world class. The first step to world class performance is to identify the peculiar needs of your business. At each stage you will need to discover your own strategies for meeting them. And by continuing to explore your needs you will gradually attain a state of continual development, out of which comes world class performance (Figure 1.3).

■ Limited improvements to date?

Perhaps you think you've done this already or perhaps you're in the middle of a quality initiative. But has your approach followed the WCM philosophy? Or

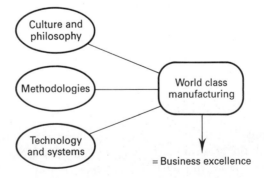

Figure 1.3   Contributors to world class performance.

were your improvements largely project-based? World class performance results from a process of continual development throughout a company, not from the completion of a single specific project.

Many companies have progressed a long way in improving their performance. But too many believed success was guaranteed if they completed a project on quality or just-in-time, or installed the best MRP2 software. Failure to match projects with strategic objectives meant benefits were only localised. And unfortunately this is still true of many firms. But those that do understand, those that can accept the full implications of the WCM philosophy, they have reaped enormous benefits.

Workers help to sustain the continual development required for WCM, but their importance often goes unrecognised. In future, workers will need to share the managers' vision of the company and be **empowered** to ensure that they can achieve targets not out of compliance but out of commitment (Figure 1.4).

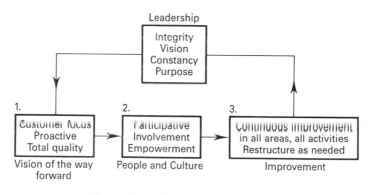

Figure 1.4 Total quality strategy.

■ Compete to survive: mean, lean and responsive

Periods of recession quickly weed out those businesses that have not become mean, lean and responsive. Shrinking markets lead to lower prices and lower profit margins. Top-heavy and inflexible companies are rapidly beaten by well-balanced and flexible competitors.

Key aims for good business practice

- be flexible and responsive
- provide products and/or services that meet or exceed customer requirements

Subsidiary aims for good business practice

- minimise costs and operating expenses

11

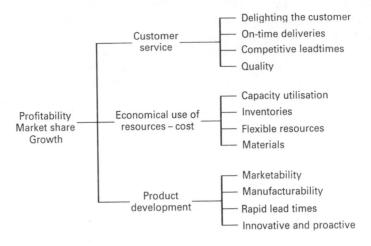

Figure 1.5 Key business objectives.

- maximise return on investment in the business
- provide an attractive environment and progression for workers

Some of these aims are interrelated. For example, reduced costs can allow price cuts or provide extras for the same price, possibly resulting in a product advantage over a competitor. Figure 1.5 expands these aims, showing their relationship with Customer Care, economic use of resources and product development.

We should *not* be trying to

- promise delivery dates that are unattainable
- provide substandard products at low cost
- cut costs without cognisance of the consequences
- stifle the business by zero investment
- demotivate staff by low involvement, poor conditions or remuneration

Few would disagree with these aims, but few would be able to plan and implement a new direction for the business to achieve them.

■ Reducing total lead time

Product lead time is not the sole responsibility of the production department. It depends on a long list of variables. Just as the company in the above example is blinkered to the concepts of internal 'customer/supplier relationships', so the latter example is a common, but ignorant belief that product lead time is a single term equation. This is most certainly not the case, with total lead time consisting of the following:

product lead time = estimate or quotation time
+ sales order processing time
+ design/modification time
+ order definition time
+ time to source special components
+ scheduling time
+ production time, assembly and test
+ warehousing time
+ shipping/delivery time

Some of these tasks may be concurrent, but the total lead time will *always* be several times the basic production time. By focusing only on the manufacturing area, more valuable improvements in other areas may be overlooked.

■ Flexibility and response

Throughout the 1990s, companies wishing to prosper need to become more flexible and responsive, to delight customers by providing what they want, when they want it, before the competition. These are acknowledged as order-winning factors by most market strategists, whereas price and quality are assumed to meet or exceed customer expectations.

Traditional views of flexibility have centred on manufacturing, the process itself and the machinery used. Important though this may be, full flexibility only comes with changes in culture, attitudes and organisational structure. From product concept to goods dispatch, every policy and procedure, system and process must have the objective of improving the flexibility and response of the whole company.

Changing culture and organisation is not easy but it is essential. If you have a business that depends on flexibility to win orders, a rigid organisation and culture will not deliver the goods, irrespective of any strategy for change. Saying is much easier than doing, and to realise sustained improvement, change needs to be initiated throughout the whole business.

■ Awareness of the way forward

Appropriate ways to work towards WCM are considered in later chapters. But before starting out, some other issues need to be understood

■ the level of awareness within the company of the need for change

13

- the nature of internal expertise in appropriate methodologies and technologies
- likely sources of resistance to change within the company

Without appropriate evaluation, investment in education and training, new computers and equipment may not produce the desired changes and tangible improvements.

Frequently companies do not evaluate or consider how such change can help to achieve strategic objectives.

Unfortunately, many managers believe new technologies and new methods can cure almost all problems, but this never happens. Imagine giving modern weapons to a medieval army, unknowledgeable and untrained in their operation or tactical use: there might be isolated success with the weapons, but many more failures and internal casualties!

The problem remains largely *management* and *workers*. Several human issues need to be understood and addressed when undertaking a major change. Communication with the workforce is crucial. Incorrectly handled, internal resistance may grow, making any changes more difficult to achieve. Chapter 8 deals with the management of change, including planning for change, internal resistance and new operating procedures.

■ WCM foundations and tools

The path to WCM draws from other well-established philosophies and methodologies. Among them are

- business re-engineering
- total quality management and control (TQM)
- just-in-time (JIT)
- a systems approach
- formal policies and procedures
- management of change
- information and automation technologies, including material requirements planning (MRP1) and enterprise resource planning (ERP)
- Underlying good practice in the above

In the future, additional elements will become part of the World Class armoury. Information technology develops by leaps and bounds. Rapid expansion of personal computers, telephones and data networks makes communication ever faster and cheaper. Fingertip access to ERP systems and customer databases will further improve the customer/supplier interface. And the expansion of computer-based expert systems for order configuration and costing will enhance their response and accuracy. Over the same period, automation and production machines may see only gradual advancement.

In essence, WCM sets out to create a new company, by business re-engineering, a company very different from the old. Its culture will be different. Its structure will be different. And its performance will be different. Figure 1.6 shows some of the changes that are needed. Each one is discussed in a later chapter, some only briefly, others at greater length. But all of them are discussed. By selecting with care, you too could have a world class company.

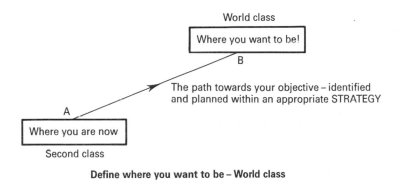

Figure 1.6 World class manufacturing: a different direction and philosophy.

■ Summary

To compete and survive in the changing global market, your company needs to be totally committed to continual development. Often difficult and far-reaching, this undertaking will significantly change your existing culture, organisation and operation, leaving nobody unaffected, workers or management. To reap the benefits requires dedication, honesty and commitment from all levels of management.

■ Further reading

Katzan, W. (1992) *Managing Uncertainty*, Van Nostrand Reinhold.
Peters, T. (1989) *Thriving on Chaos*, Pan.
Schonburger, R. J. (1982) *Japanese Manufacturing Techniques*, Free Press.
Schonburger, R. J. (1986) *World Class Manufacturing*, Maxwell McMillan.
Schonburger, R. J. (1990) *Building a Chain of Customers*, Hutchinson Business Books.
Tichy, N. M., and Sherman, S. (1993) *Control Your Destiny or Someone Else Will*, HarperCollins.
Wallace, T. (1992) *Customer Driven Strategy*, Oliver Wight.

Total Quality, World Class Performance

World Class – the journey towards excellence never ends.

■ Introduction: joggers or athletes?

The ultimate goal for any enterprise is the aspiration to become the best in their chosen field, when in competition with the other best companies from around the world – world class. At the Olympic games, only the best even get into the competition. Business will soon be the same; the dominant teams from the Far East, North America and Eastern Europe. World class athletes and world class businesses have similar qualities

- fitness
- discipline
- regular, continuous training
- continuous striving to improve all aspects of performance
- attention to detail
- comparison and analysis of competition and arena
- morale and confidence – mindset
- will to win – competitive spirit
- more fitness

Possessing only a few of the qualities – for example fitness and training – can make you a fit pedestrian, a jogger or a local club runner. But to be a world class athlete, you need them all. And once achieved, world class performance requires effort to **maintain** the very high levels of performance, competitive mindset or discipline necessary. This can be the stumbling block for many businesses, when, having attained their performance and quality goals, they relax and rapidly slip back to old practices and low performance.

Jaguar Cars

Having successfully turned around its product quality from unacceptable to more than acceptable, Jaguar Cars UK regained some market share then lost it again. After a long, hard struggle to overcome its problems, Jaguar relaxed, thinking it was secure, and quality fell again. Unfortunately, this is typical of project-based improvements that have a defined completion date; improvements that are not integral, well-defined parts of how the company should do business.

IBM

IBM was arguably one of the most world class of world class corporations. In 1992 it brought in its worst ever financial loss, shaking the corporation to its roots. For many years, IBM's size and strength derived from mainframe and minicomputer systems it sold throughout the corporate and business world. Vendor lock-in ensured that customers always came back, besides being 'safe' buying from Big Blue, as IBM was known. However, the market and competition were both changing, and IBM acknowledged these changes too late to effectively deal with them. Technology advances in computer systems, falling prices and Far East competition cut so far into its traditional customer base that IBM was left reeling. Customers downsized their computers (discussed in Chapter 11) and moved to suppliers who were able to immediately provide these more cost-effective systems. This drastically hit IBM's profitability and reduced its market share.

In summary, IBM's problems were as follows

- ignorance of or slow reaction to changing market trends and customer needs
- complacency
- late and/or ineffective development of appropriate system solutions (following, not leading)
- resources and structure inappropriate to the new customer needs

IBM's resources, capacity and technical ability have rarely been in question. But when incorrectly focused by management, they failed because they were inappropriate to the changing market needs.

Thinking you are the best is dangerous, since it can often lead to complacency, which will sooner or later result in being surpassed by the competition. This indicates the importance of a total quality business culture and mindset of **continuous improvement**, essential in creating and maintaining the motivation and will to succeed in becoming and *remaining* a world class competitor.

Figures 2.1 and 2.2 show the components of world class performance and the path towards it. Only one element needs to fail to compromise the whole enterprise or even to bring it down. Remember what happened to IBM.

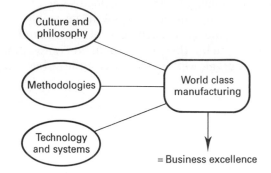

Figure 2.1   Components of WCM.

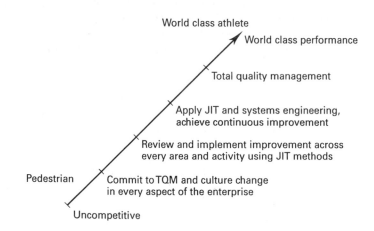

Figure 2.2   The path to world class performance.

■ A world class enterprise

To put world class manufacturing and performance into context, the author considers it to be the achievement of the highest levels of performance and quality, through the commitment to, and embarkation on a path of company-wide improvement, total quality and flexibility in every aspect of operating the business.

World class manufacturing is

■ a way to achieve and maintain the highest levels of total business performance and quality
■ a continual journey towards excellence

■ a combination of philosophy, culture, methodology and technology to continuously analyse and improve the whole business
■ a way to empower people to improve the company
■ a competitive weapon to use against non-world class competitors
■ a major, long-term undertaking for any business
■ survival

World class manufacturing is not

■ a buy-in solution
■ a technology or a method
■ a quick fix
■ a one-off project that will finish at some time in the future

Three complementary areas need to be continuously developed to meet and support the growth of the enterprise.

1. The culture and philosophy of operating and running the enterprise.
2. The methodologies that will be used to analyse and improve the enterprise.
3. The technology and systems that will be developed and/or improved to support the enterprise.

Taking these building-blocks for world class excellence, we now consider what makes up each component and how they are interrelated.

Culture and philosophy

Move to a total quality management (TQM) and operating culture that incorporates much of the just-in-time (JIT) philosophy.

Methodologies

Adopt and use appropriate approaches and methods to achieve and maintain improvement and effective working across the enterprise. Use systems engineering and JIT approaches. Use best practice in all other methods, e.g. project management. Re-engineer the business to meet specific objectives.

Technology and systems

Select, implement and make effective use of support systems and technologies where appropriate, e.g. manual and computer-based information systems, ERP, AMT, CAD/CAM, quality assurance to BS 5750/ISO 9000 and expert systems.

Definitions (*Collins English Dictionary* and others)

Methodology: *method: systematic, orderly techniques or arrangements of work for a particular field.*

Excellence: *the state or quality of excelling; to be outstandingly good, to rise up.*

■ Culture and philosophy for world class

Move to a culture that incorporates much of the JIT and Japanese management philosophies and empowers people to improve each aspect of the business.

Definitions (*Collins English Dictionary* and others)

Philosophy: *Rational investigation of knowledge or correct conduct; study of truth, fundamental principle, system or school of thought*

Culture: *the total range of activities and ideas of a group of people with shared traditions and pursuits; way of operating.* Interestingly it is also defined as: *enlightenment, refinement resulting from these pursuits.*

The most valuable asset in any business is its *people*. Systems, equipment and advanced technology can help, but the people, their approach and the culture within the enterprise are what will make or break the business. *Culture refers to the combination of company values and worth, management style and attitudes of employees and managers.*

Many improvement initiatives have failed and are currently failing because people and culture have been largely ignored. Successful improvement programmes tend to devote more time and effort to people and culture. *It is absolutely vital to establish across the whole company, a culture that understands and believes in the benefits and need for change. Only with this culture will change be supported and successful.*

This is because for virtually all major changes, the threat against success remains largely management and workers. Several human resource issues need to be addressed when undertaking a major change in the company. Communication with the workforce becomes crucial. Incorrectly handled, internal resistance may grow, making any changes more difficult to achieve (management of change is discussed in Chapter 8).

■ Just-in-time and total quality philosophies

Within our business, we therefore need to establish a set of fundamental principles to base all other plans and objectives on, and to form the basis of the new culture for change and the pursuit of excellence. There are two closely related philosophies that should be used

- just-in-time philosophy (JIT)
- total quality (TQM or TQO)

Just-in-time philosophy

This may surprise some readers who have understood JIT to be a set of methods or techniques to be applied to purchasing, manufacturing and delivery functions. JIT does contain specific methods and techniques that can certainly improve the competitiveness of most businesses, but these are only a small part of a total philosophy for the elimination of waste and unnecessary activity wherever they occur. *Waste is anything that adds cost but not value to a product.*

Better terms for JIT may be Japanese manufacturing philosophy, Japanese business system, or zero waste system. Rooted in a culture that greatly respects simplicity, quality, and people, Japanese business management founds enterprise on the very same qualities. Experience in the United States and Europe suggests this same philosophy can be adopted by Western businesses, with dramatic improvements in competitiveness.

■ JIT as a way of life

The formula for Japanese success is based on simplifying the problem. They largely reject the complex management structures and controls, the sophisticated equipment and systems found in most Western manufacturers. Instead of developing highly complex solutions, the JIT philosophy is to simplify the problem, removing the need for a complex solution.

This is not to suggest that equipment and computer systems used in Japanese companies are non-existent or basic. The JIT philosophy provides tools to make a job quicker and simpler. If that requires a flexible manufacturing cell or an MRP system, they will be fully and effectively implemented.

The origins of JIT and much of Japanese manufacturing philosophy stem from early USA mass production experiences of automobiles, including the Henry Ford principles of repetitive manufacture. Whilst Western countries moved into the development and use of more complex technology-based production, Japan concentrated on more efficient production systems. When they needed technology solutions, they simply bought them from the West!

The Japanese manufacturing and management philosophy consists of three elements

- productivity
- quality
- people

JIT deals with all these elements but is more commonly associated with productivity improvements and product quality. All three areas are tightly interwoven, so it is virtually impossible (and undesirable) to improve one area without affecting others. JIT emphasises successful **people-oriented management** and business success achieved through a culture of quality and productivity.

This approach has been dramatically successful since the 1970s, as Japan became world leader in one product after another. From TV sets to machine tools, from supertankers to silicon chips, their production superiority simply outclassed most Western countries. Productivity was climbing at over twice the rate of the USA and the UK, and much faster than the rest of Europe. Quality levels in Japan were many times higher than in the West. In 1980 Japan's GNP was number three in the world, and on current trends will be number one by the late 1990s.

Western business slowly realised the significant cultural differences between themselves and Japan: many thought the Japanese principles (developed from Henry Ford) would not work in the West because of these cultural differences. Where the East had improved, the West had stagnated. But, as several Japanese firms set up in the US, it became apparent that JIT management philosophies would work in a Western culture, provided thorough retraining and education of people was carried out. Now, many years later, many enterprises have successfully changed to JIT operation: Hewlett Packard, Hoover, GEC, to name but a few.

A large part of JIT philosophy and methodology stems from high volume, repetitive production, but JIT can still be applied to batch and job shop environments. Remember, only part of JIT addresses pure production, much of it relates to total quality and improvement in all areas of operation. Consider the following list. Though apparently confined to manufacturing, JIT objectives may have far-reaching consequences.

JIT Goal	Expansion across the business
Zero defects	Achieve excellent quality in all functions and activities
Zero set-up time	No delay between each activity, via continuous improvement
Zero inventories	Balance the system and avoid investment in wasted stock
Zero handling	Minimise the transport and handling of every item. Reorganise to create multiskilled cellular units to avoid passing between departments
Zero breakdowns	No hold-ups in processing a product, document or item of data

- **Zero lead time**   Reduce the time taken for every task
  Eliminate non-value-added-activities, to improve the response of the system
  Reorganise processes to minimise the number of activities needed
- **Zero lot size**   Produce and pass on items as individuals or as small batches to save queuing time and to improve response and inventory.

Many of these goals are in complete conflict with traditional business thinking and traditional manufacturing management. Many Western accountancy conventions do not reflect true costs as they would apply to a JIT system. Inventory generates the most obvious conflict. Traditional accounting treats inventory mainly as a valuable asset without quantifying the harmful effects of slow response, obsolescence and unnecessary investment – wasted resource.

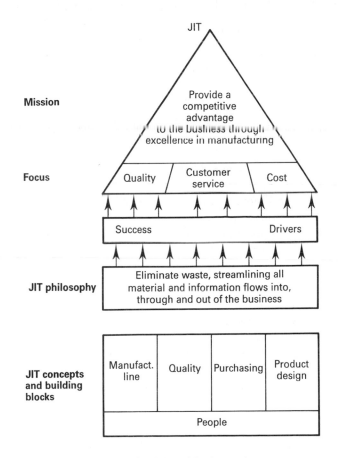

Figure 2.3   JIT building blocks and concepts.

JIT views the on-time production of goods or services as the goal, with no or minimal wasted time, effort or money. Although JIT does not acknowledge traditional overhead recovery objectives, it does excel in providing a quality product at lowest cost, in the shortest lead time. It makes money (Figure 2.3).

■ JIT philosophy, culture, methods and techniques

Not only does JIT contain a methodology for production and productivity improvement, it also encompasses a range of techniques to improve each business function right down to discrete product level.

Core elements of JIT	
People	culture, simple structure, teams, policies and procedures, quality circles
Productivity	scheduling techniques and tools, design and manufacture, modular design, set-up reduction, minimum lead time .
Control systems	pull system, levelled schedules, simple MRP control, kanban squares
Plants	group technology, focused factories, cellular units, flexibility
Purchasing	JIT suppliers, close relationships, rapid transport and delivery

There are seven productivity principles of JIT.

1. *Produce to the exact demands of the market.
2. Eliminate waste in all activities.
3. *Produce products one at a time.
4. Achieve continuous improvement in all activities.
5. Respect people at all times.
6. *Do not have contingency or back-up resources.
7. Plan and build for the future – long-term emphasis.

Those marked with an asterisk are very controversial and carry a high penalty if total quality and on-time delivery is not maintained. However, the others are immediately recognisable as good common sense that should apply to every function of all types of business.

Where failures to implement JIT have occurred, the companies have usually bought into small, narrow areas of the core, e.g. purchasing, delivery and reduced inventory. Piecemeal approaches are doomed from the outset. Several early failures of JIT were attributable to enthusiastic but inadequately educated managers aiming for zero inventory but making few changes and improvements necessary across the company to achieve it.

Many excellent texts cover JIT productivity methods. This text concentrates on the fundamental nature, breadth and depth of JIT and Japanese management cultures, and their collaborative role in creating a world class enterprise.

In order for any business to fully utilise JIT as part of a strategic improvement plan, management and workers must understand its nature, its background and its implications for people-centred quality and improvement. Without such an understanding, JIT may seem like a loose, unrelated set of methods and techniques. JIT is not just a method or set of techniques, it is a complete culture, a different way of life for a business (Figure 2.4).

JIT quality: zero defects, zero breakdowns

Although quality aspects of production and total business operation fall outside JIT and into other philosophies, these philosophies do incorporate Japanese and JIT concepts. Total quality management (TQM) is arguably the one acknowledged philosophy that embraces all aspects of running and improving any enterprise, and is the overall philosophy that is used in this text.

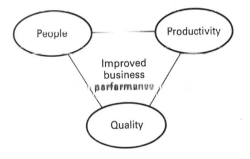

Figure 2.4 JIT relationships.

■ Total quality management

Total quality may be defined as *the optimisation of the performance of all parts and functions of an organisation – operations, procedures, systems, controls, structure and culture – to achieve conformance to the requirements and expectations of all customers*. To satisfy this definition requires a change to a total quality culture. We can think of this as a culture where everyone is committed to continuous improvement in *everything* they do as a daily routine (Figure 2.5).

Quality is never an accident, it is always the result of intelligent effort.
John Ruskin

25

Figure 2.5 Total quality organisation.

In the author's experience, a firm run as a total quality organisation is being run exceptionally well. If it has improved sufficiently through the use of appropriate methods and techniques, it will be profitable and competitive, Through appropriate and sustained use of tools and techniques, focused on defined strategic objectives within a total quality environment, a company may qualify for and compete successfully in the world class arena. It may become a world class performer.

A total quality culture is built on the following concepts

- Establishing senior management commitment
- Universal participation, involvement and ownership (empowerment)
- Focus on customer needs and expectations
- Continuous improvement of all operations and functions
- Everything is a process which must contribute to quality
- Understanding the cost of quality
- Internal customer/supplier relationships
- Improved performance at lower costs
- Growing business and profit through improved quality

As in JIT, establishment of people-centred improvement is a fundamental approach (Figure 2.6).

The TQM process

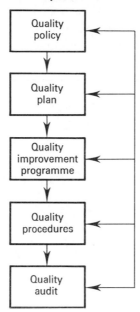

Figure 2.6 Total quality management.

TQM in turn contains many of the JIT techniques for improved productivity and elimination of waste, as these basic philosophies have a substantial overlap, and are both centred around the people and cultural approaches to business excellence. With TQM and JIT, people at all levels are fully involved and empowered with continuously improving the business, not just managers or specialists. Workers and line supervisors often make up over 80% of the labour force and can produce at least 80% of the improvements and innovations for productivity and quality. Line operators are frequently more aware than managers of poor production methods, poor quality and waste in all its forms. This makes it essential that their knowledge is harnessed and applied to company improvement.

■ *Total involvement, high motivation, continuous improvement*

Background

As a concept, total quality is not new, starting in Japan in the late 1950s and early 1960s as a major foundation to the emerging power of Japanese manufacturing, alongside other related philosophies such as JIT. The US began to awake to the potential benefits of total quality, and the now famous quality evangelists, Crosby, Deming and Duran, began to spread the gospel

27

across the Western world. Many US and UK businesses have put in place strategies for quality, but too high a proportion have not yet reached the basic levels of understanding the concepts of total quality. Far too many still hold outdated and naive views of what quality means, continuing to inspect defects out, not to build quality in.

Misconceptions of quality: inspected in

As with JIT, there have been (and still are) many firms that only perceive quality as a product and production-related concept, limited to quality inspection and control by a policing quality control function. These very basic and traditional views of quality are almost in total opposition to the TQM and JIT philosophies of building in quality at source.

Misconceptions of quality: quality assurance

An international quality assurance standard, ISO 9000/BS 5750, is rapidly growing in use across most industry and business sectors, and is becoming a prerequisite to supply services or products to several markets. ISO 9000 is a laid down set of sensible, auditable management systems, which represent good business practice and procedures.

Whilst these accreditation schemes are very useful steps in formalising the way an enterprise operates, using good business management principles, they do not necessarily lead directly to improved quality or performance within the business although many practitioners adopt a TQM approach to quality assurance. An auditable quality manual of procedures and instructions is created and used, but they may be merely a snapshot of how a poorly run company currently operates, with no review or improvement strategy to ensure such an undertaking will enhance overall performance. Such quality systems may result only in increased paperwork and non-value-added activities.

Unlike quality assurance schemes, total quality is not confined to a set of laid down, auditable procedures that define how each function and operation should be performed. It is a company-wide commitment to substantial and sustained improvement at all levels.

Fortunately, more QA experts and consultants are now actively combining the QA certification schemes with a TQM strategy, and businesses are gaining real benefits, rather than just certificates.

■ What is quality?

Popular uses of the description *quality* vary widely in the true meaning, or intended meaning of the word. 'A quality suit', 'a quality car', phrases often used by advertisements to give the impression that a product or service is better or superior in some way to another product. But these are poor uses of the word, since no standards are set, and no customer requirements are known. For example, a Jaguar XJS or BMW 735 saloon may be advertised as *quality* executive cars, implying they are superior in one or more ways to other cars. But this is only true if the customer requires the specific superior features these cars possess, and is prepared to pay the price for them. If a customer wants or needs a four-wheel drive vehicle or a seven-seater estate car, then neither the Jaguar nor the BMW will be adequate, even though the price may be similar. Similarly, a fleet buyer with a restricted budget of £15,000 per car is not able to purchase the cars because their prices are too high.

This all relates to the definition of *quality* as 'conforming to the customer requirements'. The requirements therefore include a number of facets, such as cost, performance, finish, durability, energy consumption, colour. **Conformance to customer requirements** is a definition well used in quality assurance, and QA standards such as BS 5750 address the need for businesses to possess adequate quality systems to 'assure conformance to customer requirements'. For total quality in the 1990s, the definition of quality needs to be extended.

Quality means *conforming to or exceeding customer requirements, at the lowest cost*.

■ Basic principles of total quality

A set of basic principles of quality were developed and used, stemming from the early work of Deming, Duran, Crosby and Ishikawa. These principles further demonstrate the strong overlap with the JIT philosophy.

1. Be innovative in all areas; provide the resources necessary for innovation.
2. Ensure the organisational structure will support innovation and continuous improvement.
3. Remove cumbersome or bureaucratic organisations and cultures.
4. Encourage open, involved management styles.
5. Learn and adopt the zero defects philosophy across all business functions.
6. Create multiskilled, cross-functional teams to attack and eliminate waste.
7. Utilise suitable statistical process control (SPC) and other techniques to give people tools to identify and solve problems, and hence to focus the team effort into a problem.
8. Put quality detection and prevention back onto the production line; provide ownership.
9. Reduce the number of suppliers and work with them to improve service and costs.
10. Create motivation and standards by **empowering** the workforce via

Quality maturity grid
Phases of quality development

Phase	Company attitude	Problem solving	Quality posture
I Doubt	Blame quality function for 'quality problems'	• Fire-fighting • Buck-passing • Few lessons learned	Why do we have quality problems?
II Interest	Recognise TQM would help but can't find the time	• Teams to attack major problems • Quick-fix solutions encouraged	Do we really have to have quality problems?
III Understanding	Becoming supportive and helpful	• Lessons being learned • Problems faced openly and in an orderly fashion	We are finding and beating our problems
IV Commitment	Participating to provide continuing emphasis	• Problems spotted early • All areas open to suggestion and improvement	Problem prevention is part of our normal routine
V Total quality	TQM an essential part of all company systems	Problems are anticipated and prevented	We know why we don't have quality problems

Figure 2.7 Total quality maturity table.

improvement groups; avoid the 'poster and slogan' approach to spreading the word.

As with aspects of JIT, most of these aims for improvement apply to both shop-floor and all support areas (Figure 2.7).

■ Culture, organisation and systems for total quality

Total quality: *optimisation of the performance of all parts and functions of an organisation – operations, procedures, systems, controls, structure and culture – to achieve conformance to the requirements and expectations of all customers.*

A total quality approach to business (Figure 2.8) establishes in sequence several basic elements:

- Senior management commitment to total quality concepts
- Simple, natural groups of people and skills to create effective single-office units
- Valid performance measures for each group
- Customer-oriented culture for both internal and external customers and suppliers.
- Appropriate techniques and tools to achieve continuous improvement in all functions
- Fast and reliable company-wide communications monitored for effectiveness
- Continually improving supplier relationships.

Participation and involvement

The biggest difference between a total quality business and any other is in the attitude and actions of its people. The importance and value the people place on quality and excellence of every activity, together with their continued striving for improvement in all aspects of their work, are the foundations of a total quality, world class enterprise. Culture and attitude have a direct effect on business performance. In a poorly motivated company, workers complain, 'Nobody listens to us, they don't think we matter.' This indicates poor motivation and involvement, which result in workers making mistakes and/or letting no poor workmanship pass unchecked. 'Quality of the product isn't my responsibility; I'm in sales!'

Figure 2.8 Total quality improvements.

This can result in everyone abdicating responsibility instead of checking their own work. In a TQM enterprise, similar workers might suggest, 'That product defect may be related to my late request for a design change. I'll check it out and in future I'll try to get earlier details of exact customer needs.'

Culture is the combination of company values and worth, management style and attitudes of employees and managers. This total quality organisation will be totally committed to the following objectives

■ seeking quality before profit to ensure that long-term profitability goals are achieved
■ communicating effectively throughout the business
■ making decisions based on facts and accurate information
■ continuously striving for improvement
■ developing employees through training
■ developing measurable accountability to the customer
■ striving to delight the customer
■ everyone accepting responsibility for quality

Based on a willingness to change and adapt as necessary, these objectives will be achieved through universal participation by all employees, working both as individuals and as teams. Once achieved, this will result in improved customer satisfaction and goodwill, both for internal and external customers (Figure 2.9).

Resistance to change is greater when people have not been involved in the decision process leading to the changes, so make sure all those affected by the impending changes are fully involved at an early stage. Early involvement leads to ownership, which leads to reduced resistance and easier adoption of the changes.

Participation is essential, working with people rather than controlling or confronting them. This does several things

■ it makes people part of the process
■ it motivates people
■ it brings out people's creativity
■ it provides personal development opportunities

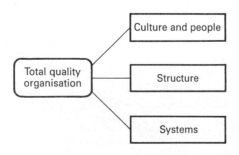

Figure 2.9 Total quality culture and people.

■   it encourages people to try new solutions to old problems

Changes for total quality

Many enterprises will become virtually unrecognisable once they are well down a fully committed total quality programme. The scale and diversity of change required to realise the full spectrum of potential improvements can be shocking, affecting virtually every aspect of the business and how it operates. For full TQM, change is massive and will be felt by everyone in the organisation. (Figure 2.10).

Area	From: Current approach	To: World Class, TQM approach
Philosophy	local competition, Just-in-case	World market, Just-In/Time
Culture	disparate, separatist, 'them and us'	empowerment, team approach
People	separate departments, specialist	multiskilled teams
Status	employees, low involvement	'partners', highly involved and motivated
Structure	hierarchical, pyramid	flat, business units / cells
Response	slow and inflexible	flexible, rapid response
Communication	vertical, poor	horizontal, improved
Methods	sporadic, piecemeal	structured, always improve
Measurements	efficiency, standard costs	throughput, effectiveness
Priorities	overhead recovery, delivery price	Customers, quality, profit, improvement
Quality	inspected in, statistical QC	Designed in, line responsibility

Figure 2.10   Total quality approaches.

Organisational structure

To achieve a people-centred approach, a simplified organisational structure is often required. This may mean the restructuring of conventional pyramid structures into flatter, horizontal matrices containing natural groupings of facilities and people around the processes that provide goods or services to a customer. *This form of organisation has a clear, overall understanding of accountability and has the aim of improving communications and response through shared ownership and involvement* (Figure 2.11).

Principles for restructuring

JIT and systems engineering philosophies are based on common sense. They are more fully described in Chapters 5 and 6, but here is a list showing how they apply to organisational structure.

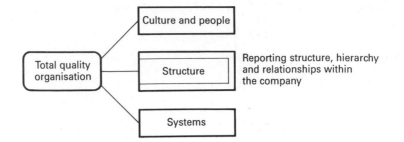

Figure 2.11 Total quality structure.

1. Break down functional boundaries and barriers; establish a first-rate communications system in every part of the company
2. Establish simple/flexible groups – modules and cells – aim to achieve a *single-office* concept, *product or service*
3. Eliminate informality and fragmentation
4. Develop simple operational procedures
5. Provide performance measures reflecting customer/supplier needs

Principles 1 to 3 are related to structure; principles 4 and 5 are related to the new procedures and systems that will be designed for more effective working.

Systems engineering practices in organisational restructuring embrace the concepts of a total quality culture, introduced earlier.

Cross-functional cells and teams

Traditional organisations tend to be functionally based, with the unavoidable break-up of a whole job or process into fragments. For example, fragmenting the production of a detailed job specification document into several parts: sales, estimating, design, bills of material. By this type of fragmentation, we divide a problem into a set of unconnected pieces, and lose a large proportion of overall understanding and ownership in the process. This in turn leads to a loss of co-operation, poor communication and more paperwork to carry the information between all functions, which results in extended internal lead times.

Functional organisations tend to become specialist and inflexible in their outlook, 'Its not my problem.' There is no 'cradle-to-grave' culture. The more functions, the more likely it will become another department's problem. Unfortunately, the functional structure tends to self-perpetuate, with more and more specialists required to deal with any growth in throughput or range of products. This structure can rapidly become large, unwieldy and slow to respond in the face of different market demands. Leaders of functions become powerful figures and may be understandably unwilling to shed their position when faced with restructuring.

To effectively address the changing market and solve the motivational problems that do exist in Western industry, the move to natural, mutiskilled organisational groupings is proposed. Based on the creation of teams, cells and modules – a single-office approach – complete tasks and problems will be processed as whole entities, with no fragmentation. Based on the major processes of product and/or information that move through an enterprise, the mutiskilled teams deal with most of the tasks within the cell, creating an environment of ownership and responsibility, and encouraging good communication. Job and task boundaries are widened, with levelling of skills where appropriate, to simplify the structure within the cell or unit (Figure 2.12).

Human relations

Training must be planned, thorough and appropriate, to assist and motivate workers in their new roles. Also, pay structures will frequently need restructuring to recognise the broader job specifications, and to recognise performance. Where bonus or incentive schemes are to be used, these should be based on groups or teams, not individuals. This encourages group and unit performance. In many traditional industries, the principles of measured day work and bonus systems are overdue for replacement by such group-based performance schemes. As internal barriers are broken down by this process, internal customer/supplier relationships are encouraged and developed in their place, within the unit, and between different units. Throughput and lead-times should also improve.

In these cells, little co-ordination is needed, and overall control is therefore simplified. Cell or unit managers have control of much more of their own destiny, and senior managers can devote more time to strategic issues and forward planning in the company

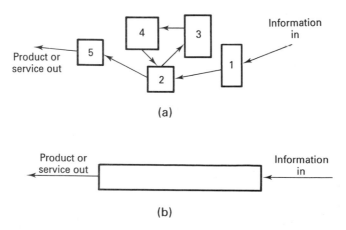

(a)

(b)

Figure 2.12 Cell and single-office concepts: (a) common internal flow between functions; (b) single office.

Examples of cross-functional units built from a traditional organisation are:

1. New product development, sales and marketing, estimating, R&D, engineering, systems engineering
2. Manufacturing, material control, production planning, cell or unit capacity planning, purchasing, stock/warehouse control, distribution, QA

Each multifunctional cell will require its own set of resources, operating policies, procedures, objectives and measurements of performance.

Because of its importance in achieving the levels of improvement required for world class performance, defining and creating a total quality structure is the subject of Chapter 5.

Support systems

Support systems comprise the operational procedures, controls and computer systems used to run the business. Many existing systems have become overcomplex and often misused, being created to deal with the large and fragmented departmental structures to be found in the traditional company. Frequently imposed from the top, they are reliant on excessive paperwork and/or computing facilities to chase information between the many departments, resulting in high operational overheads and substantial non-value-added activity (NVA). Quality, response times and accuracy regularly suffer with this type of system, due to the overlong process chain, unnecessary complexity and lack of ownership by the users.

Informal, poorly followed procedures and practices must be replaced by well-understood and formal procedures, with suitable computer system tools being created and provided wherever necessary (Figure 2.13).

Computer support systems

In recent years much emphasis has been placed on the use of computer systems to reduce manual effort, paperwork, lead times and man hours worked. These include MRP, CAD and others. Such systems can also reduce the number and

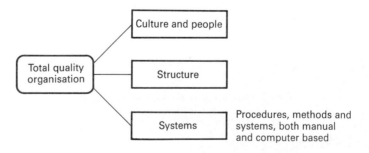

Figure 2.13   Total quality systems.

risk of errors, particularly where these systems are integrated with a common database. However, these same systems can often cause inflexibility and poor response, as they may have been designed around poor and complex manual procedures. It is often necessary to review and possibly to replace or reimplement a proportion of computer support systems when moving to a total quality operating environment.

To effectively support the total quality organisation, systems must be reviewed, analysed and developed to match the needs of the new structure at enterprise, cell and worker levels (Figure 2.14). Developed or selected in collaboration with all users, these systems and procedures must simplify operations wherever possible, be reliable in operation and reduce paperwork. Providing feedback on group performance, these systems should measure

- throughput
- lead times
- product quality
- internal customer/supplier performance
- lateness and arrears
- performance against plan at cell and enterprise levels.

Figure 2.14 Systems requirements at various enterprise levels.

■ The TQM approach

Internal customer/supplier relationships

Possibly one of the easiest and most appropriate ways to think about total quality in practice is to treat every interface *within* the enterprise as a customer/supplier relationship. One objective is to delight the customer, and this also applies to the internal customers. Internal suppliers must improve their practices and performance to meet and exceed the internal customer needs, once they have been specified. This is a key factor for improvement in any company.

In each and every type of business, there are customer/product/supplier relationships between the internal functions. For example, the sales office should correctly and unambiguously interpret the (external) customer requirements, then provide the design function with clear information. Design provides manufacturing and/or planning with drawings and data to plan and produce the product correctly, within the required lead time. Manufacturing are then responsible for delivering the product and meeting the required standards of finish, performance etc., to the warehouse and distribution functions, who will ensure its satisfactory on-time delivery to the external customer (Figure 2.15).

These interfunctional processes are carried out in every company, but are they carried out with all internal customer and supplier requirements being identified and met, every time and at least cost? Many firms' sales staff have frequently to revisit customer definitions of requirements before an accurate order can be passed to the other internal functions, wasting time and effort whenever a change or revision occurs. Although flexibility to accommodate customer change is a key aim in seeking world class performance, unnecessary change and confusion needs to be eliminated at source. How often do manufacturing have to go back to design or material planning with incorrect drawings or purchased parts? These are not examples of a total quality operation; they are symptomatic of not having the correct internal and external customer/supplier interfaces, enabling people to get it right. Each process must become 'right first time, every time'.

The internal customer/supplier concept includes a processing function at each interface. A process is a series of actions or operations that are conducted to produce a desired result, which may be information, a product or a service. Each process involves time, resources and costs, and there is always room for improvement. In a total quality organisation, every process should be the subject of continuous improvement, with any non-value-added processes being eliminated.

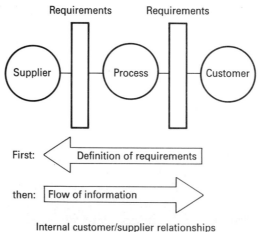

Figure 2.15 Internal customer/supplier relationships.

■ Costs of quality

Frequently underestimated, if estimated at all, quality costs are an overhead that directly reduce the bottom-line profit of the business. *The total cost of quality, nonconformance and resulting non-value-added activities will typically vary from 25 to 40% of operating costs, but can be reduced to 5 to 15% or less in a total quality organisation.*

Frequently, the costs of quality are perceived as those associated with detecting and measuring defects and nonconformances, plus rectifying defects and preventing future failures. Conventional quality costs are

- appraisal
- prevention
- internal error, within facility
- external error – once products have left facility – on route to, and in use by customers

Consider the benefits of reducing quality costs (Figure 2.16).

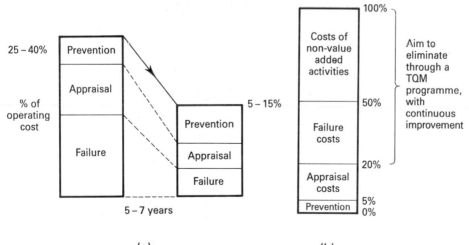

(a)   (b)

Figure 2.16 (a) Quality costs versus business operating costs;
(b) addressing NVA costs in a total quality approach.

■ Non-value-added activity costs

From a total quality perspective, there are *non-value-added activity* (NVA) costs to be added. They can be very substantial and do not normally form part of 'quality' costs. NVA costs come from all the wasted activities that people do: unnecessary tasks, duplicated tasks or repeated tasks because something went wrong. Conventional costs of quality defects and scrap are included, but will usually be a minor portion of the costs of total quality. Company directors and managers frequently think of the quality cost as the scrap, returns and operating cost of the QC department. They rarely consider the other substantial costs of concessions, rework, corrective action, re-inspection, etc. Add to this the indirect costs of noncomformance, customer dissatisfaction, premium overtime rates, reduced worker morale, etc., and the full cost of quality begins to appear.

For many firms, determining all the costs of quality and NVA will be a necessary step in preparing a case to begin the journey towards total quality. Boards of directors and shareholders often require convincing that such a major change is desirable or necessary, and a well-founded report on the full cost of quality can be the ideal instrument for this purpose.

CASE STUDY 2.1

NVA costs for a UK pharmaceutical company in 1990

Costs exclude other quality costs and are expressed as percentages of time spent on NVA tasks of every form multiplied by salary for each department*

1. Production function

- Manager (@ £30k per year) 50% × 30k = £15k
- Director (@ £45k per year) 20% × 45k = £9k
- Prod. services exec. (@ £14k per year) 50% × 14k = £7k
- Planning assistants (@ £8k per year) 50% × 8k = £4k
- Manufacturing supervisor (@ £14k p.a.) planning, paperwork, deputising
- Manufacturing technicians (7 @ £9k p.a.) paperwork 5% × 7 × 9k = £3.1k
- Packaging supervisor (@ £14k p.a.) planning, scheduling and paperwork (mainly double-entry) 20% × £14k = £2.8k
- Packing operators (9 @ £7k p.a) paperwork 5% × 9 × 7k = £3.1k

Subtotal cost £46.8k

2. Warehouse: main

- Warehouse supervisor (@ £14k p.a) 40% × £14k = £5.6k
- Warehouse deputy (@ £10k p.a) 70% × 10k = £7k
- Warehouse clerk (@ £7k p.a) 70% × 7k = £4.9k
- Warehouse staff/technicians (8 @ £8k p.a) 5% × 8 × 8k = £3.2k
- Packers (2 @ £8k) manual and terminal data entry 20% × 2 × £8k = £3.2k

Subtotal cost £24k

3. Warehouse: remote

- Group warehouse manager (@ £30k) 40% × 30k = £12k
- Warehouse supervisor (@ £14k p.a) 80% × £14k = £11.2k
- Order processing supervisor (@ £14k p.a) 80% × 14k = £11.2k
- Warehouse clerk (@ £5k p.a) 50% × 5k = £2.5k
- Warehouse staff (3 @ £8k p.a), 10% × 3 × 8k = £2.4k
- Packers (5 @ £8k) manual and terminal data entry 5% × 5 × £8k = £2k

Subtotal cost £41.3k

4. Quality control

- QC manager (@ £30k p.a.) 5% × 30k = £1.5k
- Supervisor (@ £14k p.a.) 100% = £1.44k
- QC technicians (5 @ £8k p.a) 5% × 5 × 8k = £2k

Subtotal £4.94k

5. Sales order processing

- Customer services executive (@ £14k p.a) 25% × 14k = £3.5k
- Clerks (2 @ £5k) 2 × 50% × 5k = £5k

Subtotal £10

6. Accounts

- Accounts manager (@ £30k p.a) 50% × 30k = £15k
- Assistant fin. accountant (@ £15k p.a) 30% × 15k = £5k
- Management accountant (@ £25k) 50% × 25k = £12.5k
- Clerk (@ £7k) 50% × 7k = £3.5k

Subtotal £36k

7 Management services

- Project manager (@ £20k) 50% × 20k = £10k
- Analyst/programmers (4 @ £15k) 90% × 4 × 15k = £54k

Total cost = £65.4k

Estimated total cost of time devoted to NVAs = £226k per year (salary only).
Company turnover was £5 million, giving the NVA proportion as 4.5%, which was conservatively estimated. Total salary bill was approximately 15% of turnover at £750,000. *From this the NVA portion of the salary bill was approximately 30%.*

*Appendices F and G contain details of determining the costs of quality and NVA.

■ Total quality: impact on business strategies

The commitment to move towards a TQ organisation will have profound effects on the business, both in the way it is structured and in how it operates.

Therefore these TQ-driven needs must be considered before and during the development of future business strategies.

■ Total quality strategy

To become a total quality organisation requires planning and then actioning the changes necessary to achieve this. A vision of total quality must be built into all strategies in the firm, driving all functions towards becoming a TQ organisation, striving for continuous improvement. The concepts of total quality need to be embedded in the high level strategies that will shape and direct the business for the next two to five years. Chapter 3 examines strategic aspects of total quality in relation to other business strategies.

Where do you start?

Now we have a basic grounding in what total quality is trying to achieve. But exactly where do we begin, and what do we do? Planning for total quality and JIT requires substantial amounts of time and effort to understand, plan and implement, even in a small company.

From the beginning, *total quality* has to be a team exercise, from the top management team down to the shop-floor improvement teams, led by a quality council or steering group. Chapter 5 covers the structure for managing total quality and Chapter 8 covers managing for change.

■ Methodologies and techniques for total quality

This is where the methods, tools and techniques of TQM, JIT and other best practices are brought into play.

Simplification is the key

Systems engineering uses a basic JIT approach

- examine
- analyse
- simplify
- eliminate waste and NVAs
- simplify what is left
- use appropriate tools and systems only where necessary
- continuously monitor and improve

To remove the common type of problems to be found in most businesses, the company and its systems should be reviewed and redesigned to simplify all procedures, operations and systems. Systems engineering (SE) uses this philosophy to address all aspects of a company (not just manufacturing), from strategic and organisational issues right down to individual tasks and procedures. Details of JIT and systems engineering methods are given in Chapters 6 and 7.

■ TQM, JIT and MRP2

To provide the levels of information accessibility and currency required by a modern business is a daunting task, requiring ever increasing degrees of data integration and accuracy. Manufacturing planning and control systems such as MRP2 provide integrated computer-based systems to plan, control and run each function of a business. Readers unfamiliar with the basic concepts or role of MRP2 should refer to Chapters 9, 12, and 13.

At the beginning of this chapter the philosophies for improvement were listed as total quality and JIT, with MRP2 listed under technology and systems. Why have we returned to discuss MRP2 in relation to total quality? Because, of all supporting computer-based systems, MRP2 (Manufacturing Resource Planning) and now ERP (Enterprise Resource Planning), are company-wide and systems that link into virtually every existing function. More than effective systems, they are unmatched at business planning, monitoring and control.

In the search for world class performance, companies have almost no option other than extensive computer-based data processing. This includes the successful Japanese JIT champions, which now use MRP for business planning and some control functions. MRP2 and ERP are valuable tools even after a company has restructured along simpler lines. But their appropriate implementation remains crucial.

Even Kenneth Wantuck, in his excellent book *Just-in-Time for America*, places MRP alongside JIT and total quality to effectively run a business. Good 'quality' computer-based systems remain essential platforms for business through the 1990s, but must be even more carefully matched and implemented to the business needs. Data volumes and required response times demand the use of powerful and flexible computer-based systems (Figure 2.17).

Wantuck says it all:

> MRP is a good system, but it doesn't always meet our expectations. Why?
> – because it assumes our present processes as if they were right, but are they? (No) . . . JIT attacks the outdated and wasteful practices in our manufacturing processes, simplifying the operations, eliminating all the non-value added activities and stressing quality improvement. This makes MRP easier, and really makes it pay off.

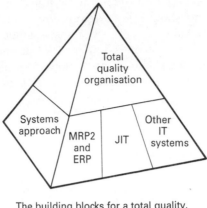

The building blocks for a total quality,
world class enterprise

Courtesy K. Wantuck, JIT for America.

Figure 2.17   The role of MRP/ERP in a total quality enterprise.

It is clear that MRP must be implemented (or re-implemented) during a company-wide improvement mission. Also, MRP and ERP should be implemented using a total quality approach for optimum benefit.

MRP2 systems are no longer confined to pure manufacturing operations, having been used for such diverse commercial fields as

- planning and managing long-term pharmaceutical trials
- project and resource management for the construction industry
- military wargame organisation

This is partly why the term enterprise resource planning (ERP) was coined; it represents the sector independency of modern business control systems. Chapters 12 and 13 are devoted to recent developments to MRP2 and ERP, and how to implement them.

■ Summary

A TQ organisation is totally committed to the following objectives and concepts

- delighting the customer
- continuous improvement
- company-wide, people-centred involvement
- understanding the cost, and benefit of quality

These objectives are achieved only through universal participation by all employees, working both as individuals and together as teams. Once achieved, TQ improves customer satisfaction and goodwill, internally and externally.

The route to total quality contains a change in philosophy and culture within the enterprise, which provides the motivation and aims for continuous improvement in all activities. To achieve these improvements requires the methods of JIT and a systems approach employing a wide range of techniques and tools to analyse performance, identify areas for change and develop new, improved practices.

■ Further reading

Crosby, P. (1979) *Quality is Free*, McGraw-Hill.
Handy, C. (1991) *The Age of Unreason*, Century.
Kanter, R. M. (1984) *The Changemasters*, Allen & Unwin.
Krajewski, L. J. and Ritzman, L. P. (1996) *Operations Management: Strategy and Analysis*, Addison-Wesley.
Schonburger, R. J. (1982) *Japanese Manufacturing Techniques*, Free Press, New York.
Schonburger, R. J. (1986) *World Class Manufacturing*, Free Press, New York.
Schonburger, R. J. (1990) *Building a Chain of Customers*, Hutchinson Business Books.
Shingo, S. (1985) *A Revolution in Manufacturing*, Productivity Press.

3	# Strategy for the Whole Business

If senior management do not have or take the time to properly evaluate and devise strategies for the future, there may be no future for the business.

■ Introduction

To produce an effective and competitive business, it is essential to address the overall business goals and financial targets. Piecemeal development and improvements in any one area of the business will give only limited benefits, and it is by developing strategic plans that we can ensure all areas are moving in the 'right' direction.

This chapter defines the interlinking strategic plans for each area of the business and outlines their development.

Several important points are discussed

■ An overall business strategy must exist to guide the development of underlying strategies.

■ A disciplined approach is necessary for the effective development, use and maintenance of strategies.

You are unlikely to achieve the best overall company performance if you allow independent, local improvement programmes to develop without regard to overall aims and objectives. Global optima rarely equal the sum of local optima. Production may achieve very high levels of efficiency, but at the cost of large batch runs, high inventory and slow response times, all detrimental to overall performance.

We have to develop long- and short-term plans for the company, and all major functions within it. These plans must be complementary, contributing to overall competitiveness. This attention to the highest aspects of company operations and objectives is necessary to set the framework and priorities for detailed improvements. The required strategies are established using a systematic approach to simplify structures, reduce overheads and encourage continuous improvement.

■ Strategies

Strategy describes the planning and execution of a war or a business campaign. Plans list strategic objectives and how to attain them. Objectives may be attained using tactics. And tactics control resources, plant and people.

For example:

Objective*: to increase profit margin on product A, a white good.*

Strategy*: address sales price and cost of manufacture. Following market research, devise a sales strategy to realise a price increase of n%, then advertise, repackage and reposition. Break down cost of manufacture into labour, materials and overheads. Work on all three but concentrate on the highest, typically materials. Develop cost reduction programmes for product A, including alternative materials, methods review, redesign and weight reduction. Where appropriate change the manufacturing process to reduce cycle time or labour content. This may be strategically important by helping the overall business to cope with changes in market volume and/or flexibility.*

Action plans*: by following the strategy, draw up action plans to achieve the overall objective, which affects sales and manufacturing.*

■ Steps in business planning

Business strategies are usually developed in five interrelated steps (Figure 3.1), which lead to and shape the supporting strategies. These are well-established steps in business planning, but have frequently been seen as interactively looping through steps 1 to 3, and non-interactive or passive in steps 4 and 5. Manufacturing is often not a major player in business strategy development. What is required is full interaction between all steps, particularly between

---

Reactive

1. Define the main business objectives
2. Create the marketing and sales strategies to support these objectives
3. Review how each product or product range wins orders against competition

Passive

4. Determine the most suitable manufacturing systems to produce the products
5. Determine the most appropriate support systems for manufacturing

---

Figure 3.1 Passive and reactive steps in business strategy development.

marketing and manufacturing. Only if this is done can the full potential of the business be realised, with manufacturing employed as a strategic weapon.

■ Manufacturing: a traditional strategic misconception

Traditionally, corporate business strategists and marketing strategists decided what they wanted to do, then expected production to blindly follow their plans. Manufacturing was not considered strategic, seen simply as an operating function to produce the goods that sales and marketing had wanted. This form of narrow thinking led to the creation of inflexible production systems in a large number of firms, because the manufacturing managers were not involved in defining and developing overall business objectives. Therefore they could not offer the senior management guidance on the options or impact of various manufacturing scenarios.

Manufacturing can be a strategic weapon

The aim is to create a manufacturing system that is capable of rapidly responding to changing market conditions, in the appropriate manner. Once achieved, the power and flexibility of your manufacturing can and should be used as both a tool and a *weapon* for gaining business, a weapon to attack and defend against competitors.

This is very different from the previous traditional view of manufacturing, where it was seen as a relatively unimportant function with the sole brief of producing. For most of this century, business and industrial strategists have been largely driven by financial and/or marketing objectives, with the firm's operation and production function aligned with these primary aims. Already we are seeing manufacturing strategies becoming dominant over the way businesses are led and operated, and this must become the norm over the next decade. Manufacturing can and does provide a flexible, strategic weapon for the business, because there is now the essential culture, methodology and technology that allow us to revolutionise the speed, range and methods of production.

■ What drives the business?

Depending on the company, emphasis and drive will come from different quarters, but most frequently from sales. Sales-driven companies are both common and understandable, but must be carefully controlled to avoid becoming imbalanced and unrealistic. For example, sales staff bring in the orders and money, so they may consider everyone in the company should be driven mainly by their requirements.

The over-simplistic 'no sales, no business' argument has proven very attractive to many chief executives, leading them to dispense with proper balance and direction in the company. Most sales executives are very aware of the financial aspects of their sales, in terms of profit margin, contribution, etc. They often succeed at focusing the company into a 'we sold it, you'll need to make it' way of operating. In this scenario, sales tend to dictate much more than they should, in terms of product variety, pricing, delivery dates, etc. Meeting and exceeding the customer's requirements is certainly a reliable way to maintain and grow a business, but must be derived from well-founded, effective operational practices and systems. Too often the excessively sales-driven company will aim for these goals when the business is not capable of delivering. The result is often failure.

One effect of this will be the gradual demoralising of the rest of the company. A more serious outcome can be the erosion of any ability to compete in the marketplace.

Break down barriers

The need for flexible organisation and the breaking down of departmental barriers is obvious. Reviews and discussions on flexibility can often be used to bring marketing and operations together, identifying areas that need attention if 'flexibility and response' is to be achieved.

Strive for realistic balance

Achieving correct balance is the key, and it would be equally wrong for excessive drive to come from design, manufacturing or finance. The chief executive has a duty to maintain a balance, since it may affect the business plan now or in the future. Once the problem is recognised, solutions are comparatively easy. The hard part is recognition.

CASE STUDY 3.1

A sales-driven company

The author was asked into a manufacturing company in Southern England to carry out a review of production. The company produces a range of electrical switchgear cabinets for home and export markets.

The managing director was very concerned about his plant's low levels of output and frequent parts shortages. It soon became apparent this was a typical sales-driven company. Orders were taken for products that were not yet ready for production. Delivery dates were given with little or no consultation with production or planning. Changes to orders up to and beyond the production stage were readily accepted and no related costs were recovered from the customer. Fairly easy to accommodate at long product lead times, orders became a serious problem at short lead times (which are within purchased or manufactured part lead times). *The problem wasn't with production; it was at the front end – sales and customer services. And with the managing director.*

Sales were blaming production and planning for not being able to satisfy their customers. In reality, it was sales who were not satisfying their customers within the company – design, planning, and production – who in turn could not satisfy the final external customer (Figure 3.2). The results were shortages, line delays and stoppages, high running costs and late deliveries. Morale was very low, as you might expect.

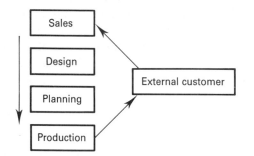

Figure 3.2 Internal and external customer relationship problems.

The solution was to establish formal procedures for order definition and acceptance that involved all key functions.

■ Flexibility

The ultimate aim is for the whole company to become sufficiently flexible and responsive to win business and deliver on time. We must therefore ensure that each and every function within the company is improved, through the systematic development of the integrated areas that make up and support the business strategy. This is particularly relevant to production processes, which may well have to become much more flexible to deal with market and product variations.

Who develops the strategies?

It is wrong to assume that strategic issues concern only the director and senior managers. Middle managers need to be fully aware of the company objectives and any plans to achieve them. This is essential because middle managers are responsible for executing strategy even if they do not devise it. Once running, systems engineering task forces should also be used to contribute to relevant strategic planning sessions. In many cases the systems engineering professional can actively assist in the development of strategies and action plans. Where strategic planning is to be a new or revisited undertaking for a business, their assistance will be invaluable. Good communication and the sharing of problems and plans on a wider basis is essential and is part of the foundation for motivation and project success.

■ A new generation of manufacturing leaders

Many top managers continue to give inadequate attention or thought to long-term issues, despite the continual call for strategic planning by all leading authorities on business excellence. However, there has been significant progress in both academic and professional education for business and manufacturing over the last ten years, producing a new breed of professional manager.

Almost all universities and business schools now run good MBA programmes and postgraduate degrees or courses on modern business administration, systems engineering and related subjects. Also widely available are seminars and courses run by professional institutions, consultancies and other experts. Staff training and retraining at the highest levels is therefore well provided and has equipped many managers with the necessary understanding and skills for improving business competitiveness.

Today's manufacturing leaders appear far better acquainted with business operations and planning at the highest levels. They understand market and business issues, both in terms of customer requirements and financial implications. By employing their talents to the full perhaps we can achieve boardroom advances in strategic thinking, advances that are long overdue. It is also essential for established mangers to retrain and update their knowledge, especially senior managers responsible for driving change and improvement. So, senior management, having been reinvigorated and reinforced, begin the task of developing strategies.

■ Strategy foundations

Total quality is the *optimisation of the performance of all parts and functions of an organisation – operations, procedures, systems, controls, structure and culture – to achieve conformance to the requirements and expectations of all customers*.

In the author's experience, a firm run as a total quality organisation is being run well, and if it has improved sufficiently through the use of systems engineering methods, it will be profitable and competitive. It is the aim of us all to run our firms as well as possible but what does this really mean? If we assume *well run* to mean efficient, this does not answer the question, because a business can be profitable without being efficient. Or, it may be efficient without making a profit.

The route to total quality and world class performance begins with planning, followed by appropriate action to achieve the planned changes. The vision of total quality (TQ) needs to be built into all strategies in the firm, driving it to become a TQ organisation, an organisation that strives for continuous improvement. The concepts of total quality must therefore be embedded in the strategies that will shape and direct the business for the next two to five years.

As described in Chapter 2, total quality is built on the following concepts

- Establishing senior management commitment
- Universal participation and ownership
- Focus on customers needs and expectations
- Continuous improvement of all operations and functions
- Everything is a process which must contribute to quality
- Improved performance at lower costs
- Growing business through improved quality

On these concepts we should construct our strategic business plans.

■ Total quality and business strategies

These TQ-driven needs should always be considered before and during the development of business strategies, being a company-wide commitment to substantial and sustained improvement at all levels. From strategy to detailed action plans, the philosophies and aims of total quality and systems engineering must be incorporated.

Systems engineering in strategy development

Systems engineering considers business, its market and then the organisational structure required to deal with the market. This means taking a fresh look at how the market and business are moving, with no predefined notions of how the firm should be structured or operated.

- Search out the NVAs and eliminate them
- Look for and create natural groupings of people, tasks and products; do *not* automate existing ineffective and fragmented systems
- Review, simplify and improve then apply automation if necessary
- Regroup service functions into the areas that use them
- Establish performance measurements for *all* and relate targets to business performance
- Continue improvement in all areas

These should be translated into appropriate aims and objectives within each separate strategy: sales, manufacturing, R&D, etc.

Organisational structure

Organisational structure often limits, cripples or constrains improvements in communication. For total quality, the creation of a simplified organisational structure is essential, containing natural groupings of facilities and people around processes that provide goods or services to a customer.

Strategy then structure

Strategies are implemented within an organisational structure. A strategy is developed then the optimal structure is chosen. The organisation and its functional responsibilities must be shaped to create the framework for implementing the strategy. Again, change in local areas of a structure should be avoided, as they will not serve to advance the overall objectives. Local optima do not lead to global optima again. Structure and organisational change are further covered in Chapter 6.

Summary

The full implications of total quality and systems engineering need to be understood and adopted at the strategic level; they should not be confined to manufacturing. Systems engineering is equally applicable in sales, design, materials and distribution. The company-wide application of systems engineering requires an understanding of background aims, objectives and constraints. This leads to the development of appropriate strategies for all areas and provides competitive advantage in an ever changing market.

■ Systems engineering strategy

Systems engineering needs to be a major influence when planning a business strategy and its supporting strategies, otherwise improvements will be confined to lower levels in the company. Best practice in systems engineering, itself requires a strategy – a systems engineering strategy – integrated with all the others. To achieve integration, aspects of systems engineering should be included in all other strategies (Figure 3.3). The systems engineering objectives should include

1. Detailed definitions of
   - review procedures and intervals between reviews
   - procedures for changing organisation and/or operations when required to meet new market or business needs
   - procedures and targets for performance measures, including quality of internal and external services and cost reduction.
2. Formal procedures and practices for operating effective multidisciplinary improvement teams, including scope, authority and responsibility, reporting, performance targets and timescales.
3. Training and support facilities and materials for systems engineering practitioners, including reference materials and audiovisual aids.
4. Definition and publication of total quality aims and objectives, other quality and procedures and standards (e.g. BS5750/ISO9000) and supporting quality practices and techniques.

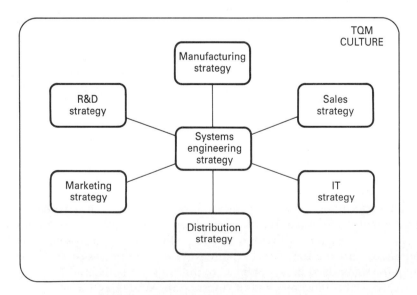

Figure 3.3 Systems engineering used in all other business strategies.

5. The definition of the simplest and most effective arrangement and combination of all company resources, machines, people, systems and buildings that can provide a competitive service to the market.
6. The definition of the simplest cell-based shop or office layout and operation achievable, including flows of product and information.

As with the other business strategies the systems engineering strategy (Figure 3.4) requires regular review to embrace trends in the market or the competition. It is also affected by in-house changes to products, materials, procedures and systems.

Strategic view of manufacturing development

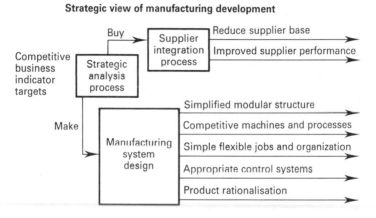

Figure 3.4 Strategy components for systems engineering.

■ Business objectives

Each company requires a plan of overall direction, a mission statement, defining its purpose, aims and objectives. Prepared and agreed by the senior management team, it needs to cover medium-term or long-term business issues, including market trends and shifts, investment plans, personnel policies and manufacturing plans.

The statements covering these plans should be as explicit as possible, with no margin for misinterpretation. They should include measurable targets and realistic deadlines. This document will drive the company for at least a year, so realism is essential. It's not a rod intended for future beatings. Statements at this level will shape and direct the resulting lower level strategies. They will be the driving force for creating a company-wide commitment to a common set of objectives. That is why realism is paramount.

The business objectives of each company will be different, because of size, market sector and state of the economy. However, most will include measurable objectives relating to growth, market share, profit and expenses. Other performance measures and policies may also be included.

The business objectives of the company need to be supported by several other integrated strategies. Figure 3.5 shows how they are related.

Each of the business planning steps needs to be defined in terms related to your own business, in order to identify and promote discussion on the options available to your firm. High level debate on the viable options is essential, involving all functional leaders. Typically, the activity and discussion in each of the planning steps will be as shown in Figure 3.6.

When carried out competently, the investigation and review of the market and product needs (steps 2 and 3) will lead logically to questions of manufacturing and business capability. This should expand into consideration of what manufacturing and support systems will need, and more importantly

Task	Strategy area
1. Define the main business objectives	Business objectives
2. Create the marketing and sales strategies to support these objectives	Marketing strategy
3. Review how products win orders against the competition	Market / R&D
4. Determine the most suitable manufacturing systems to produce the products	Manufacturing
5. Determine the most appropriate support systems for manufacturing	All support and IT and manufacturing

Figure 3.5   New steps in business strategy.

1	2	3	4	5
Business Objectives	Marketing Strategy	Order winning criteria in Market	Manufacturing Process	Strategy Support
Growth Profit Market share Financial ratios Major projects e.g. TQM, MRP2, BS5750/ ISO9000**	Target markets and sectors Product range Market size and needs – standard or make/customise to order need for change and innovation current and future position – leader or follower in market?	quality response – speed and flexibility reliability product range options customisation innovation and leadership in R&D service	process options flexibility future market/ capital equipment skilled/de-skilled tasks	systems engineering business units control systems organisation culture procedures IT systems

** Major projects can and should be used as vehicles and tools to achieve other business objectives

Figure 3.6   Planning objectives, strategies and support implications.

what they can offer in terms of advantage to the market. These topics are expanded as we consider each strategy in turn

- Marketing strategy
- Product research and development strategy
- Manufacturing systems engineering strategy
- Business systems engineering and organisational strategy
- Financial control strategy
- Information systems strategy

Surrounding and supporting them is an equally important strategy for personnel development and training.

■ How do you measure up?

Does your company have such strategies? Many companies prepare a business plan that is purely financial, with little or no consideration of how these financial goals will be achieved. Alternatively, objectives relating to the business plan may be set for each division or profit centre, but again these frequently tend to be inadequately thought out from a tactical, executable standpoint. All too often a business plan will be constructed with a superficial strategy to speed it past the board of directors, never to be seen again.

It is the responsibility of the company managing director and senior management to ensure the development and pursuit of these strategies in the fullest sense of the word, to allow the company to gain competitive advantage.

CASE STUDY 3.2

A purely financial plan divides a company

A European branch of a large US company had developed a business plan purely in consultation with sales, marketing and finance. This produced business targets for all functions, but with little real common sense as regards production and distribution levels. Sales forecasts were based on last year's sales plus a percentage. Virtually no market research was carried out.

Production staff were tied into levels of output and overtime that offered inadequate room for manoeuvre when the sales demand changed its bias from one line to another. To deal with the sudden and unforeseen shift in demand, overtime rose significantly and efficiency fell. The US parent was understandably concerned over the worsening monthly reports and it began to doubt the credibility of the business and production plans. The root causes of the trouble were the lack of time and expertise devoted to the plan. It could all have been avoided.

Clearly all functions need to be involved in developing the long-term business strategy and the detailed plans, and *all* must be accountable for

57

achieving it. Internal politics and personalities will undoubtedly figure in the equation, but it is the responsibility of the chief executive to ensure the plan is both balanced and achievable. He or she must not let any persuasive individual sway the company from its path.

Set up for line and batch production, the company was geared for volume, requiring this to return a reasonable margin on sales. Shifting trends in the market led to increased demand for modified and customised products, and sales began to sell more and more product that had not been fully tested or prototyped, leading to difficulties in producing the products, quality problems and also increasing customer dissatisfaction. Late deliveries became common, and failures in the field increased, due to inadequate testing in the compressed R&D times.

This situation was the result of a shifting market, where order-winning criteria had moved away from quality and price to include flexibility and customisation. The poor market intelligence of this firm gave no warning of the shift, and hence no review of the marketing and manufacturing strategies took place. Even as it hit, the shift triggered no high level action to review the situation and plan contingency measures. Several basic faults show up:

- Lack of firm management and agreed direction concerning policy of dealing with and selling heavily customised product.
- Sales promising delivery dates with little or no discussion with R&D or manufacturing.
- Manufacturing and R&D overstretched and unprepared for heavy customisation loads – incorrect process and support systems.
- Lack of market intelligence and forward vision.

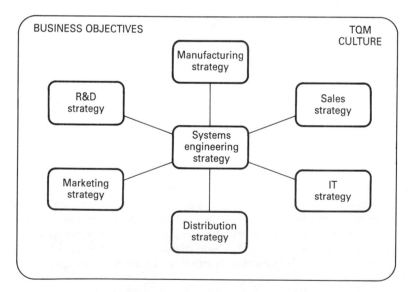

Figure 3.7 (a) Strategy diagram or wheel.

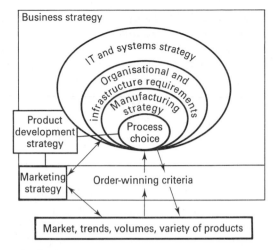

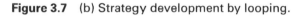

Our business

Figure 3.7 (b) Strategy development by looping.

This example shows what can happen when the five business planning steps in Figure 3.1 are only partially used. Any significant shift in objectives, market or order-winning criteria must rapidly be discussed and reviewed at the corporate level, reworking the business plan and component strategies as necessary (Figure 3.7).

■ Strategy development timescales

When planning the strategy development sessions, ensure at least several days are put aside for this purpose, well before the deadline for completion. *Failure to plan is planning to fail.* The preferred activity sequence is as follows, where each activity is fed by information from the previous activity.

Session 1: brainstorm all major areas

Session 2: research market and competitors
- Preliminary marketing strategy (largely capacity independent); open discussion on all relevant intelligence information
- Preliminary manufacturing strategy (capacity dependent)
- What advantages could be gained over the competition by adopting various manufacturing strategies? Focused on major proposed or required changes in process/product type
- Review and align marketing and manufacturing
- Revisit alternative manufacturing versus market strategies as required

Session 3: set business strategy
- Firm up marketing

- Firm up manufacturing strategy to reflect marketing needs
- Consider and amend any organisation and infrastructure (personnel/ resource implications)
- Develop information systems strategy to meet needs of business and manufacturing
- Feedback into business strategy

Session 4: firm up all strategies before issue

Detailed development of underlying action plans can now be commenced by each function. There will be a certain amount of movement in each preliminary strategy, as all tangible and intangible opportunities and constraints are explored. For example, manufacturing input will be considered with respect to shop capacity, staffing, flexibility and the options available to deal with variations in market demand. This will shape the information systems requirements and will be a key factor in forming the final version of the business plan.

Similarly, manufacturing will consider the IT options and resources available to perform these options. This will affect the final manufacturing strategy, as the availability of specific systems may significantly impact planned capacities and response time.

■ Marketing strategy

Marketing strategy defines the market sectors we currently target and wish to target, volumes and types of product. Assuming a concurrent engineering approach (design/sales/marketing closely integrated), marketing strategy will be developed and maintained by the business and new product development directors and the sales directors. It considers the following questions:

- Why market in these particular areas?
- What areas offer greatest potential profit/business? Why?
- Barriers and competition
- Fit of existing and new products
- Pricing and discount strategy
- Product life and lead time implications
- Life-cycle management
- Market and consumer trends
- Potential volumes and specifications versus operational capacity
- Rates of market change
- Approach and tactics to be used by sales staff and other customer contact methods
- Public communication to get the best from the internal improvements and benefits

The marketing strategy must spell out where the company can go and how to get there. It must be developed in parallel with the interrelated product development strategy and manufacturing strategy.

Market movement in terms of volume changes, variety or other aspects of products and services to be supplied by the business can be of vital importance in overall business strategy. This can be both changing customer needs, and/or proposed business refocusing to address other areas more effectively.

Current or future variations in market needs may well require changes in product volume, specification, lead time, etc. These may not be optimally dealt with by existing production process facilities or current organisational support structures. A more flexible or appropriate process or organisational structure may well provide short- and long-term competitive advantage.

Order-winning factors

Order-winning factors include

- quality
- response – speed and flexibility
- reliability
- product range
- options
- customisation
- innovation
- leadership and innovation in R&D
- after sales service or other services
- price

Different markets have different emphasis placed on each of these factors, with certain factors moving up and down the scale of importance. For example:

Market	Prime order-winning criteria
Petrol	Price, and other services (shop facilities)
Supermarket goods	Price, quality, freshness, choice
Automobiles	Price, quality, delivery, innovation
Commercial refrigeration	Price, customisation/flexibility, delivery
Aerospace products	Quality/reliability, price, delivery
Custom manufacturing	Response, delivery, price and quality

In many situations, certain factors may become accepted by the customer as always being there, for example, Sony TV and hi-fi are seen as conforming to or exceeding customer requirements. Here the quality criterion is met, so a different factor becomes prime, either price or delivery. This is a continuing trend in all industry sectors, with product quality being the requirement of the

61

1980s. Now, in the 1990s, quality is expected to be there, and if not, you will not normally be considered as a supplier, irrespective of price, delivery or other factors. *Quality is a competitive essential, not a feature or option.*

The previous case study illustrated one firm's poor handling of customised products, causing problems in both quality and delivery. Once these basic criteria are not met, the customer is unlikely to be interested in any future business, irrespective of price. If the firm is to prosper and grow, it is vitally important to ensure that each order-winning factor is understood, planned for and continuously achieved.

It is worthwhile considering the role that sales and marketing have to play in the new flexible and responsive firm of the 1990s. This is discussed below.

■ Sales and marketing

Marketing

> . . . the management process responsible for identifying, anticipating and satisfying customer requirements profitably (Institute of Marketing).

We can think of marketing as the longer-term view of customer and market trends and requirements, involving market research, input to R&D, and providing direction for future product and market needs. Marketing is the *planning* aspect of business development.

Sales

Sales deliver the company's marketing mix to the customer. Sales is the *execution* aspect of business development. The sales function tends to be relatively short term; marketing tends to be more long term. The two functions are required if optimum coverage is to be given to business growth. Rarely can you be forward planning while engaged in sales activity – running not fighting.

Many companies largely ignore longer-term marketing aspects. Instead they pay lip service to future customer needs and direction.

Changing trends

Sales itself is a vitally important function in any company, being the link between customer and product. Current trends are for customers to seek longer-term relationships with fewer suppliers. In return, suppliers are expected to provide more and better services than before. Failure to recognise and provide this higher level of service will result in loss of customers.

Almost all functions in the company must respond to this requirement, including sales and marketing. A new kind of sales executive is needed to succeed and then excel in this environment, being better trained and more professional in approach.

Recessionary conditions only make the sales task harder, and most sales functions will need to undergo fairly drastic organisational and job changes over the next few years to cope with the shrinking and more demanding market trends

■ falling orders
■ smaller customer base
■ increased competition
■ closer links required with key customers/accounts
■ increased service levels required
■ more selling per salesperson needed
■ new and improved selling techniques needed
■ greatly improved feedback and communication of customer requirements and perceptions

Improved feedback focuses continuous improvement efforts within your business. Surveys have also indicated that the average salesperson spends only 5% of their time actually selling the product. This offers considerable scope to increase sales activity.

Sales responses

Various approaches to increase sales force efficiency and effectiveness have been adopted to deal with some of these challenges

■ key account managers
■ pre- and post-sales support
■ computer aids such as mailshot databases
■ call management
■ forecasting aids
■ sales teams
■ telesales

However, the result of these and other moves is to make selling a much more complicated and sophisticated process, often requiring several different experts to combine their talents at the various stages of a sale.

Team selling

The use of a mixed discipline team has obvious benefits when attempting to provide full, expert support to a customer call or presentation, particularly when your customers may field their own expert panel. Be prepared.

Team selling (Figure 3.8), involving staff from appropriate company functions as required, can provide a powerful and effective sales mechanism. Staff from R&D, engineering/technical support, finance, operations, quality and sales can and should be used in the team approach. Obviously this should be planned, with appropriate training and preparation for staff before they commence team selling. Role play is very useful and will familiarise backroom staff with what to expect on a sales visit.

Mix and match

The team selling approach should be matched to customer needs as they develop. Group-to-group sessions may break into specialist subgroups for detailed discussions. R&D or quality directors of each company may arrange to meet and co-ordinate future plans.

Sales professionals

The quality and sophistication of sales executives must rise to meet these new challenges, with higher qualified and trained staff becoming the norm, although fewer in number. The number of generalists will fall, with a corresponding rise in the number of key account managers to deal with the increased needs of the remaining customers.

Continuous training

Internal training must continue to include product and technical aspects, but expand to give much greater knowledge of

- customers
- market trends
- products
- interpersonal skills
- versatile communications
- new technology aids
- analysis of customer needs
- time management

Many companies already using similar approaches have seen sales effectiveness growing steadily, proving the worth of this form of investment.

Whilst some traditional sales tasks remain (tasks dependent on less skill or knowledge) they can be carried out by a more cost-effective resource,

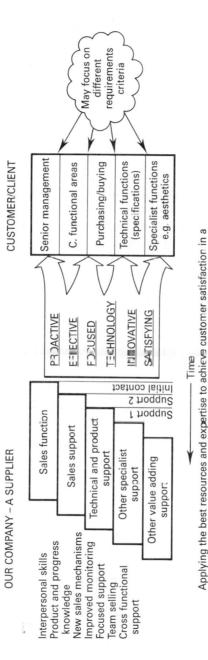

Figure 3.8 Phased team selling.

together with the use of new technology such as computer databases, mail merges, etc. Certain functions may also be better subcontracted outside the company, such as market research studies.

Forecasting

Most sales departments are required to provide regular forecasts of sales volume mix and value, to allow the execution side of the business to produce, or prepare to produce against these expectations. The nature and quality of these often essential forecasts can vary enormously, and frequently offer significant scope for improvement. *The only definite quality of a forecast is that it will be wrong, but if they are frequently reviewed the error will be reduced.*

A wide variety of powerful forecasting aids and computer tools (Figure 3.9) have been available for many years, offering greatly improved facilities and user-friendly operation. However, in the experience of the author, the adoption of these tools into the sales function remains low. Part of the reason for this is the heavily statistical nature of forecasting software, relatively new ground for many sales staff.

Through training and education, sales personnel can quickly understand and use such tools, leading to improved personal and company effectiveness. Once mastered, people will often discover other valuable uses for such tools. Alternatively, specialists can be trained or recruited into the marketing team to perform this function.

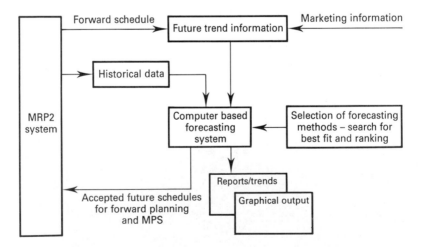

Figure 3.9 Forecasting tools: operation and dataflow.

Product development: R&D

There is a vitally important interface between sales and marketing and R&D. Development of new and existing products is essential to ensure continued company survival, but this development must be well guided to match and anticipate customer and market needs. This will not always happen naturally, and development can become internally driven, rather than market driven. Finding the correct balance for a particular industry is not easy, as it will reflect the personalities and resources involved.

If left to their own devices, R&D staff may produce what they want to build or what they believe the customer will want. Liaison with sales and marketing allows both groups to decide the content and order of the design task list by pooling all relevant knowledge and analysing the available options.

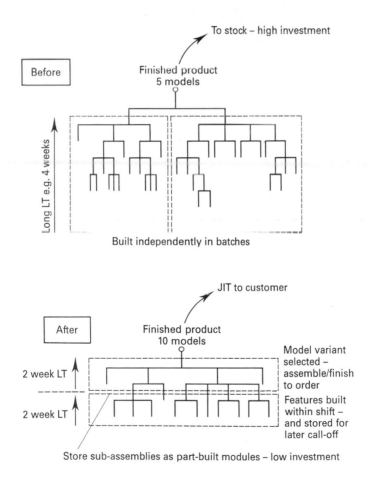

Figure 3.10 Product and process redesign for improved structure and shorter lead time.

Areas for improvement and strategic consideration include

- product range
- degree of product flexibility and modularity
- product structure and build stages

A wide product range may be desirable for the customers but, due to the lack of modularity of product structure, be expensive to manufacture and stock. Product variety is likely to be a key selling factor in many sectors, and R&D input to this objective will be worthwhile investment. Redesign of products can frequently result in greater product options, but with more effective and efficient building and stocking methods (Figure 3.10).

Product development strategy

As described above, the product development strategy will be developed in parallel with the business development strategy, because of the interrelationships between them. Fundamental points include

- Is product development driven by customer or supplier?
- How do our competitors deal with development?
- What identifiable trends exist for types of development requested?
- How do we calculate the cost of manufacture?
- Design for manufacture/simultaneous engineering
- Formal design review procedures and committee
- Continuous improvement programme in all areas
- CAD strategy and effective operation

Because of the importance of close liaison between marketing, business development and product development, in defining where the company could go, the next section discusses the interaction between these areas in some detail.

Market research

The source for most of the information needed by marketing and sales should be provided by market research. This can be either an internal or external function, but it must be done well and continuously. For some companies one full-time market researcher is ample; for others large internal resources are supplemented by one or more external agencies. It depends on the size and nature of the business. If you leave it as a part-time task for an inexperienced or unqualified person, then don't be surprised when your competition start taking market share away from you. If your market intelligence is poor, you may have no way of detecting the loss of share, other than reduced business; by then it's too late.

The use of modern PC spreadsheet packages can prove to be ideal in the analysis of market and share information. The example in Figure 3.11 shows the build-up of several sources of market data for a UK manufacturer, and

Shakey Ladder Manufacturing Ltd			Sales compared to various surveys 1992					
BUILT-IN LADDER SALES					Estimated	Actuals		
Total market estimates	1989	1990	1991	1992		1992	1993 est	1994 est
A competitors's published figures	16000	17000	18700	21000		19000	21000	21500
Chairman's figures	17500	18000	22000	22000		18500	22000	21500
DTI figures	8190	22000	20625	18000		18000	18000	19000
DTI trend		171.30%	-6.25%	-12.73%		-12.73%	0.00%	5.56%
Independent Association figures (IAF)	17000	18648	22984	20000		25000	27000	28000
IAF trends		9.69%	23.25%	-12.98%		8.77%	35.00%	12.00%
Average DTI, Independent Chairman	14203	19549	21870	20000		20500	22333	22667
Shakey figures	5312	4910	4064	3863		3863	3678	3678
Shakey internal trend		-7.57%	-17.23%	-4.95%		-4.95%	-4.79%	-4.79%
Shakey % share of IAF	31.25%	26.33%	17.68%	19.32%		15.45%	13.62%	13.14%
Shakey share trend vs IAF		-15.74%	-32.84%	9.24%		-12.61%	-29.47%	-14.99%
Shakey share vs average (built-in)	37.40%	25.12%	18.58%	19.32%		18.84%	16.47%	16.23%
Shakey share trend vs average		-32.85%	-26.01%	3.94%		1.40%	-14.74%	-13.89%
FREE STANDING LADDER UNITS					estimated	actuals		
	1989	1990	1991	1992		1992	1993	1994
Chairman's figures	12000	12900	14000	15000		15000	15000	16000
DTI figures	7838	19114	16673	16000		15500	16000	17000
DTI trend		143.86%	-12.77%	-4.04%		-7.04%	0.00%	9.68%
Independent Association figures (IAF)	15000	16302	21685	22811		22000	23000	24000
IAF trends		8.68%	33.02%	5.19%		1.45%	0.83%	9.09%
less custom units	9000	10902	14895	15283		15000	15000	15000
IAF trend less custom units		21.13%	36.63%	2.60%		0.70%	-1.85%	0.00%
Average DTI, IAF, Chairman	9613	14172	15189	15428		15167	15333	16000
Shakey figures	5544	3875	4029	3251		3251	3351	3351
Shakey trends internally		-30.10%	3.97%	-19.31%		-19.31%	3.08%	3.08%
Skakey % share of non-custom	61.60%	35.54%	27.05%	21.27%		21.67%	22.34%	22.34%
Shakey trend vs IAF		-42.30%	-23.90%	-21.36%		-19.87%	5.02%	3.08%
Shakey share vs average (free std.)	57.67%	27.34%	26.53%	21.07%		21.44%	21.85%	20.94%
Shakey share trend vs average		-52.59%	-2.99%	-20.56%		-19.19%	3.71%	-2.29%

Figure 3.11 Market share movements.

derives trend data on both overall market movement and own company market share trends.

Market research must also cover your competition: you are not just aiming to maintain or increase market share, but also to be in front of what your competitors are doing, or planning to do. Although obvious, it is surprising how many businesses are unaware of competitive movement. By the time they realise, it is likely to be too late. Market movement, where significant changes in quantity or type of supplied product is occurring, can render existing manufacturing or process facilities less effective or even inappropriate. This must therefore be part of any market research carried out, with findings fed back to the strategic team.

One London-based chief executive considered his secretary could provide all the market intelligence he needed. Mistakenly believing his company was the leader in the UK, and possessing no data to the contrary, he was eventually surprised by a subordinate's report showing his three main competitors had passed him by, growing their market share at his expense. The competitors had invested prudently in reorganisation and manufacturing equipment, improving quality, reducing cost and giving better standards of customer care.

Information sources

Once one realises its worth, information is often readily available from many sources. These sources can be many and varied but are rarely used to the full. Your customers and suppliers are most likely shared by the competition. They are a ready source of information, as are the trade press. Staff within your own company may gain information through informal relationships with competitors' staff. Use it. You can never have too much market information. Ensure you provide enough time and resources for acquisition and analysis.

Company profiles and accounts

In the UK, Companies House provides detailed reports on individual companies and groups of companies by market sector. These are the submissions required by law from each limited company in the UK and they contain a wealth of information on financial performance, investment, staffing, growth, etc. *Business Monitor* makes available reports from Companies House on most sectors in the UK; other publications cover Europe and the US, including Dunn & Bradstreet. The type of comparative data available can be used for trend analysis coupled with additional intelligence gathered on the competition.

Similarly, it is worthwhile defining the strengths and weaknesses of your own company in respect of your competitors and your customers. Strength, weakness. opportunity and threat (SWOT) charts are useful in laying out such

STRENGTHS	OPPORTUNITIES
Our products meet spec.	Increase model range
Good customer knowledge	Increase market share
	Imports
	To grow the refit programme!
	To look after the right people
	To improve customer knowledge
WEAKNESSES	**THREATS**
No direct access to factory	Cheap competition
Too many mark-ups	Contracting 'deals'
One product sitework	Limited products approved
intensive (redesign?)	
Contract manager??	
Lack of technical info and	
understanding	
'Them and us attitude'	
Need further tightening of	
contract into the Company	
Pricing structure	
Energy consumption	
Distance from decision makers	
Not looking after the 'right' people	

ACCOUNT A

STRENGTHS	OPPORTUNITIES
Good customer liaison	To improve cust. liaison
Regular customer visits to factory	To increase visits to customer
Customer input to product design	To increase such input
Full range if cases to customer spec.	
Good recent track record	For offering single source supply
WEAKNESSES	**THREATS**
Slow response	Cheap imports
Price sensitive	Price sensitive
Over reliance on previous business	Competition influence – increasing mkt share
	price cutting
	– close to main group
	– offers single source
	– easier to deal with

ACCOUNT B

Figure 3.12   SWOT charts for business versus two customer accounts.

data, as shown in Figure 3.12. They help to maximise areas of advantage and to minimise threats.

■ Manufacturing strategy

This is a particularly important segment of the overall business strategy, as the concepts and techniques employed within manufacturing can and should impact on all other functions in the company. The operations or manufacturing director and process/production managers will be responsible for this segment.

The overall aims are simplification, flexibility and quality, to ensure the lowest manufacturing cost base consistent with high quality of performance and service. Invariably the core elements and philosophies of JIT form the basis of the improvement objectives in manufacturing.

Process types

Virtually all businesses can be considered as suppliers or customers of services or products; the actual nature of the business being frequently immaterial. The market characteristics and volumes of product or service are the main factors in the choice or evolution of the *production process* – the way in which a business produces or supplies its goods. There are several distinct types of process, with a degree of overlap between the 'classic' types:

1. Continuous
2. Line
3. Batch
4. Jobbing and bespoke units
5. Project

Line process

Line processes produce or deliver a limited range of standard products to a short lead time, relative to the design/development time. Examples are cars, package holidays, fast food, furniture, pharmaceuticals and fridges.

Batch/hybrid

In batch/hybrid processes there will be a significant amount of finishing or assembling to order. Examples are commercial refrigeration, housing estates and (some) commercial vehicles.

Job shop/bespoke/project

Job shop products are custom one-offs. Examples are ships, films, racing cars, power stations, buildings, one-off holidays.

Process choice

The choice of manufacturing or business process is strategically important as it can expand or contract a business's capacity to cater for changes in market volume and/or flexibility needs. It is likely that the existing process facilities were installed to deal with market and product requirements that are now different in various ways, and may be less well suited to providing optimal response to current or future business needs. Hill's *Manufacturing Strategy* provides an easily understood discussion and explanation of the importance of process choice to manufacturing strategy, which the reader is encouraged to obtain.

Some products can be produced by all types of operation, for example houses can be mass produced, made in unique batches or architect designed as bespoke units. Process type is normally selected to suit the product and business sector, but it can have a major effect on its future operation and competitiveness.

The process choice must effectively support the market needs, as they relate to products and/or services. The process should be measured to see how well it supports order-winning criteria:

- How capable is the current process at dealing with the range of products requested?
- For a given product, is the current lead time longer/same/shorter than the best competitors?
- What options are there for changing the process to meet changed market needs?

Depending on where the expected markets are moving, processes may be too inflexible to respond to changing customer and market needs. The process-related infrastructure and skills may cease to be optimal. Process type has a major effect on most aspects of the business, so considerable attention is required during strategic planning.

1. Market trends vary
2. Order-winning criteria vary
3. Marketing strategy may change; product/process span and flexibility may need to change
4. Manufacturing strategy may need to adapt
5. Business support infrastructure may need to adapt, ideally being flexible enough to change easily
6. IT and systems support strategy must be flexible enough to cope with the above

Figure 3.7 illustrated the importance of process choice in overall business strategy, and the interactive effects involved. Figure 3.13(b) shows how variations inside and outside the business can affect the market area.

CASE STUDY 3.3

Focused plant problems

A batch production company with a single factory unit services the commercial air-conditioning market with a range of self-contained and remote AC systems.

These were built on dedicated lines in batches, both to order and to stock. There was a unionised workforce of skilled and semi-skilled employees, with limited flexibility between jobs. Market demand was for sale from stock of self-contained units, and delivery on a four-week lead time for made-to-order systems.

From market information, the market for integral units was expected to increase in volume over the next few years, with little change in product range. From this, a capital plan for £3 million was prepared and approved to create a second factory, providing two focused plants for producing the two product ranges. The plan was to provide each focused plant with its own facilities on the shop-floor, which involved dividing the press shop into two. All service functions were to remain centralised.

The capital was used for the construction of an additional building, new production machinery and the plant rearrangement. As part of the rearrangement, much of the floor space used for work in process (WIP) was eliminated, as the additional capacity of the plants was expected to remove the need for such buffer stock. The basic floor layout was production lines per product family, but with little flexibility between lines.

Unfortunately no investment was made in detailed analysis of the proposed plant, such as simulation. Neither was there any organisational review or analysis of operating procedures, nor investment in improved IT and MRP2 systems to run the focused plants. In short, there was a substantial change in manufacturing process and potential capability, without due consideration of the overall business needs, or the practical effects of these changes; an incomplete strategy was adopted, and failed.

The project took several months to complete, with a proportional loss of production capability during the changeover. Once up and running, it became apparent that operating problems were not confined to start-up bugs.

The loss of WIP space, and movement onto significantly smaller batch sizes of manufactured components, led to capacity restrictions on the assembly lines. No set-up reduction work had been done, with management assuming that JIT production would just happen. It didn't.

To compound the problems, the market for both product groups began to shift, both through changing customer requirements and general market recession. This gave a reduced demand for the self-contained units, meaning the lines were run at reduced capacity and small batches, to avoid excessive finished stock. Conversely, demand for the remote units grew, with greater requirements for product customisation, and for reduced lead times. The focused plant for remote units was soon running at full capacity, and not coping with demand. It could not pass or sub-contract work to the other plant because of the differences in equipment and skill base, only able to move a small proportion of the personnel from the integral plant to assist in the remote plant. WIP could be stored in the integral plant area, but effectively destroyed the flow of work through the plant.

Conclusions

- No detailed analysis of the proposed plant, such as simulation.
- No organisational review or analysis of operating procedures.
- No investment in improved IT and MRP2 systems to run the focused plants.

- Inadequate consideration of the overall business needs and possible market trends.
- Substantial change in manufacturing facilities and capabilities, without due consideration of the practical effects of these changes.
- Assembly line mentality, with little ownership for product, resulting in quality problems.
- JIT and focused plant approaches were only superficial.
- An incomplete strategy was adopted, and failed.

Remedies

- Market research to be more focused and receptive to shifting product and volume needs.
- Review and redesign lines to focus on flexibility, rather than pure volume of a limited product range.
- Analyse WIP needs to provide just enough desirable inventory (JEDI).
- Determine requirements, and invest in improved IT and MRP2 systems to run the plants.
- Train and organise workforce for flexibility and ease of movement between areas.
- Create cell-based units within the business module, to provide product accountability and improved quality.

Process- or product-based structures?

With few exceptions It Is certainly wrong to generalise about improvements to any type of manufacturing process. However, the JIT and manufacturing systems methodologies are universally applicable and should form the basis for the majority of improvements in the manufacturing area. Where high product volume is required, the production processes will tend to be more inflexible as they are product-centred, designed to meet the needs of a narrow range of products. For batch and hybrid businesses, many conventional manufacturing systems tend toward higher volumes, giving low flexibility; they are more process-centred than product-centred. Even in job shop or bespoke production, manufacturing facilities often tend to be grouped as process areas, not product centres.

Manufacturing systems engineering (MSE) methodology attacks this process-based approach for a wide variety of businesses, proposing instead the reorganisation of manufacturing (and support) resources into product-based units that have complete autonomy and responsibility for the products and/or services of that unit. This re-emphasises the need to fit the process to the product, not the reverse. Also, flexibility must be built into the product-based units to allow for product and market movement over time.

Focused plant concepts are now commonplace but can often be too inflexible in the face of such changes in market and product trends. The case above shows an extreme example of a focused plant failure. MSE advocates smaller business units to deal with products or product families, with flexible

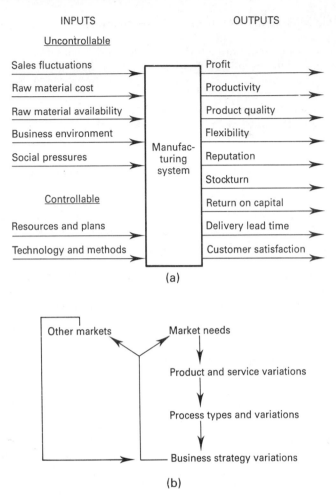

Figure 3.13 Manufacturing system: (a) inputs and outputs;
(b) relationship between various elements of the system.

resources and procedures to deal with such variations. *In many situations, careful consideration should be given to the creation of business units, business modules and product-based cells as an alternative to existing inflexible facilities and resources.* Chapters 6 and 7 discuss these concepts in greater detail.

Developing a new manufacturing strategy

■ Define a manufacturing strategy, approach and process to provide the maximum competitive advantage to the business (see Hill, *Manufacturing Strategy*).

■ Investigate whether a change of process or approach is desirable. If so, are the cellular/business unit concepts appropriate?

If yes:

- Design each operational cell as a product-based unit for simplicity, to operate as a small business unit, often as part of a larger module.
- Simplify and restructure shop-floor systems and introduce new flexible job functions, within clear product units/cellular structures.
- Use disciplined procedures and practices to ensure effective teams – a total quality approach.
- Provide all such units with aims and measurements related to achieving business ratios and customer service.
- Offer clear leadership and hands-on management.
- Devolve service functions, such as quality, maintenance and local planning/scheduling, down to direct operating staff.
- Provide staff with support via training, computer tools and simple control techniques such as kanban.
- Develop continuous improvement teams in *all* areas, with the aim of improving measured performance indicators. Review targets regularly.
- Aim for very short changeovers on machinery and processes by effective production engineering.

Detailed improvement to both manufacturing and other support areas is described in Chapters 5 to 8 on the design and development of cellular units. Internal reorganisation and improvement is a prerequisite to improving the external links to customers and suppliers. Only when the **internal** capacities and performances are established can we accurately and honestly consider those external to our business. This is shown in Figure 3.14.

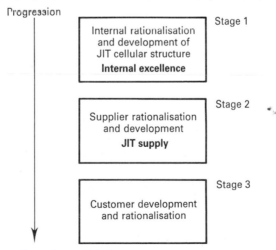

Figure 3.14 Strategic applications of JIT to internal and external functions.

■ Purchasing strategy

Make or buy: strategic decisions

We examine business targets and performance indicators and consider the potential of manufacturing systems and supplier/subcontract base. Our aim is to compare the two options to discover which is best for the business, in terms of cost, response and quality. What may have been sensible to buy in before improving the facilities may now be advantageous to make in-house, and vice versa. The in-house capabilities after cellular improvements will create a simplified, modular structure that has more effective processes and machines, known capacities and performances, and flexible jobs and organisation. Determine guideline internal costs and capacities, then make or buy according to response, cost and capacity.

Supplier integration

The cells and modules developed in the manufacturing and office areas of a business cannot exist in isolation from the outside world; they must interface with customers and suppliers of information and materials. To reach optimal integration with external agents, we must undertake a series of rationalisation and development stages.

Stage 1: internal rationalisation

Stage 1 begins internally, based on the company's manufacturing and functional strategies, with the development and rationalisation of the facilities into a modular and cellular structure that provides effective performance and response to the market. The contents of this chapter have described the steps and methods that can be used to achieve this first stage; Chapters 5 to 7 describe the detailed application of this internal improvement in all areas.

If we do not work systematically through stages, much of the effort that is applied will be diluted and wasted, due to the inevitable duplication and repetition of development work that will be done on relationships with customers and suppliers before the internal needs and abilities are fully known. Only when we have established comprehensive customer/supplier relationships between cells and modules should we progress to the next stage, supplier integration. Only then can JIT deliveries be properly effective, without excess storage and material handling.

Stage 2: supplier integration

Stage 2 develops relationships with suppliers so they understand and support the total business operation requirements. This is often in the form of a partnership between the enterprise and its suppliers, aligning and collaborating on many aspects of their shared business:

- product development
- quality and reliability
- cost and functional requirements
- delivery and response requirements

In many cases the partnership extends to cover a wide range of topics, from linked information and MRP systems via electronic data interchange (EDI) through to shared product development resources and knowledge. Transfer of JIT and systems engineering skills from customer to supplier is now becoming a common improvement and training path, and specific objectives need to be recognised:

1. Reduce supplier base to less than four but more than one per item
2. Improve supplier performance to at least meet your own requirements of quality and delivery

Reducing the supplier base is required to build up the above relationships and high quality levels with a core group, all of whom should be capable of becoming and remaining key JIT suppliers. Guidelines here include reducing to less than four, but more than one, supplier per component. Many JIT experts will advocate single sourcing, one supplier per component or service. In the author's experience, this can lead to overdependence on a sole supplier, making the enterprise vulnerable to supply failures or price increases. It may be better to create a small number of capable JIT suppliers, each of whom takes a large portion of the component business. For a total parts contract of £350k, two suppliers could be selected, one taking £200k (57%) and the other £150k (43%) of the total spend. Make sure this does not exceed a specific percentage of the supplier's turnover. This will limit your exposure in case of any problems. Also, each supplier should be capable of taking on the complete order quantity in the event of problems with the other supplier.

Improving supplier performance aims to achieve zero defects and zero lateness. This means that goods-in inspection can often be eliminated or drastically reduced; it is carried out by the supplier before shipment. One hundred percent on-time delivery allows minimum possible levels of inventory, with complete confidence.

Stage 3: rationalisation and development of customers

This stage is commenced once all the stages 1 and 2 activities are working and performing on a Just-in-Time, Total Quality basis. In stage 3 customers are introduced to the new facilities set up in stages 1 and 2. Customers are encouraged to become part of the JIT process by the joint creation of an optimal interface:

- mutually understood and agreed terms, definitions and capabilities
- specified procedures, deadlines and timescales
- specified products, variations and change procedures
- specified payment terms and conditions
- agreed performance measures

As with suppliers, customers should normally collaborate on product development and forward planning for market trends, through shared development costs, pooled ideas and a general two-way flow of non-sales-specific information between the two enterprises. Shared product scheduling, flexible and bulk ordering with planned call-offs, are all potential benefits from growing this partnership.

Customers will require education and training in the philosophy and methods of JIT and total quality, if they are not already going down this path. Like our own business, customers will be moving towards a smaller group of top quality suppliers, and we must ensure that our company is the best of these suppliers, and remains so. Only this will ensure our continued survival. According to our earlier guidelines, companies will reduce their supplier base to less than four per item. *If you aren't in the top three, you won't be there at all!*

Other customers must be expected to follow suit, and, in the majority of business sectors, will more than likely wish to use the same small top group of suppliers. Therefore, membership of an elite group becomes essential, not just desirable. Any business not at this level of competitiveness and integration will quickly fall by the wayside.

■ Distribution strategy

Both service and manufacturing business may require comprehensive and complex distribution systems to deliver product and/or services to a customer, on time and at low internal cost. The planning and management of distribution is virtually a complete business within itself, but must also be linked to business strategy for optimal overall performance. It is worthless developing the most effective production systems in the industry if their products cannot be distributed to the correct place, at the correct time.

Distribution systems can require a significant proportion of the operating costs of any business, typically representing up to 10% of turnover in a capital products sector. Over the last five years there has been a trend towards subcontracting of transport activities to third-party specialists. There has also been a massive growth in the number and quality of express delivery and distribution providers. What this provides is a wide range of choice for our distribution needs, allowing selection of different agents for specific price and delivery situations.

Over the last few years, warehousing and stock storage requirements have generally reduced under the effects of JIT and other inventory procedures. Some sectors continue to require high levels of finished stock for immediate call-off and delivery, but in general the improvements in production response and lead time will place increasing demands on distribution systems, which must therefore improve at least as quickly.

All the underlying approaches for improvement and elimination of waste apply equally to distribution systems:

- Follow a total quality approach.
- Eliminate waste.
- Simplify and restructure systems and procedures.
- Introduce new flexible job functions within clear unitary/cellular structures.
- Use disciplined procedures and practices to ensure effective teams.
- Provide all units with aims and measurements related to achieving business ratios and customer service.
- Offer clear leadership and hands-on management.
- Provide staff with support via training, computer tools and simple control techniques.
- Develop continuous improvement teams in *all* areas, with the aim of improving measured performance indicators. Review targets regularly.
- Eliminate *all* NVAs by simplification of functions and removal of non-essential overheads.

However, distribution frequently involves working with outside agents and suppliers of services, requiring the full adoption of close supplier relationships, as described in the previous sections. Working with the supplier, the aim is to achieve zero defects and zero lateness.

■ Business systems and IT strategy

Using the same basic principles as the manufacturing systems approach, business systems and IT strategy encompasses organisational structure, support functions, business and computer systems. By systems, I refer to manual procedures, paperwork and computer-based systems. Again, the aims are simplification, flexibility and quality, to ensure the lowest manufacturing cost base consistent with high quality of performance and service.

Parts of this will be covered by an information systems/technology strategy, but this can be narrow in its outlook. The term preferred by the author is a business systems approach, encompassing all aspects of information flow. Information technology must only be thought of as a tool to help run the business, and must service the company's requirements. Key elements are

- Create an environment that encourages team approaches to problem solving and removes blame.
- Only apply new technology where it can measurably improve performance indicators.
- Install core system skills with updated training procedures and communications around the company.
- Define in detail the mix of people and control systems necessary to meet the needs of the market and product strategies.
- Regular six-monthly reviews of strategy, performance measures and market/product trends.

A procedure or method of working in an office or service function is essentially no different than a manual task in a production environment. Both involve people in carrying out an activity to process input information and/or physical items, producing output of a physical or informative nature. They can both therefore be effective or ineffective, and can always be improved. This approach to developing a new business systems strategy for company systems includes several elements common to the previous operational strategies (as listed on the previous page)

- Follow a total quality approach.
- Eliminate waste.
- Simplify and restructure systems and procedures.
- Introduce new flexible job functions within clear unitary/cellular structures.
- Use disciplined procedures and practices to ensure effective teams.
- Provide all such units with aims and measurements related to achieving business ratios and customer service.
- Offer clear leadership and hands-on management.
- Provide staff with support via training, computer tools and simple control techniques.
- Develop continuous improvement teams in *all* areas, with the aim of improving measured performance indicators. Review targets regularly.
- Eliminate *all* NVAs by simplification of functions and removal of non-essential overheads.

plus

- Design each functional group as a product unit for simplicity and to operate as a small business.
- Devolve service functions, such as quality assurance of clerical jobs, down to line staff.

Information technology (IT) strategy

The IT strategy should then be developed to provide for the requirements of the business systems, not as a separate function or activity. It can be worthwhile to place the manufacturing information systems (MIS) function fully within the operational structure of the business. The objective of this is a closer alignment and improved focus of computer MIS services to the actual needs of the business. When retained as a separate function, MIS will naturally tend to have different priorities and perceptions, caused by distanced non-involvement with the day to day running of the operation.

Integrated business and manufacturing

Computer-based information systems form the life-blood of 99% of businesses, providing information, planning and control systems for every function, across all business sectors. Whilst the 1980s concepts of full computer integrated

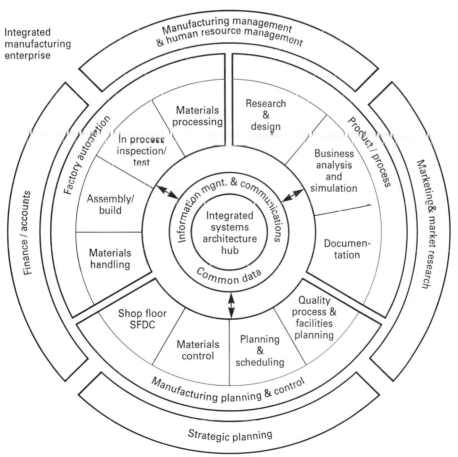

Figure 3.15 Business integration wheel.

manufacturing (CIM) and lights-out operations have been successfully applied in a few businesses, they are not ideal or cost-effective for the majority.

However, a high degree of business integration *is* both desirable and cost-effective, with common data and information management providing the hub of business operations. This is illustrated by the Integration wheel in Figure 3.15, showing the need for an integrated systems architecture at the core, allowing data sharing between all functions.

From mainframe to microcomputer: downsizing

Information systems and data processing are in the middle of a period of extreme change, in the wake of a world-wide move to downsizing of computer systems, moving away from minicomputers to smaller, faster and lower cost microcomputer and network systems. With the global move to open systems – hardware and software that conforms to agreed international standards – there has been a growing trend of updating old, costly systems, with replacements that are portable between various hardware platforms. This reduces dependency on a single computer supplier, offering greatly improved competition and service. It can also involve substantial change in the MIS function, with different hardware and software expertise being required.

As a result, many firms have been able to reduce their MIS resources substantially, as the effort and manning required for servicing the incoming hardware and software is much less. Greater standardisation makes it more viable for data processing to be contracted out. Information systems and computer hardware are discussed in detail in Chapter 11.

CASE STUDY 3.4

Strategy and actions for a UK £26 million turnover manufacturer in 1990–92

Following a company-wide review of procedures, systems and ability to compete in the European market, management developed a new business strategy for the next five years.

Overall aim
To reclaim and grow the number one position for market share in the UK and European market, by committing to company-wide improvement and the achievement of a total quality business.

Major objectives in pursuit of this aim
> Supply quality products fully complete and on time
> Reduce product and service costs to below competitor levels

Achieve these objectives by
■ Introduction of new formalised operational procedures across the company (resulting from the review)

- Introduction of effective, measurable use of material requirements planning (MRP1) and manufacturing resource planning (MRP2)
- Establish personnel awareness and training programmes
- Measure company performance against a set of defined cross-business measurements and targets, and achieve class A performance within three years.

Major objective 1: supply quality products fully complete and on time

Strategy
To achieve this we have to ensure that all functions are communicating properly and formally, using valid and accurate data to derive achievable plans and schedules for all stages of procurement, manufacture and delivery. This relies heavily on the previous strategies to create the procedures and systems necessary.

Continuous improvement will be sought in all areas, commencing with full, unambiguous order definition, allowing rapid investigation of materials and capacity constraints. This will produce agreed and achievable build and ship dates, which will be held firm. Changes to specification or delivery date will be by negotiation only, with amendments to lead time and cost to the customer enforced as standard practice.

Improved planning and scheduling will be achieved through the use of improved policies, procedures and systems. This strategy will ensure that production is provided with the right schedules and materials at the correct time, with a substantial reduction in the amount of rescheduling and redundant material that currently occurs.

A detailed review and work study of production is now under way. This will identify areas for improvement to the processing and flow of work through the factory.

Final goods inspection will identify and remedy all shortage and quality problems prior to a case being passed to the warehouse. This will act as an interim catcher for quality problems until the improvements and quality circle benefits eliminate product nonconformances.

Major objective 2: introduce new formalised operational procedures across the company.

Strategy
Establishment of improved and formal procedures will be initiated by providing education and awareness training on total quality to all personnel. Procedures will be developed by small groups of personnel involved, assisted by in-house experts. This will be commenced in areas identified as critical, then expand across the plant. When agreed, each procedure will be signed off by those who will use it, and entered in the procedures manual.

Major objective 3: introduction of material requirements planning (MRP1) and manufacturing resource planning (MRP2)

Strategy
To create a solid foundation for the successful introduction of MRP by improving

and formalising our business policies and procedures, and through the provision of appropriate training and support from in-house and external experts. Selection of the system will be carried out against a formal definition of requirements document. Outline and detailed implementation plans will be maintained.

The project will be managed by a steering committee comprising the managing director, functional directors, financial controller and the project manager. This group will meet fortnightly.

The full-time project manager will lead the project team, consisting of full- and part-time members as required. Each member will lead a local user group, which will identify the requirements within MRP in terms of functionality and training required.

The implementation will be phased, providing the core modules of MRP initially, followed by additional functions and areas as shown on the plan. Prototype and Pilot studies will be carried out, to prove the new systems alongside those existing. Following the successful completion of these studies, we will go live on the new systems. Reviews will be held regularly.

Major objective 4: establish personnel awareness and training programmes

Strategy
All directors and managers will identify the training needs of their staff, both for specific and general topics. With the director of training, they will develop appropriate training programmes, using in-house, external and programmed learning assistance.

Detailed actions from the review: planning division

Improvement programme
Inventory control policies review
Formal policies and procedures
Order change control – delivery dates and specifications
Full order specification
MRP1
MRP2
Purchasing work study services
Costing system development
Capacity planning

Action plan: supply quality products complete and on time, responsibility of managing director

Decription of task	Date to start and complete	Person responsible
Provide final goods inspection	Oct 90	Quality director Production director
Provide goods-in inspection	Nov 90	Quality director and planning director

Continued on page 87

Continued from page 86

Decription of task	Date to start and complete	Person responsible
Carry out operator training on self-inspection	March 91	Training manager and production director
Review and improve inventory policies	March 91	Planning director
Review and improvements in purchasing policies and procedures	March 91	Purchasing director
Introduce formal vendor appraisal and quality assurance	May 91	Purchasing director
Provide improved capacity planning and scheduling methods and tools	Oct 90 to May 91	Planning director and project team

Action plan: introduce new formalised operational procedures across the company, responsibility of managing director

Decription of task	Date to start and complete	Person responsible
Formalise and adopt order acceptance and change control procedure	Sept 90	Engineering director Planning director
Development of new policies and procedures in all functions	Dec 90	All directors
Set up continuous improvement teams (CIT)	Oct 90	All directors
Complete work study in specific factory areas	Nov 90	Production director
Development of job costing system	Dec 90	Project team, finance director and MIS director
Provide in-house training for CITs	Nov 90	Quality director Selected staff as appropriate

Action plan: introduction of MRP, responsibility of planning director

Decription of task	Date to start and complete	Person responsible	Dependency
Establish project steering group and project team membership	Sept 90 to Oct 90	Managing director planning director	
Provide computer hardware and software for MRP1 and MRP2	Oct 90 to June 91	MIS director and project team	
Training and education for MRP personnel	Oct 90 to July 91	Project team and training manager	
Implementation of material requirements planning (MRP1)	First quarter of 1991	Project team Planning director	New policies and procedures
Implementation of manufacturing resource planning (MRP2)	Fourth quarter of 1991	Project team and all directors	Achieving MRP1

Action plan: establish personnel awareness and training programmes, responsibility of training manager

Decription of task	Date to start and complete	Person responsible
Define training needs of all personnel and develop programmes	Sept 90 to Nov 91	All managers
Product training for order entry/sales staff	Oct 90	Sales director Training manager
Provide in-house training for CITs	Nov 90 to June 91	Quality director All directors
Training and education for MRP personnel	Oct 90 to July 91	Project team and planning director
Provide education and training on total quality to all personnel	Sept 90 to Aug 91	Quality director

Specific actions

From the above action plans and objectives, a more detailed set of actions are created by each function. In turn, each of these actions requires a full description or procedure to carry out the task, a mechanism. For example, actions in planning division with target completion dates to be agreed:

Review and improve inventory management policies

1. Use sales forecasting as a guiding factor in inventory investment planning
2. Use ABC analysis for inventory classification and control
3. Define safety stock, order quantities and reorder control related to ABC classification
4. Establish monitoring and cycle counting by classification
5. Establish inventory value and turnover targets based on type, product line and ABC code
6. Improve accuracy of inventory transactions and records in relation to ABC classification
7. Provide training to reduce keying errors
8. Establish measurements and frequency of measurements for the above
9. Identify and carry out training and education required by staff
10. Specific MRP related training

Review and improve purchasing

1. Introduce formal vendor appraisal and quality assurance
2. Define relative importance of price, delivery and quality
3. Provide forecasts of demand to vendors and call-off on a single purchase order
4. Identify alternative suppliers for key parts, also reduce overall number of suppliers
5. Identify potential restrictions by design/engineering to specified components
6. Carry out a comprehensive make-or-buy analysis
7. Hold annualised price increases on top fifty suppliers to no more than 6% during 1991
8. Identify annualised material cost savings of £250,000 during 1991

Warehousing and distribution

1. Identify and dispose of all finished stock built prior to January 1990
2. Control transport costs by reducing shipping costs per case
3. Reduce packaging costs

Production and material control

1. Form the medium-term planning group to provide master scheduling and validation of proposed orders
2. Put in place a standard document for order confirmation
3. Include refit line work in plans and schedule loads
4. Bring policies and procedures up to speed for MRP (see Chapter 13)
5. Establish measurements of performance, e.g. number of orders/number of changes (Figure 13.7)
   - number of changes/order/build per week
   - number of rescheduling activities
   - floor shortages
6. Investigate scheduling mill and paint shop and/or use of simplified flow control kanban

■ Summary

For attainment of world class performance levels, strategy development within the business must be founded on the basic building blocks of WCM, TQM and JIT philosophies. These are applicable beyond the manufacturing function, indeed they are essential improvement mechanisms for *all* business areas. Strategic planning is not difficult but it can be a worrying prospect to businesses that have only considered local aspects of their development. Following the straightforward steps outlined in this chapter will allow most businesses to take the important step into strategic planning. For greater depth of treatment, the further reading provides breadth and specific guidance on strategic planning and management.

■ Further reading

Business Monitor (1992) Central Statistical Publications.

Glueck, W. F. and Jauch, L. (1988) *Business Policy and Strategic Management*, McGraw-Hill.

Gunn, T. (1987) *Manufacturing for Competitive Advantage*, Ballinger.

Hill, T. (1992) *Manufacturing Strategy*, McMillan, particularly chapters 3, 4 and 5.

Krajewski, L. J. and Ritzman, L. P. (1996) *Operations Management: Strategy and Analysis*, Addison-Wesley.

Lucas Engineering & Systems (1989) *Manufacturing Systems Engineers Handbook*, The Lucas Miniguides.

Parnaby, J. (1986) 'The design of competitive manufacturing systems', Int. Journal Tech. Management, 1.

Parnaby, J. (1988) 'A systems approach to the implementation of JIT in Lucas Industries', Int. Journal Prod. Res., 26.

Schonburger, R. J. (1982) *Japanese Manufacturing Techniques*, Free Press.

Schonburger, R. J. (1990) *Building a Chain of Customers*, Hutchinson Business Books.

Review of the Company

4

To measure is to know
Lord Kelvin

■ Introduction

Many companies have approached problems and improvement areas in a local or piecemeal fashion, without fully considering the overall operation and performance of the business. In many instances a local improvement can have a negative knock-on effect on other functions, which should have been anticipated at the time of analysis.

In order to ensure that we fully understand the scope and relative importance of each area of potential improvement, it is much more worthwhile to commence with a structured review programme of all business and production areas, looking for potential improvements and opportunities in every area. To remove the common type of problems described in Chapter 1, the company and its systems should be reviewed and redesigned to simplify all procedures, operations and systems.

In many cases there are substantial benefits to the company that can be achieved without the high capital spend involved with advanced technology applications. So, before embarking on improvements or projects that may be of limited overall benefit, take the opportunity to conduct such a review: it will certainly pay off in both the short term and the long term.

■ Review what?

What exactly are we looking for in a review? We are seeking to identify not only problems, inefficiencies and areas for improvement, but to identify their root causes. For example, a company may intend to install plant to increase the capacity of a certain function. But is it a genuine bottleneck? Perhaps, or it could be an apparent bottleneck, highlighting problems that lie further back.

visible effect: overloaded machine
    caused by unbalanced workload
        caused by poor/unbalanced scheduling
            caused by erratic orders/lack of training

In many cases the effects can be traced to more basic causes, which should then be addressed. It would be wrong (but not uncommon) to merely invest in an additional plant or extra staff. *A good management consultant is reputed to ask why at least five times in such situations.*

Q1: Why do you want to install an additional machine here?
A: Because we cannot cope with the workload with the existing ones.
Q2: Why is that so? Have you currently got a heavier schedule?
A: No, but the scheduled workload is very erratic and changes frequently so we can't get on top of it.
Q3: Have the schedulers always given you this form of erratic schedule?
A: No, they used to give us level loadings in advance that were much easier to plan for. But now the sales mix is different and level schedules are impossible.
Q4: Have you discussed your problems with the schedulers?
A: Yes, but the new scheduling manager still maintains that sales won't sufficiently define our forward build to provide a better schedule.
Q5: What about flexible staffing on this section? And have you considered the batching alternatives for orders and for stock?
A5 (No answer)
Q6 Set-up times are high on this machine. Have you carried out any set-up reduction programmes?
A6 (No answer)

Here we now have cause to investigate several other areas:

1. Scheduling methods
2. Sales orders and forecasting
3. Management of the existing plant to improve throughput, e.g. set-up reduction, operator training

Root causes will probably lie in areas 1 and 2, but item 3 should also be addressed. Cause and effect analysis is a very useful tool for identifying problems and is further described later in the book.

■ Areas for review

When aiming for a total quality operation, no area or function should escape review but some areas may require more urgent attention than others. In all cases, this should take the form of a systematic analysis of current operating procedures and practices, which would typically involve the following:

- company business objectives
- the manufacturing strategy
- current and future production requirements
- production routes and methods, including process planning
- set-up methods and production batch sizes
- job design for cell autonomy
- support services and systems
- scheduling and control systems
- full costs and benefits

■ Basic steps in a review

It may appear a daunting task to even consider carrying out a company-wide review, but providing the correct approach is adopted, it will break down into manageable sections, each preparing for and leading onto the next. *Simplification is the key*.

JIT and systems engineering use this simplification philosophy to address all aspects of a company from strategic and organisational issues right down to individual tasks and procedures. This can be considered in several stages:

1. Examine
2. Analyse
3. Define improvements required
4. Simplify
5. Eliminate waste and NVAs
6. Simplify what is left
7. Use appropriate tools and systems only where necessary
8. Continuously monitor and improve

Examination and analysis form the basis of a review, normally including a cost and benefits study to define in detail how proposed improvements will assist in practice, both from operational and financial perspectives. Case studies are given later in this chapter and in Appendix B.

Throughout this approach, we should use the aims and practices from other well-established philosophies and methodologies as desirable objectives and benchmarks to compare our business with other similar and appropriate organisations. These include TQM and JIT, underlying good management practice, and the effective use of information and automation technologies, including manufacturing resource planning (MRP2) and executive information systems (EIS). Each of these topics are discussed in the text.

■ Planning and carrying out a review

Several aspects must first be defined:

- What are the background, objectives and scope of the review?
- What outcome or deliverables are wanted and in what form?
- How is the review to be carried out, in what timescale and by whom?
- Who is responsible for the review?
- What resources will be used?
- What methods and techniques will be used?
- What training and skills are required?
- What will happen after the review?

These and other questions should be answered by a formal brief or document prepared by the senior management of the company, to be used by the internal or external personnel tasked with conducting the review. From the previous discussions on world class performance and pressures on every company to improve, the review is ideally triggered by a top-level commitment to achieving a total quality, world class business. From this, the objectives and scope of the review become apparent. We are setting out to turn the company around, and require to identify all areas of improvement.

■ Implementing change following the review

This chapter describes how to conduct a comprehensive review of a business and the type of questions to be asked. Following a review, we define what changes are necessary across the company. Also, the relative importance and payback of each improvement must be defined and understood, together with the preferred sequence of addressing each area. If improvements are required in sales because there are knock-on effects in manufacturing, the sales area should be tackled first.

■ Resources for the review

The comprehensive review of a business is a full-time engagement for one or more competent analysts. Part-time working rarely provides sufficient continuity and takes much longer to gather as much information. Review analysts need

- interpersonal skills
- analytical skills data and activity flow analysis
- experience of overall general business operation and management
- knowledge of best practice business procedures

Where specific skills are weak or absent, suitable training may provide acceptable levels of competence.

In-house resources

In some cases there may be company staff who have the required skills and experience to carry out such a review, particularly where a large organisation is involved. Otherwise outside experts – consultants – should be sought. Remember that detailed reviews can often expose weakness and/or inefficiency, potentially a source of conflict. People need to be treated sensitively, so consultants are often preferable (Figure 4.1).

Consultants

Consultants offer an outside view of the business, free from internal politics. This can be crucial to a review's success.

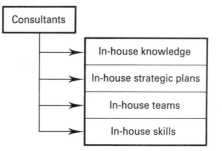

Figure 4.1   Effective use of consultants.

■ Guidelines for using consultants

Although relatively expensive, consultants can provide significant analysis, provided the appropriate people are identified and effectively used. Most consultants operate at the levels of middle or senior management, where major decisions are taken. But a growing number look at lower levels: team leaders or members of an internal systems engineering team.

Careful selection of consultants helps to avoid disappointment. Perhaps they have encouraged their image as gurus, but they are no more than suppliers of services to you, the customer. Specify a wide or narrow brief to suit your requirements. The consultant will act accordingly.

If the brief is basically 'we need help', then you can expect a good consultant to conduct a review which analyses the company at various levels, assisting in determining where consultancy assistance is required. *Success with consultants largely depends on the client.*

Consultancy support is normally initiated by senior management, who have to be realistic and honest about their aims and objectives. These aims can be very political, where an external opinion is thought necessary to sway a decision or justify a major change within the client company. Whilst this remains a necessary form of consultancy, the most effective use of their resource is arguably in working with the client in developing business improvements that will then be implemented within the business.

A good consultant won't make recommendations a client won't want to implement (Geoffrey Kitt, Touche Ross Consultants).

Many clients do not follow all the consultant's recommendations and fail to experience the benefits they hoped for. Incomplete adherence may be because some parts are too expensive or seem inappropriate, but it usually signifies a poor relationship between client and consultant and maybe an indequate briefing. Although there are situations where recommendations will be painful or disappointing to the client, it is up to the consultant to prepare the ground for such potentially difficult situations. He or she would be expected to build up the scenario by regular reviews with the client, preventing a big surprise at the final presentation. *Both client and consultant have the responsibility of completing the job, and of delivering satisfactory solutions.*

Making the best use of consultants therefore depends on several factors:

- Understanding how to use consultants effectively
- Developing an appropriate and measurable brief for the consultants
- Selecting suitable consultant(s) for the job
- Participating in a relationship with the consultants
- Implementing the consultants' recommendations

Obtaining consultancy assistance

Getting good value from a consultancy requires careful consideration. Some of them may not offer the approach, style and levels of understanding you require. Consultancies vary widely in their approach and in the type of people they employ. Similar to accountancy firms, many of the larger consultancies have a high proportion of younger staff with good qualifications but maybe not much experience. Experience and guidance comes from a small number of managing consultants, who may have held senior positions in business. Through their size, larger consultancies can normally put forward specific experts in the desired subject area, although they may have little operational experience. Smaller consultancies tend to have a more mature portfolio of staff, with a mixture of experienced and well-qualified consultants who may be both generalists and specific subject experts. Choose according to the size and nature of your requirement.

Guidelines for effective use of consultants

- Fully specify the assignment brief, timescale and deliverables
- Avoid 'rookie' consultants; demand specific details of previous experience
- Ensure the knowledge and experience is relevant
- Interview proposed consultant(s) to ensure a good fit of approach and personality
- Arrange period and attendance pattern
- Monitor progress with regular reviews

Team approach

The best of both worlds can be obtained by creating a team of internal and external experts. This can combine internal company knowledge with external experience and consultancy skills, resulting in a higher standard of review overall. In this type of role, a consultant will work alongside the team members on detailed analysis and improvements, offering direction and assistance in all aspects of business review. The consultant can also provide the unbiased appraisal that may be difficult or politically sensitive for the internal team to put forward. Finally, the consultant can liaise with the senior management on progress, internal problems and solutions that may fall outside the remit of the improvement team.

Basic review skills: systems analysis techniques

Definition

System: 1: a group or combination of interrelated or interactive elements forming a collective entity. 2: a method or complex of methods 3: orderliness.

The term *system* can also be used to describe a business, a wrist-watch, a space shuttle, a production line, a method used to process a sales order, etc.

In the context of business operations, systems are used in every part of the company, both manual and computer-based, to obtain, process and deliver information and/or physical items.

Office and factory systems are often fairly complex, large and difficult to describe in words. This is also true for many other systems, including computer software design. Systems analysis and its tools were developed to assist in the orderly description of these complex systems and are widely applicable.

The underlying philosophy of structured systems analysis is to break down large, complex problems into smaller, more easily understood elements. This allows us to deal with any size and type of system, describing it as an orderly combination of elements. Graphical tools are used extensively to reveal the system's workings at a glance. Repeated decomposition achieves the desired simplicity. And the simple elements are what we try to improve.

Decompose (break down) complex systems and information into smaller blocks.

There are several common techniques used for systems analysis, most of them using diagrams. We will look at

- flow charts
- input/output diagrams
- structured analysis and design techniques (SADT)

In practice, a combination of graphical techniques is frequently required to understand the system under review.

■ Graphical analysis techniques

There are several significant advantages to graphical charts:

- they show interrelationships
- they allow easy movement from top-level descriptions down to detailed levels in a well-mapped, referenced fashion
- most people find them easy to use

Flow charting

Well known among graphical methods, flow charting is best suited to detailing operations and steps in a sequence. It uses a standard set of symbols easily accessible to system reviewers and software designers.

In a review, flow charts are useful for identifying process steps in most areas of a business, offering insight into potential NVAs, poor work or information flow and weaknesses in the operation sequence.

Flow charts are not so good for large systems or when several systems interact. Their step-by-step nature makes them cumbersome to use and difficult to update. In general, they are good for overviews or detailed analyses, not for top-to-bottom descriptions.

There are several types of flow chart, designed for different applications. The most common is used for information flow processing (Figure 4.2) as used by systems analysts.

Another form of flow chart is designed for representing physical actions and movements, rather than just information. The standard symbols are shown in Figure 4.3 and an example in Figure 4.4.

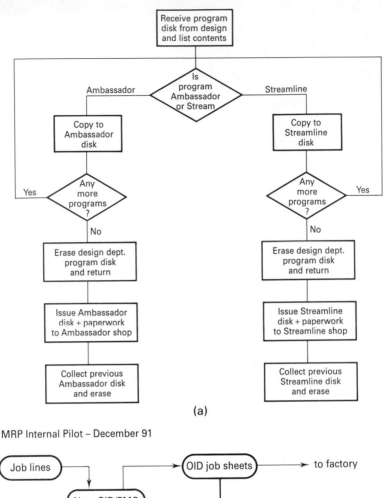

(a)

MRP Internal Pilot – December 91

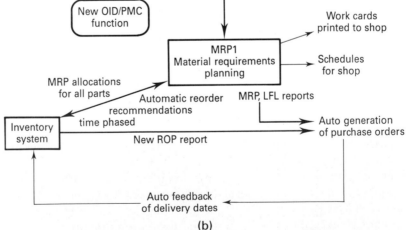

(b)

Figure 4.2 Flow chart examples: (a) file update procedure for product lines A and S; (b) company-specific MRP operations.

Symbol	Meaning	Example
 ◯	Operation	Drill hole Fill out form Design a part
 ⇨	Movement/ transportation	Deliver a document from A to B
 ◇	Decision point	Send to person A or person B
 ▭	Inspection	Forms audited Buy-off
 D	Delay	For signature From supplier From stores
 ▽	Storage	Filed documents To/from stores
 →	Direction of flow	Document Product Process output
 ⟿	Transmission	Data transmission
◯	Connector	To continue flow to next line or page

Figure 4.3 Flow chart symbols for physical processes.

IDEF analysis of information flow and activity

First developed and used by the US Air Force to describe their large and complex operational systems, IDEF (Information DEFinition) contains several levels of diagram to deal with top-to-bottom coverage of system operation and information flow (Figures 4.5 to 4.7). At the top level, IDEF 0, the diagrams are used to describe high level company operations. This shows a basic business operating with three main activities:

1. Commercial activities
2. Planning and control
3. Implementing plan

All blocks describe activities, rather than titles or functions. The blocks are

99

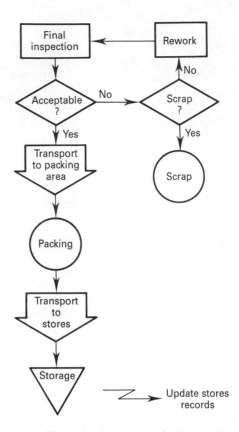

Figure 4.4 Flow chart example for inspection and rework of products.

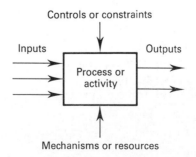

Figure 4.5 Basic IDEF building blocks.

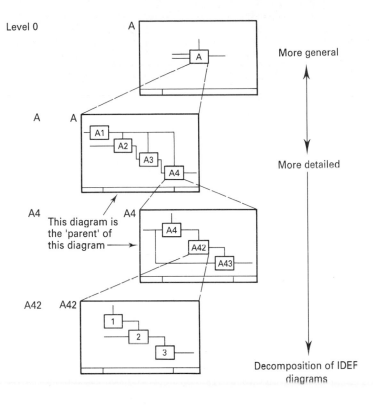

Figure 4.6 IDEF structure and decomposition.

constructed to show main input and output information, together with resources and constraints on the activity. This is similar to input/output charts, but it allows easier grouping and display at any level.

Once the IDEF 0 charts are defined, each of its activity areas is exploded down to the next level. In this example, commercial activities, IDEF 1, could be the subject of up to five activity blocks. And each of these five blocks is itself exploded down to another level. This process continues until it reaches the desired level of decomposition. (See Figure 4.7.)

Inputs and outputs need to be consistent between each level; there should be no untraceable elements. Written text is also used to further describe each chart and its operations. The IDEF numbering system allows easy referencing of each level and area of the whole IDEF study.

Computer analysis tools based on IDEF and other SADT methods automatically screen for illegal entries at every level (Figure 4.8). For a summary, see Appendix B.

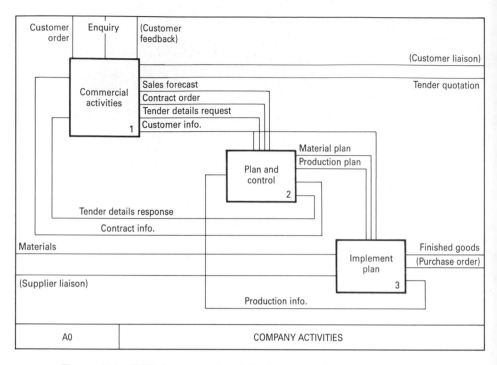

Figure 4.7 IDEF diagram of activity in a manufacturing company.

Input/output analysis charts

Input/output (I/O) analysis is a method used to display processes with the required inputs and the generated outputs. Normally used for much smaller and more localised descriptions than IDEF or similar approaches, I/O charts also form the basis of other techniques such as job design (Chapter 5). They may be linked together to show larger systems, but like flow charts can become unwieldy unless carefully constructed. Feedback between processes can be shown.

Essentially simple descriptors, I/O charts show inputs, process and outputs. They assist when looking for missing or superfluous inputs or outputs, and when defining boundaries between processes and functions within a business (Figure 4.9).

When constructing the I/O diagram, the process is listed first, then the desired outputs are added. Only necessary outputs should be present, with any loose, redundant or duplicated outputs eliminated. Inputs are added to produce the defined outputs. Similar to IDEFs, I/O charts can be decomposed to show lower level activities. They can be applied to a wide variety of areas,

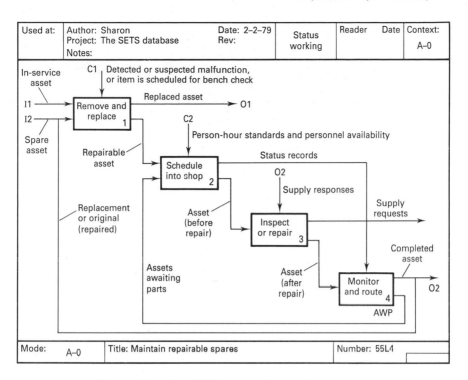

Figure 4.8 AutoIDEF generation tool output

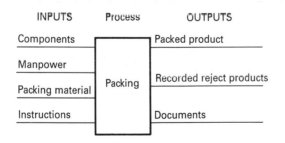

Figure 4.9 Basic I/O chart.

from job analysis to company representation. Figures 4.10 and 4.11 show two different examples. When the target area is large or complex, divide it into subsections, each with its own I/O chart. Then examine the links between each chart. A large I/O analysis is given in Chapter 7, showing the overall cell design process.

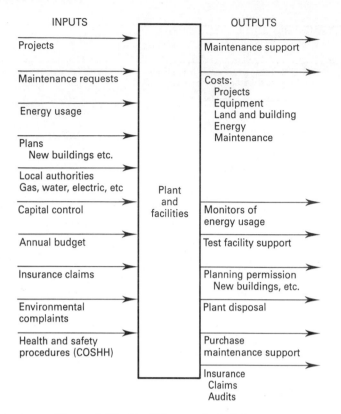

Figure 4.10 Facilities analysis I/O chart.

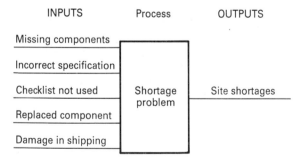

Figure 4.11 I/O analysis on a problem area.

■ Gathering information: data collection

Before collecting any data, we must define our brief or terms of reference. The review will concentrate on

1. Examination of current operations
2. Analysis

Of these tasks, examination involves the identification and collection of relevant information and data. In the early stages, it is more likely that operational information (practices, problems, etc.) will be required. Actual data and statistics will become more important as the review progresses, and detailed data for analysis is needed. A suitable programme for data collection is as follows:

- problem/task definition
- data selection
- data collection and storage
- data analysis
- revision of all stages

Group brainstorming sessions can help to discover what information is required and where to obtain it. The majority of the information will be collected face-to-face, from interviews. Interpersonal skills will be needed to conduct interviews and other parts of the review.

Interpersonal skills

Personal behaviour, speech and body language all play a part in encouraging your interviewee or audience. Consider how your mannerisms and presentation influence others. Are they encouraging or discouraging?

Discouraging	Encouraging
continued talking	listening
autocratic, dictatorial manner	discussion and interested approach
not giving full attention	being attentive and interested
unprepared and disorganised	well prepared and organised
giving few or no comment/feedback	commenting, giving feedback and discussion
criticising	encouraging, agreeing
several interviewers	one-to-one interviews

Systems engineering practitioners need to be good listeners. Good listeners require a focused mind, little or no distraction and no preconceived judgements of the topic under discussion. When involved in the gathering and analysis of information, it is often necessary to listen, relate and analyse information very rapidly, particularly when attempting to seek out root causes for NVAs.

Interviewing

The majority of interviews are relaxed and amicable, but inevitably some will become more brisk, particularly when it becomes necessary to discuss items of conflict, such as alternative approaches or deficiencies in previous plans (Figure 4.12).

In all interview situations, being prepared with the right information and questions is essential. It is normally useful to prepare specific questions for your interviewee, to ensure all necessary items are covered. Depending on the situation, it can be more productive to let your interviewee have the questions in advance, in order to prepare for the meeting.

All of the interpersonal skills and perceptions are required for effective interviewing, especially listening. The list includes:

- Structuring the discussion
- Having a friendly and open attitude
- Asking appropriate questions
- Listening
- Analysing
- Note taking
- Giving feedback

With a one-to-one interview, eye contact becomes more important, indicating your interest and attention to what the interviewee is saying. Avoid staring out of the window or at your notepad; they can suggest lack of interest or something to hide. If you find it hard to maintain eye contact, try concentrating on the forehead/nose of your subject. He or she will not notice the difference, but it may help you. Alternate between eye contact and note taking, getting some

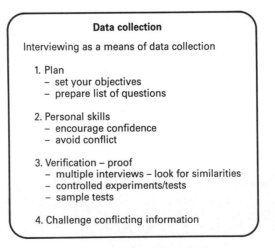

Figure 4.12 Interviewing for data collection.

relief during the latter. When a negative or unexpected response is given by the interviewee, try repeating it back to them, allowing them a second chance to consider.

Note taking

Take plenty of notes, but make sure you can read them later. Developing a summary after the interview can be very helpful, before you forget the details. Dictaphones can be useful, but ensure the interviewee agrees to their use.

During the interview, don't be afraid to ask for an answer to be repeated or rephrased, or to ask the same question in a different form. This may be to

- confirm what you thought you heard or didn't hear
- let the interviewee change or rephrase a dubious answer

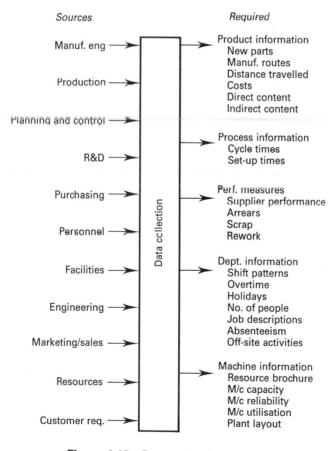

Figure 4.13 Data collection I/O chart.

You may want to revisit the interviewee to clarify certain points. Explain this to them at the first session. It may save them anxiety about future meetings.

Practice

Good interviewing skills are only developed with practice, and by the end of a systems engineering project you will have had plenty. To bring out any potential problems, It is worthwhile taking part in role plays with your team members.

Information and data collection is likely to be required from all functions within the business. The I/O chart in Figure 4.13 illustrates the common information obtained from each function. Further decomposition will be needed within each function.

■ General checklists for business

General lists of questions and ratios help to ensure complete coverage of every function in a review. The lists are useful in interview situations as checklists that should be amended and added to as appropriate to each business situation.

Where are we now?

The aim of using such a list is to assist in preparing a report card on the target company, identifying areas or activities that may offer varying scope for improvement. These would then be the subject of more detailed investigation and action. In some cases it may be preferable to extract questions and prompts to address one specific area, such as distribution or production, fleshing out the questionnaire as necessary. Appendix A contains a comprehensive list of review questions and performance ratios.

Business Control

- Is an integrated management information system used?
- How fully is the information used to run the business?
- Describe the system of control and the purpose of information produced.
- Gauge competence of personnel responsible.
- How much effort is needed to collect information?
- What are results compared with and how?
- Are reasons for variations examined on a regular basis?

Organisational issues

- Is an organisation chart available? Obtain organisational charts and examine them.

- Is there a reasonably flat structure or are there too many layers and few direct reports?
- How many levels of responsibility/accountability
- Does the organisation conform to the current needs of the business?
- Is centralisation or decentralisation planned or required?

Management ratios can often provide a good indication of business performance against competition, where information from Companies House or other sources can show how you fare in certain measurable financial and general areas. Other performance ratios can be examined or measured as part of the review. Here we would be looking for trend information, rather than specific values.

- Profit per employee
- Profit margin
- Stock turns
- Investment levels
- Staffing

CASE STUDY 4.1

Summary report on an integrated business control system for the Shakey Ladder Corporation, March 1990

The following report was the result of several weeks review of a £25 million turnover manufacturing company in Northern England.

1. Executive summary

The company is in need of substantial and rapid improvements in its operational procedures and systems. Existing procedures and systems are wholly inadequate for the effective operation of the business, and are directly contributing to current poor performance and high cost of sales. The recently commenced TQM programme will be rendered largely impotent if additional measures are not quickly taken to provide appropriate business planning and control systems. The company requires

- A full review of business and operational methods and procedures
- The determination of the manufacturing strategy
- Application of systems engineering to the company, providing quick fixes and improvements in all areas, e.g. kanban control systems on the shop-floor
- Establishment of formal procedures and disciplines in *all* areas and departments
- Selection and implementation of the company-wide business planning and control system, based on a package solution

If the company does not follow such a path, what will it do instead? We believe there is no viable alternative.

Is the company ready for the level of control and discipline required by such systems? Not at present, but it will rapidly become ready through the adoption of the above recommendations. It needs these disciplines and systems and must take the appropriate steps to review and prepare for them. This should be viewed as a unified improvement and implementation plan, as described in the report. Commitment to the new manufacturing strategy will provide the driving force.

2. Findings

- Recent major changes in top management
- Low staff morale
- No overall manufacturing or business strategies exist
- No cost and benefits analysis for recent capital investment in new plant
- Manual, paper intensive systems predominate, grossly inadequate for the volume and nature of the business
- Lack of formal procedures across the company
- Computer systems are minimal and disparate, little or no integration
- Crisis management with massive amounts of effort expended on expediting, chasing and fire-fighting
- Absence of essential business planning and control systems is responsible for NVAs and related costs of approximately £2.1 million per year
- Recent investment in new plant and TQM programme will derive minimal returns without the selection and implementation of essential business systems
- Significant scope to control and reduce inventory

2.1. Sales ordering

- No knowledge of profit or loss per sale; no data on cost of case specified
- Customer order narrative produced does not fully define the product requirements; order interpretation department then produces a full bill of materials/works order – duplication of effort
- No formal procedures or discipline for change control in order specification or due date creates significant replanning and excessive difficulties in manufacturing

2.2. Operations and production control

- Production line environment but job shop efficiencies
- Cycle of sale/design/manufacture is informal and unreliable
- Poor ratio of lead time/process time – 25 days/4 days – significant scope for improvement
- No measurements exist for production/process times
- No capacity planning/loading carried out, apart from press shop (unrealistic to carry out manually)
- Management of change control on customer order and specification is sales driven and very ineffective, causes significant disruption to production and purchasing
- No agreed measurements of scrap and rework levels or customer satisfaction, so there is no quantitative knowledge of performance or costs
- Parts shortages cause significant delays in product completion; in 1990 first quarter 39% of cases into warehouse have shortages (mainly due to made

parts) thus very high levels of chasing and expediting exist; shortage list in use
- Massive knock-on effect of shortages on number of site visits and additional delivery costs
- Purchasing requirements calculated and processed manually, grossly ineffective and unreactive

2.3. Financial

- Networked accounts systems, but no links to sales ordering or purchase ordering
- Manual data transfer and entry required
- Costing system exists but is not used; no job times available
- Should be integrated to company-wide business system

2.4. Management services

- Responsibility for several sites, with four staff
- Developing effective site-wide and multisite facilities
- Implement and maintain package and bespoke software
- Not a feasible option to generate an in-house integrated system in terms of resource, time and functionality; in our experience this is not a successful route

3. Proposals summary

From the investigation of the operational systems and procedures to date, it is apparent that the majority of problems exist at the front end of the business, i e not in the production areas but in the suppliers of information and services to production. In order these are sales, design/order definition group, production and materials control, purchasing. Problems include

- No planning, only scheduling and rescheduling
- No formal or effective procedures for order change control
- Inadequate communication between functions
- Informal systems used to supplement inadequacies in the formal systems
- Scope for improved support in the materials planning area and development of inventory policy for short and long term.
- Lack of accurate job times for production, so no realistic capacity planning

3.1. Change control

A substantial amount of shop disruption, shortages and expediting can be attributed to the lack of control on order changes. These are largely sales driven, with only token consideration of the resulting effect on material and production. There is a blanket assumption that everything can and will be accommodated. The SOP system doesn't take account of such changes. Changes also bypass credit control, which is unacceptable. The actual cost to the company of a change is not estimated or used. A guestimate by sales may be used in requote discussions with a customer, but this is informal.

Sales only use a price list for basic carcasses and common options. This does not provide adequate pricing information for the variety of cases being sold. The work currently under way on the price list is addressing this, but should be seen as a high priority.

A formal procedure for change control is essential, and must be put in place immediately. This procedure is to be followed by *all* company staff.

Sales cannot issue changes to the shop. All requests for changes must be formally passed and signed off by order processing before being passed to production control (PC) for investigation and approval/rejection.

Production control will check all necessary aspects of material availability, order definition group/design work and shop loading, to determine the impact on lead time and delivery date. PC will determine when the requested change order can be delivered, and advise sales of this. This will be logged. They may also derive a *cost* for implementing the change within the initial lead time. This is to be used for situations where the customer cannot/will not move on delivery date. Sales may not offer a customer a delivery date that is inside that advised by PC.

If the delivery date is altered but is acceptable to the customer, PC will be advised and will amend all relevant schedules and dates in the system. If the date is not acceptable, the change request must be passed to the PC director for consideration (using the cost estimate from above). He will use the *cost/authority matrix* to determine sign-off level, i.e. disruption cost and implications for materials and production. Where senior management wish to overrule or expedite, they must become aware of the effects and costs of their action, and sign off the overrule.

It is also recommended that a set of *costs of change* be created for use by sales and change control in determining the cost and impact of any change. This would use factors including type of change, type of case and progress point of order. This should supplement the aforementioned price list.

It is proposed that the company issues a statement to all customers and dealers detailing the procedures for change control and the cut-off dates. A suggested format is given below.

As part of the programme for improved customer service and delivery, we are firming up on the procedures and timings for changes to orders. We believe that this will provide the following advantages to you:

- Guaranteed delivery on agreed dates
- Guaranteed complete cases delivered

Position of order	Changes allowed
Before week 5 (preceding delivery)	Provided no material –
At or during week 5	sourcing problems
During week 4	Otherwise by negotiation
During week 2 or 3	No changes accepted on existing delivery date, amended delivery date and cost by negotiation
During week 1	Full cost of order

3.2. Planning and scheduling

Virtually all the effort of production control and management appears to go on short-term scheduling and fire-fighting. Medium- and long-term planning is ignored. Forecasting is primarily for financial reasons, rather than for use in long-term planning. This effort must be redressed to provide solid planning that will remove undue pressure from the short-term scheduling function, and hence from production. Planning and scheduling should be dealt with as follows:

- Long-term (one year and over)
- Medium term (eight to twelve weeks out)
- Short term scheduling (up to eight weeks out)

Long-term/strategic planning

It is noted that forecasting has degraded significantly over several years. This activity should involve all sales managers, finance and a tempering influence. This group will use information on market trends, historical data and product marketing strategy to produce regular forecasts. Statistical forecasting tools should be used as appropriate. Forecasts produced must be agreed by the group and have a measure of accountability from those involved. This is to guard against over- or under-forecasting. Forecasts will be used for capacity planning design focusing.

There is a requirement to focus on the trends of the outside market and the plans of the competition, as far as they are known, and any other factors, such as the pending sale of the company or any PR requirements. It is likely that staff at the highest levels will be involved. This group will also plan the medium- and long-term capacity against the forecast. Where proposed systems or initiatives are programmed within the company, this group will be responsible for ensuring that planning and communication remain adequate.

Medium-term planning

There is presently little activity in this area, which we believe is critical if the company is to move towards a well-planned, effective environment. The aim is to do more planning and less scheduling. Using order information and available forecasts, the intention is to investigate material requirements and capacity/load, drawing office load and warehouse/delivery implications.

Although communicating with scheduling and materials, the planning team should remain largely independent. Its first focus is likely to cover problems on or within the five-week lead time, e.g. orders with a delivery date shorter than the particular lead time.

Scheduling

The primary aim is to reduce the need for the pressurised scheduling and rescheduling that may currently occur. This should be provided by improved planning of load and mix, coupled with the change control system. Any informal or illegal production/requisition that occurs must also be eliminated or formally controlled, e.g. overseers having a private stock.

Scheduling should be for up to approximately seven or eight weeks out, with continuous liaison with the planning group. It is anticipated that the current staff will carry out this function with assistance from appropriate managers.

Inventory policy

Our understanding of the current procedures is as follows. No formal policy exists. EOQs are established by buyers; safety stock levels are set and act as prompts for attention. ABC/Pareto analysis is provided by the system. Corporate moves to reduce inventory may be unstructured, causing knock-on problems for material supply to shop/schedule demand. There needs to be a compromise between job delivery date and material lead time (LT). If a sourced part has an LT of eight weeks, you cannot and should not expect to get it in less. The LT for this job is eight weeks. If some customers don't like this, then

- sweet-talk them
- offer alternatives
- consider stocking the part(s) if the order is continuing
- ask for commitment to your supplying/stocking these customers, preferably in writing
- decline the order

Inventory needs to be reduced as part of a comprehensive improvement programme. Doing it in isolation is usually a disaster.

Internal services should be treated like suppliers. Although LTs may be slightly varied, only rarely should they be expedited. If design or the order definition group quote two weeks to provide drawings, this time must be respected and built into the delivery date and plan.

3.3. Interfunction communication

There is scope to improve the levels of communication and effectiveness from order acceptance through to production. Presently there is unnecessary overlap between sales and the order definition group. In time the pending order definition group system should provide a significant improvement in order definition and planning. But there is scope to offer improved order processing by removing departmental boundaries and duplication of effort. It is recommended these be as follows:

- Rationalisation of sales and order definition group functions, creating a single team with greater knowledge and ability to fully specify an enquiry or order in a shorter time.

It is anticipated that once this function is operating effectively, there will be scope to reduce or redeploy a proportion of the staff. In creating this focused team, try to avoid creating barriers between different groups.

The current developments in the order definition group system offer greatly improved jobsheet descriptions. It is essential that all potential users are involved as fully as possible. There appears to be scope to improve this, through the use of a comprehensive communications programme. All systems implementations should have one. This is a priority, due to the closeness of the live date.

3.4. Staffing and organisational issues

Certain communication recommendations are likely to require changes in the structure of the existing departments. This requires further discussion. The present management support in the production and materials control functions offers scope for improvement. Perhaps due to inexperience, little guidance seems to be provided, particularly in the materials planning function. The situation could be improved by injecting an experienced manager into the area or by a management reshuffle.

Action list for subsequent activities

A comprehensive list of necessary actions was also developed to improve several areas and functions, as part of a comprehensive strategy for improvement. This includes

1. Carry out full review of existing systems
2. Complete merger of customer services and order definition groups
3. Implement change control procedures
4. Establish market research function
5. Confirm establishment of medium-term planning group
6. Define training requirements and schedules for
   – estimating and sales staff
   – cross-company total quality approach
7. Plan and commence company-wide communications regarding TQM and continuous improvement
8. Initiate development of specific strategies by directors

Task allocation

    Production
        – Develop production manual
        – Communicate review to supervisors/leaders and workforce
        – Initiate improvement groups, measurements and required training
    Training and quality
        – Undertake training needs analysis for all areas
        – Draw up action plan and costs/resources
        – Make measurements
    Production and material control
        – Review methods used and identify training needs
        – Commence change control procedure
        – Make relevant measurements
        – Perform ABC parts analysis
        – Check on items with long lead times
        – Is there master scheduling of sub-assemblies?
        – Inspect goods-in
        – Investigate vendor relationships and alternative suppliers
        – Report exceptions
        – Produce parts release, kitting
        – Check stores data accuracy and miskeying
        – Supply forecasts to vendors
        – Set up medium-term planning group
    Production Control
        – Rough CRP tools to be used
        – Schedule levelling
        – MPS
        – Runner, repeater, stranger aspects
        – Exception reporting
        – Order/work in process tracking

3.5. Existing and future computer-based information systems

Some areas are completely missing, areas critical to improving the overall materials and order processing performance:

- Full and accurate order description (order definition group)
- Purchase order system
- Materials requirements planning (MRP1)

In-house software could be created to perform these functions in three or four months, possibly better than sweeping clean with all new systems. This requires further discussion. On initial inspection it would appear that many of the existing systems are basically adequate, lacking only minor features and full integration.

4. Costs and benefits analysis

The following analysis has been based on figures and costs obtained from company directors and management. They are as realistic as possible from this brief review. For comparison, we have prepared three separate scenarios, giving potential cost benefits of 25%, 50% and 75%. Potential benefits are little short of massive; 25% savings seem easily achievable; 50% should be taken as an obtainable target; and 75% savings are a longer-term objective.

4.1. Lead time, inventory and WIP reductions

Through improved planning and scheduling of production and purchases, we expect a reduction in the average LT/process time ratio, currently 8:1. This reduction in LT will allow stocks of finished goods to be reduced, since the production facility will be able to respond to demand much more quickly. Improved scheduling of raw material (RM) will reduce this inventory, with work in progress (WIP) being more precisely controlled. Less work in process will be pushed into the system, hence reducing the WIP level. Conversely, the reduction in WIP will also reduce queue sizes and queue time, hence reducing product lead time. This one-off saving is the cost of carrying the current level of inventory at 15%, £850k per year.

Scenario	LT Ratio (% reduction)	RM (£m)	WIP (£m)	FGS (£m)	Total saving (£m)
Now	8:1	1.38	1.70	2.60	0
1	6:1 (25)	0.34	0.42	0.65	1.40
2	4:1 (50)	0.69	0.85	1.30	2.84
3	2:1 (75)	1.03	1.27	1.95	4.20

4.2. Intangible benefits

In many cases the intangible benefits can be more rewarding than the tangible benefits.

- Ability to grow without a proportional cost increase
- Reduced management fire fighting therefore more time to plan
- Improved accuracy and currency of management information
- Improved quality of service between departments
- Improved employee morale and job satisfaction

- Improved customer service
- Ability to react quickly and effectively to change

5. Proposed improvement implementation plan

The following plan outlines our recommendations on how the company should proceed towards the goal of maximised profit through effective, disciplined operating procedures and appropriate business planning and control systems. No elements are stand-alone, since the goal can only be fully achieved by satisfying all stages of the plan.

Our plan proposes that the first five months be devoted to the review and improvement of procedures and methods within the company. Only once this is well established will the requirements for the business planning and control system be formalised. Selection and implementation of the system follows, over a period of seven months, with reviews at regular intervals. Education and involvement of people at all stages is essential. This is not a system solution, it is a business and people solution.

6. Project structure and resources

For successful review and implementation of improvements across the company, we recommend that a project team be established, reporting to a high level steering committee. The proposed structure is shown below.

STEERING COMMITTEE

PROJECT MANAGER   CONSULTANT

PROJECT TEAM

FULL AND PART TIME   AS REQUIRED

Production planning	Sales and marketing	Key users
Production	Information systems	Vendor
Purchasing	Personnel	
Finance and accounts	Quality control	Design

6.1. Structure: Steering committee
Managing director (chair)
Operations director
Project manager
Quality director
Purchasing and materials director
MIS director
Consultant

The role and responsibility of the steering committee is to

- Establish overall policies and procedures
- Provide adequate staffing and funds
- Monitor progress frequently and regularly

■ Recommend hourly meetings every fortnight
■ Resolve conflicts
■ Encourage acceptance and support within the company
■ Participate

6.2. Project team

The project team has the following responsibilities:

■ Develop and maintain a detailed project plan
■ Provide programme direction
■ Plan, organise, direct, co-ordinate, measure and report on all project efforts
■ Carry out the decisions of the steering committee
■ Co-ordinate the education programme
■ Work closely with users during all phases of the project
■ Develop new policies and procedures
■ Act as departmental champions of the project

As shown in the above chart, the project team will ideally have a core of full-time members, plus part-time and as-required members.

The importance of creating and sustaining the correct full-time core team cannot be overstated. Dilution of the team will result in dilution of the project, making the end goals much harder, if not impossible to achieve.

This case study illustrates appropriate ways of deciding what is good practice and how to measure it effectively. A significant portion of review and improvement will be the correct application of common sense, and this should always be remembered. Appendix A contains further indicators and ratios for business measurement, which should be selected and used depending on the company and the brief.

■ Revision of review stages

Once the data and information has been collected, analysed and is thought to be complete, it is necessary to review progress:

■ Is the analysed information and data complete and accurate?
■ Is the data in a transferable form, such as computer databases?
■ Is further analysis required?
■ Define and carry out any additional requirements
■ Review situation again

By this revision of the review, we can reduce the possibility of missing important information or operational practices that would affect the validity of the review.

■ Summary

From a detailed business review must come solid and well-thought-out recommendations for future improvement, recommendations that both fit the market needs and the business capabilities. Possibly the most difficult aspect of any review is trying to grasp the overall operation of the business. High level flow charts and IDEF definitions are invaluable when focusing on problems and drawing out solutions.

It can take substantial resources and time to decide what information is needed for the review then to collect and validate the data. Ideally, this information should be entered into personal computer systems for ease of storage, analysis and presentation.

The basis for reviewing and improving any business must fit the overall business strategies discussed in Chapter 3. Following acceptance of the review, planning and implementation begin. Business improvements are described in Chapters 5 to 7.

■ Further reading

Bititci, U. (1993) 'Measuring your way to profit'. Paper presented at FAIM 1993.
Gane, C. and Sarson, T. (1979) *Structured Systems Analysis*, Prentice Hall.
Lucas Engineering & Systems (1989) *Manufacturing Systems Engineers Handbook*, The Lucas Miniguides.
Model, M. (1992) *Data Analysis, Data Modelling and Classification*, McGraw-Hill.
Tucker, S. (1961) *Successful Management Control by Ratio Analysis*, McGraw-Hill.

5 | Methods and Techniques for Improvement

> *There will be more changes within manufacturing over the next ten years than there have been over the last fifty.*
>
> Hal Mather, 1988

■ Introduction

Changing the performance of a business can seem daunting to the inexperienced. But however, by using what is basically a straightforward and common-sense approach, we can achieve significant, often enormous benefits in the performance and competitiveness of almost any business.

This change in performance will be achieved by using a systems engineering methodology, employing concepts and techniques drawn from virtually all other proven world class philosophies. This will be the basis of the strategy for both manufacturing and all other support functions within the business.

■ Technology or methodology?

Methodology: *method, systematic, orderly techniques or arrangements of work for a particular field.*

Studies of a variety of industries show that 80% of the savings and benefits of high technology solutions frequently come from the improved methodology, the methods and techniques used in all areas of the business. This is not to say that the appropriate and correct application of advanced manufacturing technology (AMT) will not benefit the company, but it illustrates the potential for achieving substantial benefits without major capital investment. The potential benefits are even greater benefits if methodology and technology are effectively employed together.

■ Procedures: get the basics correct

The basic operating methods within the business are the foundation for any further improvements. Problems in the foundations will create more serious problems later on. Make sure the foundations are solid; use systems that are formal, not informal.

When correctly designed and operated, manual and paper-based systems can work well in a large proportion of businesses. Formal procedures are the basis for any later developments into computer-based systems, and are an essential prerequisite. Appendix D contains examples of several common procedures.

CASE STUDY 5.1

A small batch operation

A fabrication company producing small batch and one-off steel enclosures for six major end-users was having problems with late changes to job specifications and delivery dates. The customer services function received orders, issuing them to a production planning function for material planning and shop scheduling.

This involved the information flow in Figure 5.1. However, when customers were pressing for earlier deliveries than planning had offered, or when changes to an order specification were asked for, there was a tendency for the customer services staff to bypass the planning function, who were seen as blocking customer needs and hence being non-commercial. Unfortunately, the production staff were allowing this to occur by reacting to the informal requests from customer services, leaving planning outside the information flow. This resulted in production missing other order dates, causing friction between all functions, who saw the problem as lying in the other areas. Planning also maintained all order specification and progress data, which became gradually out of synchronisation with actual requirements and status, compounding the problem.

This is not an uncommon problem, and has been observed in many different businesses and sectors. Basically a procedures and disciplines problem, there was also a lack of understanding of the total business and functional objectives by most staff. What was required was a review and improvements to the ways these functioned, and how they related to one another. There was also need for some reorganisation, described in Chapter 7.

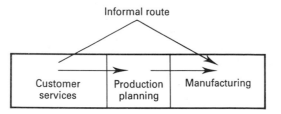

Figure 5.1 Formal and informal information flows in a fabrication company.

■ Key skills from Japan

Much of JIT and systems engineering is good, practical business sense employed in a systematic way that cannot fail to impact on the business. Many systems engineering guidelines are borrowed directly from Japanese manufacturing and JIT philosophies:

- Total quality of performance in each and every function
- Communication of objectives and performance measures to all
- Planning to a detailed, meticulous level, covering all requirements
- Economic achievement of quality conformance
- Flexible manufacturing systems and personnel
- Economic production of a high variety of products

Reproducing the seven Zeros of JIT from Chapter 2 to provide us with a set of targets:

JIT Goal	Expansion across the business
■ Zero defects	Achieve excellent quality in all functions and activities
■ Zero set-up time	No delay between each activity, via continuous improvement
■ Zero inventories	Balance the system and avoid investment in wasted stock
■ Zero handling	Minimise the transport and handling of every item. Reorganise to create multiskilled cellular units to avoid passing between departments
■ Zero breakdowns	No hold-ups in processing a product, document or item of data
■ Zero lead time	Reduce the time taken for every task Eliminate non-value-added-activities, to improve the response of the system Reorganise processes to minimise the number of activities needed
■ Zero lot size	Produce and pass on items as individuals or as small batches to save queuing time and to improve response and inventory.

We need to consider the detailed methods and mechanisms for achieving the JIT goals listed in Chapter 2. And we need to apply them to the actual facilities, operations and procedures in each area of our company.

■ People and productivity

This chapter covers people and productivity aspects of elements from the lower level structure, and on procedures to operate the business effectively with greatly improved productivity.

People

Culture, simple structure, teams, policies and procedures, quality circles

Productivity

Plants: group technology, focused factories, cellular units, flexibility
Products and services: Design for manufacture, modular design, set-up reduction, minimum lead time

The methodology improvements potentially apply to all areas of a business, not only to the manufacturing system. In fact manufacturing often needs fewest improvements; most problems and ineffective practices are found in other areas of the business. Business and manufacturing systems consist of value-added and non-value-added activities across direct, indirect and staff functions

■ Non-value-added activities

Non-value-added activities (NVAs) include any and all activities that do not assist in achieving delivery of a product or service to customer requirements. Translation of people and productivity into tangible improvements across the

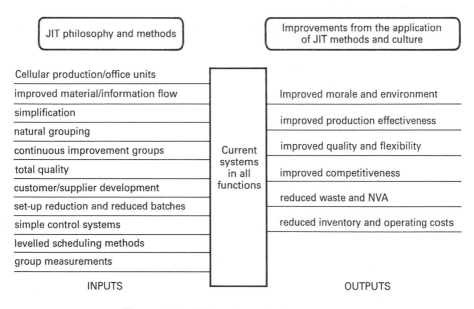

Figure 5.2 JIT business improvements.

business is wholly based on the elimination of waste. To a large extent, waste and NVAs are the same. Some examples of NVAs are

- where two departments duplicate or rekey information.
- long distances travelled by products between work centres in poor layouts
- repetitive information processing by people instead of computers
- mistakes and inaccuracies

So, our overall objective is to eliminate this waste and NVA from all activities within the business. But, since the existing arrangements of people, departments and plant facilities are unlikely to be ideal, it also requires us to investigate alternative layouts, arrangements and facilities to see whether they would provide more effective operations (Figure 5.2). Should a given production line be rearranged to form several cells to manufacture distinct families of product? Should service functions be redeployed to each of these cells to give better response? *First define the problem, only then define the solution.*

■ Key elements of systems engineering

There are no sacred cows or restricted areas in this type of business review, though it is common to encounter them in each firm that undertakes a systems engineering review. Systems engineers follow basic guidelines appropriate to all areas of the business from high tech computer firms to cleaning services.

- Consider *everything* as a process with its own lead time and reliability. Processes consist of value-added and non-value-added activities.
- Search out and eliminate the NVAs.
- Look for and create natural groupings of people, tasks and products. Do not automate existing ineffective and fragmented systems.
- Review, simplify and improve before introducing appropriate automation.
- Regroup service functions into the areas that use them.
- Establish performance measurements for *everyone*; relate targets to business performance.
- Use analytical tools and techniques in the review, e.g. cause and effect analysis.

We can break problem solving into simple stages. This makes the task easier because *simple is efficient and simple is effective.*

■ Problem clarification, analysis, solution and prevention

Not only should we break down business areas and operations into smaller, identifiable units, we should consider exactly how to approach a problem. There are four basic aspects to solving and eradicating any problem: clarification, analysis, solution and prevention.

1. **Clarification**: identify the problem.
2. **Analysis**: define and examine the problem.
3. **Solution**: solve the problem.
4. **Prevention**: ensure the problem does not recur.

In small areas, the first three steps tend to be pretty straightforward. We all solve many different problems every day, but they tend to be the easy ones! *Larger-scale problems often need a more careful and deliberate approach.*

Here are some simple tools for expressing various aspects of business operations. Most are general management and analysis tools but a few are drawn from statistics and operational research. They are described in the *MSE Handbook* of the Institute of Production Engineers.

For clarification and analysis

- Ideas generation, including brainstorming and lateral thinking
- Flow charts (Chapter 4)
- IDEF and data flow diagrams (Chapter 4)
- Input/output analysis (Chapter 4)
- Critical path analysis (CPA)
- Cause and effect analysis and relationship diagrams
- Tree diagrams
- From/to matrices
- Problem rating
- Strength, weakness, opportunity, threat (SWOT) charts
- Pareto analysis
- Failure mode and effect analysis (FMEA)
- Deeming circles

For solution

- Foolproofing (POKA YOKE)
- Decision analysis

For prevention

- Statistical process control (SPC)
- Taguchi methods

■ Problem solving

Ideas generation

When faced with one or more problems to solve, whatever the size and nature

125

of the problems, it is frequently valuable to have group participation to stimulate and generate ideas for causes and solutions. This occurs before applying more structured problem-solving tools such as cause and effect analysis. Brainstorming, the nominal group technique and other techniques are used to obtain as many ideas for future discussion and shortlisting. Brainstorming sessions should

- obtain a large number of ideas quickly
- accept all ideas without question
- combine and build on previous ideas

Similar to brainstorming, the nominal group technique uses lists of ideas submitted by participants before the session. Used by individuals or groups of any size, lateral thinking can help to find innovative solutions to problems of product, design, production, servicing and communication. It can

- generate alternative ideas and approaches
- challenge previous assumptions and rules
- break down problems into components
- turn problems round, e.g work brought to employees instead of employees brought to work

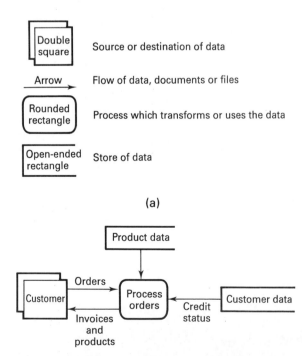

Figure 5.3 DFD diagrams: (a) symbols; (b) information flow including activities between customers and suppliers.

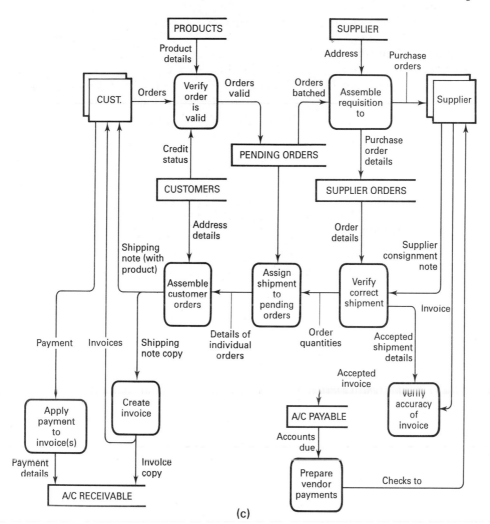

Figure 5.3 (c) Fully exploded version of (b).

Problem clarification: IDEF and data flow diagrams

IDEF is described in Chapter 4. Data flow diagrams (DFDs) are another structured design technique used to analyse information and activity processes and/or systems. Figure 5.3 (a) explains the symbols and Figure 5.3 (b) shows an example. This DFD is depicting logical flows of data and information in the form of paper, computer disk, telephone call, fax, etc. The processes may be manual or automated and take place in an office, factory, building site or any other environment. Data stores can be filing cabinets, computer memory, paper cards, and so on.

These four symbols are used to represent all systems, simple or complex, and can decompose each process box down to a further level of detail, if required. Figure 5.3 (c) shows a fully exploded version of the DFD in Figure 5.3 (b).

We do not show flow of products or services. For the purposes of this diagram, products are not data so they are not included. A materials flow diagram can be constructed when required as a variation of DFD and flow chart methods.

Data flow diagrams and lower level IDEF diagrams work at low levels of information flow and activity. They are better suited to defining data and database requirements at a detailed level. This can be a useful alternative to the common flow chart when recording data movement and/or activities at a very low level. And when there may be a need to create or modify a computer based application.

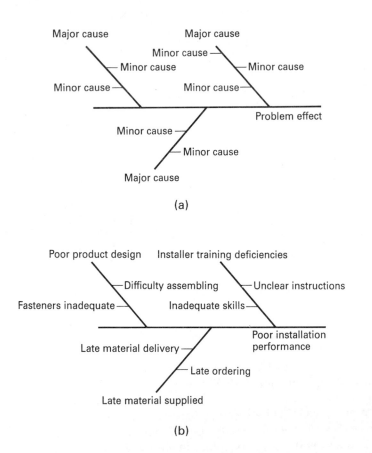

Figure 5.4   Cause and effect diagrams: (a) basic structure;
(b) for analysing installation problems.

Clarification and analysis of problems: cause and effect analysis

A cause and effect diagram is a method of representing the contributing elements within an analysis. Starting with a problem, the effect, the aim is to trace all contributing causes back to their beginnings, eliminate these causes and solve the problem.

The diagram has a tree structure, (also known as fish-bone or Ishikawa); with the causes branching out from a spine representing the problem (Figure 5.4).

A different way of describing causes and effects is the relationship diagram (Figure 5.5). This is of particular use with complicated situations, developing the diagram from the hub problem and expanding to primary, secondary and further causes. It can show relationships between the different levels of causes.

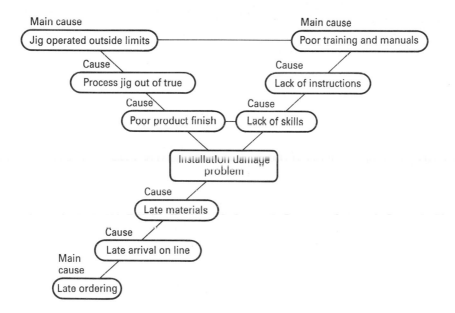

Figure 5.5 Relationship diagram.

Problem clarification: SWOT charts

SWOT arranges strengths, weaknesses, opportunities and threats in a standard, single-page layout, giving a clear summary of the area under examination. It aims to

■ consolidate and maximise the strengths
■ maximise the opportunities, converting them to strengths where possible

- minimise and overcome weaknesses, converting them to opportunities or strengths
- minimise or nullify the threats, possibly by converting them to weaknesses or opportunities

The layout of the SWOT diagram (Figure 5.6) allows positioning of corresponding strengths and weaknesses, opportunities and threats. In this way it can aid focusing attention on improvements, with possible relationships across the diagram. For example, a business strength may be a proactive design department, but a weakness may exist in the current product range. This strength can then be related to the weakness for action.

SWOT diagrams were used in Chapter 3 to illustrate various customer accounts. A further example is Figure 5.7, the application of a total quality, multimethodology approach.

Strengths	Weaknesses
Opportunities	Threats

Figure 5.6   SWOT diagram for visual impact.

STRENGTHS	WEAKNESSES
Uses proven methods and techniques Low capital costs Improves business in all areas Improves morale and team approach	Possibly little in-house expertise Resources to carry out the mission
Applicable across most business types Integrates most aspects of improvement and good business management Improve staff awareness and abilities	Market competition Competitors also using TQM Failure to complete and achieve world class levels Internal commitment Time
OPPORTUNITIES	THREATS

Figure 5.7   SWOT diagram for proposing total quality improvements.

■ Improvements: where do you begin?

Investigative analysis usually uncovers a wide range of problems and ineffective practices. If it hasn't, it needs to be more rigorous. *Everything is capable of improvement.* Faced with a long list of problems, and ineffective working practices, which do you address first? The answer is simple: those nearest the 'front' of the company.

Similar to many situations in business, there is most to be gained by improving operations at or close to the input or start of any process, then progressively moving through to the end or output of the process. Benefits gained at the front end are available early, and help all succeeding stages (Figure 5.8). This may seem obvious, but many businesses forget this concept, even working from final to initial stages.

Sometimes it may be more productive to deal with the largest problem early on. If this problem occurs at the back end of the company, it makes sense to work from back to front. But there need to be very good reasons for this.

Where several problems and areas of potential improvement are of approximately the same magnitude, 'front end first' is a sound policy. For

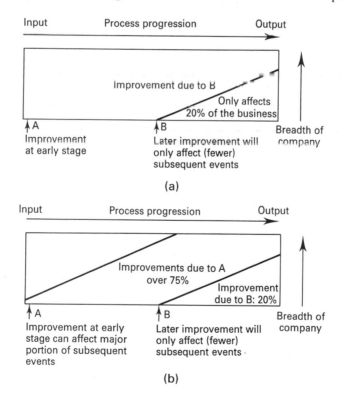

(a)

(b)

Figure 5.8 Sequences of improvements within a business.

131

example, where a conventional functional structure exists, a mid-process improvement may benefit only a proportion of the production, distribution and after-sales functions. (This will also apply to the positional effect of improvements within self-contained cells or business units. But an improvement early in the process, e.g. improved order definition in sales, will immediately benefit most downstream functions (Figure 5.9).

Where a large number of improvements or areas for improvement have been identified, it is worthwhile to display these on a matrix or table to show their relative positions, potential effects and values. Several techniques can be used.

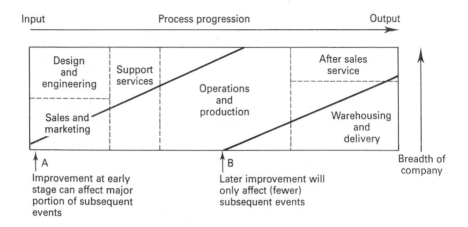

Figure 5.9 Business functions related to the sequence in Figure 5.8 (b).

■ Problem analysis and rating: Pareto analysis

Widely used in all walks of life, the 80/20 rule or Pareto principle is a simple way to prioritise problems and data. Working in the nineteenth century on mass/elite interactions, the Italian economist, Pareto, found that 80% of the wealth was held by only 20% of the people, still largely true to this day. The principle is largely common sense, as we can see from several examples:

- The majority of crimes are committed by a very small proportion of the population, e.g. 95:5.
- Eighty per cent of inventory value is likely to be in 20% of the part numbers.
- Eighty-five per cent of the nonconforming product may come from 20% of the feeder areas.

Pareto is not a hard and fast 80/20 split distribution; it is a more general guide to proportions. For example, data being analysed may give a ratio of 70/30, 80/20, 87/13, 92/8, or any other ratio that generally conforms to the Pareto principle.

Pareto is frequently used for data series analysis and for prioritising or rating problems. Progress is graphically displayed by plotting Pareto diagrams before and after improvements. Choose an appropriate format from the wide range available, to suit your application.

CASE STUDY 5.2

Raw material inventory

Figure 5.10 analyses the cumulative value (£) of raw material inventory of a £25 million turnover company and compares it to the proportion of the total 2,676 part numbers held. This found that 80% of the value was held in only 11.4% of the parts. This will vary by company and industry sector. This type of analysis has long been used for inventory analysis into class A, B, C, etc. for focusing control onto part groups by value.

Data for this type of analysis is commonly available as transferable files from inventory or MRP2 computer systems, easily loaded into PC spreadsheets for analysis and graphing.

Steps In Pareto analysis for problem clarification

1. Identify all problems or data available in the particular area. Collect the relevant data, discover how frequently the problems occur and find the costs associated with the problems.
2. Create a table listing all problems then sort them by magnitude, cost, or other criteria. For problem ranking, a comparable financial value is preferred.
3. Construct a graph or histogram using a spreadsheet with suitable axes.

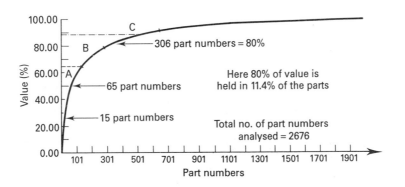

Figure 5.10 Pareto diagram of purchased inventory versus cumulative parts usage.

CASE STUDY 5.3

Air-conditioning audit

The installation of air-conditioning plant into office premises was not conforming to the customers' requirements; dissatisfaction was growing. A quality audit was carried out and the results were analysed (Figure 5.11).

Reason for nonconformance	Cost of scrap and repairs (£)
Faulty materials	80,000
Incorrect instructions and diagrams	16,000
Operator errors in assembly	4,000
Operator errors in commissioning	1,500
Instrument calibration errors	1,200

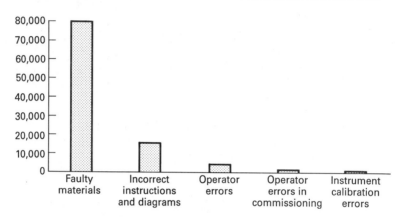

Figure 5.11 Pareto analysis histogram for costs of nonconformance.

■ Analysis and rating: impact changeability

Impact changeability analysis ranks problems according to their impact on the business and on how easily improvements can be made.

It is a useful technique for summarising and shortlisting problems, focusing effort on those with the greatest impact or benefit. Each problem is described as a short written statement, then ranked 1, 2 or 3 for impact/benefit and ease of change.

Ranking table for impact tangibility

Amount of impact/benefit	Impact/benefit ranking	Ease of change	Change ranking
Little	1	Very difficult	1
Some	2	Moderately difficult	2
Significant	3	Easy	3

A problem that has significant potential benefit and is easy to change ranks 3 for impact and 3 for change. These entries in the ranking table are expressed as 3:3 in the priority table. A problem with some potential benefit that is very difficult to change is expressed as 2:1.

Priority table for impact tangibility

Shorthand ranking	Priority	
3:3	1	Maximum benefit, minimum effort
3:2	2	
2:3	3	
2:2	4	
2:1	5	
1:2	6	
1:1	7	Minimum benefit, maximum effort

CASE STUDY 5.2

Two potential areas for improvement are identified

An ineffective manufacturing jig is causing a degree of wastage in component materials during product injection moulding. Curing the problem will require removal of the jig from service for 48 hours, resulting in lost production worth £6,000. The material loss is approximately 2%, with a value of £40 per day. This is given an impact ranking of 2, since there is some benefit, and a change ranking of 1, due to the disruption and cost factors. Overall 2:1 gives a priority of 5.

A methods revision has been identified in the CAD drawing office which will allow drawings and CNC data to be prepared 10% faster, a reduction in lead time of up to four days. Training is required at a cost of £2,000 and one lost day in the drawing office. Due to its considerable impact on the overall response lead time, this is ranked 3 for benefit. The training day makes the change ranking a 2. Overall 3:2 gives a priority of 2.

Conclusion: improve the drawing office before improving the jig.

■ Analysis and rating: FMEA

Failure mode and effect analysis (FMEA) is a more specific assessment technique for defining and rating the products or methods of production in terms of their likely failure and its consequences. By analysing and rating these characteristics and weaknesses, attention is focused on the more important aspects to be addressed. *Failure* should be taken as any nonconformance to a specification or other aspect of customer dissatisfaction.

135

Rating	Frequency of occurrence	Severity	Detection of problem
1	Virtually never	Negligible	Completely obvious
:	:	:	:
10	Frequently	Serious effects	Virtually undetectable

From this form of table, each problem is assigned a rating. For example, Figure 5.12(a) shows an FMEA table for condenser failures, with data analysis on the failure modes and effects as shown. Here the fitting of too small a bolt has an occurrence rating of 6, severity of 7, and detection of 6. These ratings are multiplied to give a risk priority number RPN = 252, a greater risk than the omission of a bolt (RPN = 56). The recommendations aim to fit all bolts of the correct size.

An alternative layout is shown in Figure 5.12(b), reported field service calls and failures of fitted antiradiation screens to VDUs. Modes and causes of possible failure are examined, together with the existing controls for reducing them. From the FMEA rating, problems and causes are targeted for suitable prevention and/or solution, e.g. by statistical process control, and for focused effort on contributing functions, e.g. supplied material quality assurance, design for manufacture and reliability.

Potential failure mode Potential effects of failure	Becomes detached Pipe fractures and case fails	
Potential cause of failure	Bolts too small diameter	One nut and bolt omitted
Existing conditions Current controls Occurrence Severity Detection Risk-priority number (RPN)	Advise special bolts to use 6 7 6 252	Inspection 2 7 4 56
Recommended action and status	Standardise on next bolt size, discontinue smaller size and add final inspection check	
Resulting conditions Revised controls Occurrence Severity Detection Risk-priority number (RPN)	Revised final inspection. Operator in process inspection. 1 7 7 49	
Responsible engineer/area	Fred Bloggs/Engineering	

FMEA analysis for detached condenser

Figure 5.12   (a) FMEA table.

Example of an FMEA

A company provides a service to computer users, fitting anti-radiation screens to VDUs. These are attached with two brackets and adhesive pads. The potential failure modes might be as follows:

Problem 1: breaking of the clips holding the anti-rad screen to the VDU

Problem 2: adhesive pads failing to stick permanently

Both problems result in a similar failure effect, but with differing occurrences and detection factors. Both had a high resulting RPN, and were therefore both subject to revised action. A new set of statistics was collected, and as shown, gave improved quality and reduced failure.

FMEA can be used in any circumstances where failure and failure modes can be defined, including processes, procedures, within offices or a factory situation.

Potential failure mode	Potential effect of failure	Potential causes of failure	Current conditions				
			Existing controls	Occurrence a	Severity b	Detection c	RPN (axbxc)
Falls off VDU	Breakage of screen	Clips break	Check operation after fitting	6	6	5	180
	Return visit to site for repairs	Adhesive pads not sticking	Visual check	6	6	3	108

Recommended actions	Current conditions					Area(s) responsible for actions
	Revised controls	Occurrence a	Severity b	Detection c	RPN (axbxc)	
Redesign clips	None	1	6	5	30	Design department
Source and fit stronger adhesive pads	Check firmness of installation	1	6	3	18	Purchasing dept.

Figure 5.12 (b) FMEA table – alternative layout.

■ Analysis and rating: matrix diagrams

Matrix diagrams are useful for displaying relationships between problem types and associated areas or functions. For foam injection, Figure 5.13 tabulates work areas in the top row of the matrix and potential problems in the left-hand column. Relationships are entered in the body of the matrix. In this example there are entries for improvement of waste, labour and materials, addressable through improved jig design and better working practices. Analysis can be

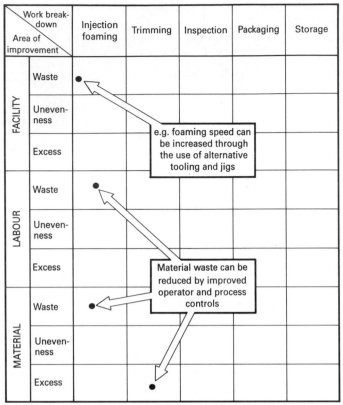

NB: Each of the intersection points should be checked as a means of logically looking at all possible combinations of problems

Figure 5.13 Matrix diagram of problem relationships.

completed by using cause and effect analysis, ranking tables and other techniques. Ranking is used to prioritise solutions.

■ Solution: the Deming circle

A common sense approach to problem solving lies in the four steps of the Deming circle (Figure 5.14):

> **Plan**: describe scope of project, aims, requirements, performance measures
>
> **Do**: define why problem occurred, determine solution or countermeasures then implement
>
> **Check**: look at solution effectiveness, any shortfalls or nonconformances and/or side effects

Act: make further control system modifications to eliminate problems and investigate whether other applications of this solution may be beneficial

Figure 5.14 Deming circle approach.

■ Solution: decision analysis

When faced with problems to be solved or choices to be made that have far-reaching implications, it can often be worthwhile to analyse the problem and/ or options available against the known and unknown effects and constraints surrounding the subject in question. This is a decision-making process and model for any business, determining the degree of risk present in any solution or strategy. The components can be described as

- Objectives: What is to be achieved? Short- or long-term plans? Narrow or broadening market? Performance measures required?
- Decision set: What are the alternatives? Do any others exist? Is there a specific or predetermined outcome?
- Constraints: Limits on freedom to choose? Timescales? Resources? Company policy?
- Objective function: How to decide on the optimal choice? What can be measured? Costs and benefits comparison? Intangibles?
- Uncontrolled variables: What are the risks involved? What may go wrong? How accurate and appropriate is the information? What data is not available?

In order to make decisions using this process, we should ideally be able to make quantitative answers to several key questions:

- What are the aims and objectives? How do we rate them?
- Which objective is the most important?
- Which options are available? How are they rated?
- How do the options compare rated against each of the objectives and overall?
- Which is the overall preferred option?

139

©

CASE STUDY 5.4

Choosing computer hardware for a new IT strategy

- IBM AS400
- IBM RS 6000
- IBM 486 Server
- DEC Microvax
- Tricord Superserver
- Compaq 486 Server
- AST 486 Server

Relevant criteria for comparing the options are defined and given relative weightings that sum to 1.0.

Price	0.3
Processing speed	0.3
Vendor support	0.15
Upgradability and cost	0.25
TOTAL	1.0

You may wish to break down each of these criteria further, for example processing speed into clock speed, millions of instructions per second (MIPs), disk I/O speed, RAM access time, etc. These local criteria are then given local weightings summing to 1.0. Once each option is properly defined, find the *product of the local and main criteria weightings*. This gives the overall rating of the option. Each hardware system could then be compared quantitatively with the others, against our defined criteria.

You can then decide on how to interpret the results. It could be to select the option with the highest overall rating (sum of products of criteria); this is a medium risk. However, you may wish to select the *maximin*, the option with the highest *low* outcome and therefore a low risk, or the *maximax*, the option with the best *high* outcome and therefore a high risk.

This form of decision analysis can be very useful when comparing the benefits and weaknesses of MRP2 systems, using a spreadsheet to calculate ratings for each system against a wide range of criteria.

■ Organisational review: DPA

Departmental purpose analysis (DPA) analyses the mission or goals of a business, and determines how its constituent functions interact as internal or external **customers and suppliers**. Intended to create and measure improved organisational structures, DPA focuses on the customer/supplier interfaces, and on the ability of each supplier function to satisfy its customer. It is basically for identifying the purpose and amount of added value contributed to each department. *In customer/supplier relationships, view every department or function as a business unit* (Figure 5.15).

DPA is normally carried out during the review of an existing structure to determine the strengths and weaknesses of the current department-to-department relationships and service levels, and often after restructuring to define performance measures for a new department or grouping. DPA is then used repeatedly over several years when an enterprise is committed to a mission of continuous improvement.

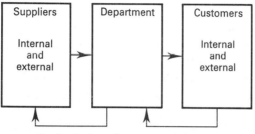

Figure 5.15   Customer/process/supplier relationships.

Begin by defining and agreeing the customer needs then consider supplier organisation and requirements. Next, establish the internal groups, procedures and systems to reorganise the tasks and set up suitable performance measures. Eight steps are involved.

1. **Define and agree mission and responsibilities of the department:** Why does this department exist? What are its goals? This assists in aligning the objectives with the overall business mission.
2. **Identify all internal and external customers of the department:** First step in defining customer/supplier relationships.
3. **Define customer requirements and standards of performance:** Meet the customers and agree their requirements. Regular reviews should be established, with feedback to measure how well a department is meeting customer needs. *Agree appropriate and practical measurements.* Customers may be requiring products, services, response times and information. Each of the different needs must be understood and provided for in the department processes and internal priority system. Detecting changing aspects of customer needs is part of this ongoing review.
4. **Decide and define what functions and activities the department or unit needs to perform to meet customer requirements:** The previous steps define and prioritise all customer requirements. Now we determine the specific activities needed to provide the necessary service. What functions is it to perform? Use a task-by-task analysis of what is necessary. Final checking of these activities should be carried out by referring to the customers. Exclude NVAs. Identify the skills and resource capacity needed in the unit.

141

Find out how rapidly you need to respond and what level of reliability is required.

5. **Identify all internal and external suppliers:** External suppliers correspond to all companies and individuals outside the business who provide goods and/or services. Internal suppliers are the other groups, departments and individuals that have to provide the department in question with goods or services. We decide the needs of our department then tell our suppliers, we are now their customer. A full listing of all internal and external customers should be produced.

6. **Define our requirements to these suppliers:** State performance requirements with agreed measurements to define when nonconformance occurs.

7. **Define, design and develop an effective feedback system:** Key to successful customer/supplier relationships, it develops a discipline for informing both parties of successes and failures, and warns of the need to change a service or priority.

8. **Review regularly and identify any defects – continuous improvement:** With the initial sound definition of customer and supplier needs, reviewing the feedback information allows early detection of any problems or defects in the services (Figure 5.16). When problems are found, carry out analysis by the appropriate problem-solving techniques.

In a given department, regularly answer the following questions:

- What are the main priority tasks?
- What levels of service are we giving in these?
- Are they meeting customer requirements?
- If not, why not?
- What are the causes of the problems?
- What are the solutions?
- Will this prevent the problem happening again?
- Is our detection and feedback system working properly?
- How can we further improve each activity?
- Are there any NVAs due to changes in customer needs?

Only by doing so will continuous improvement be maintained.

In turn, each department, group, cell and office is part of the total business system, and each will be the customer and supplier to several other functions.

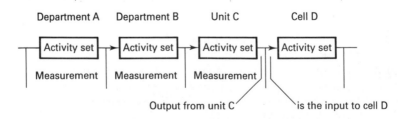

Figure 5.16 Business department analysis.

Every group must be set up and run in the manner described, continually striving to improve, analyse its own performance against the requirements of the customers it serves, and informing the supplier groups of amendments to requirements or falling performance levels. Instead of a blame culture, everyone now concentrates on quality and added value both inside and outside their unit. Figure 5.16 shows the chain of groups within an enterprise all working towards world class performance.

■ Organisational review: from/to matrices – used for forming natural groups and single offices

The 'from/to' matrix helps to determine what business functions should be involved in forming natural groups. Where a business has complex or large numbers of functions involved in various processes, this tool simplifies the plotting of information between all constituents. The basic matrix lists all functions on both row (to) and column (from) headings, as shown in Figure 5.17.

Each matrix cell is then labelled with information and entries on the number and type of communication that passes between the various functions. Obviously, the more communication between functions, the greater the need to include these functions in a natural group or cell.

We also give a rating assessment to each cell entry to describe its frequency (f) and importance (i), placing each in a standard position in the cell. An overall cell weighting is then calculated from a weighting product table, allowing all cells to be readily compared and ranked.

FROM: \ TO:	Design	Engineering	Order definition	Cust. services	BOM	Planning
Design	×	3 6 3	3 5 2			
Engineer.	3 5 2	×				
Order def.			×			2 5 3
Cust. serv.				×		
BOMs					×	
Planning			3 6 3			×

Importance Total rating Frequency

Figure 5.17 From/to matrix plotting interfunctional relationships.

Weightings	Importance	Frequency	Frequency per week
1	Low	Almost zero	1 to 5
2	Some	Low	6 to 20
3	Medium	Medium	20 to 50
4	Very high	Very high	51 to 100
5	Highest	Highest	over 100

Frequency may also have relevant quantitative measurements per unit time to give comparable meaning to the weightings.

Product of weightings ($w_f \times w_i$)	1	2	3	4	6	9
Overall weightings	1	2	3	4	5	6

A related technique for factory floor reorganisation is the measurement and comparison of physical parts and documents sent between resources and/or sections, together with the distances covered. Here the aim is also to minimise distance travelled.

The ratings in each row or column can be summed to produce a row sum or a column sum. The from/to value for each functional *area* is its row sum added to its column sum. These figures are used to decide which business functions should move to become full members of a natural group or single office. Commonly, those functions with a total rating of 4 or more would become part of the single office.

Finally, the information from the matrix is displayed on a **pie chart** (Figure 5.18) with each functional area within the group shown as a segment of the pie. Connections between areas with total rating of 4 or more are shown with a connecting arc drawn between them, and a solid dot added where the from rating is 4 or more. From/to values (row sum plus column sum) are shown outside the pie chart.

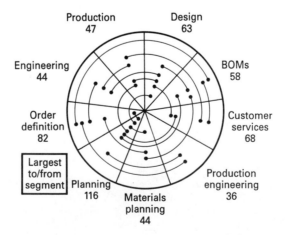

Figure 5.18 Pie chart compiled using the from/to matrix in Figure 5.17.

144

■ Summary

This chapter described how to approach improvement systematically. Several of the techniques are amenable to immediate application and offer a chance to gain valuable experience before commencing a large-scale programme. But further reading is strongly recommended for full details and examples. The basics of cell and single-office definition were also described, and their next stages are covered in the following chapter.

■ Further reading

de Bono, E. (1990) *A Textbook to Creativity*, Penguin, Harmondsworth.
Gane, C. and Sarson, T. (1979) *Structured Systems Analysis*, Prentice Hall.
Hirano, H. (1989) *JIT Factory Revolution*, Productivity Press.
Ishiwata, J. (1991) *Industrial Engineering for the Shop Floor*, Productivity Press.
Johnson *et al.* (1972) *Operations Management: A Systems Concept*, Houghton-Mifflin.
Lucas Engineering & Systems (1989) *Manufacturing Systems Engineers Handbook*, The Lucas Miniguides.
Silver, E. A. and Peterson, R. (1991) *Decision Systems for Inventory Management and Production Planning*, Wiley.

<table>
<tr><td>**6**</td><td># The Total Quality Organisation</td></tr>
</table>

If anything should be excellent,
it should be management.
Akio Murita

■ Organisational structure issues

The traditional company is structured like a pyramid, a hierarchy of heads, departments and functions that contain specialist staff and skill groups (Figure 6.1). This tends to result in a departmentalised attitude of isolation and separation, discouraging communication and creating barriers between people. This in turn leads to reduced response and more frequent errors in everyday activities. Pyramid organisations may have reasonably good communication up and down (vertical communication) but poor communication between functions (horizontal communication). And fast, flexible responses require good **horizontal** communication.

Even within a relatively small company, we frequently find identified functional units being used at different stages of the process. In practice, there

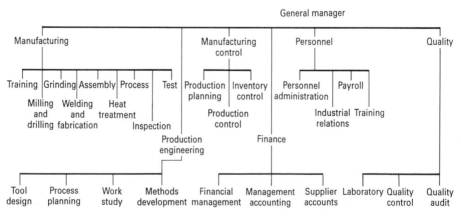

Continued ▶

Figure 6.1 ABC Engineering – The current factory organisation.

are obviously many other functions, and this places ever growing reliance on good communications between them.

The larger and more complex the organisation becomes, the harder it is to maintain rapid and efficient communication. Special measures may be required, but they can create inefficiency and turn into NVAs.

■ The greenfield site approach

If we start any improvement project with preconceived ideas of the areas or functions that cannot be changed, we will not achieve the maximum benefits that are possible. A change in mindset is required, and to assist with this, the factory or business should be considered as a greenfield site. It should have:

- No trade boundaries
- No union boundaries
- No functional boundaries
- No management boundaries
- No personal boundaries

Starting from a clean sheet, we can more easily identify desirable aims and objectives for any existing organisation.

■ Break down functional boundaries and barriers

The total quality, systems engineering approach to achieving a more effective organisation is to move away from the complex, multilevel structures that are commonplace, moving instead to the establishment of simple and flexible groups – modules and work cells based on a single-office concept. Fragmentation and poor communication are attacked at source, by placing all necessary

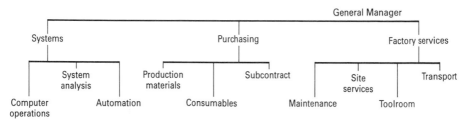

skills and resources under one module or cell. The organisational structure must always be reviewed against the business needs and objectives, not purely from the perspective of staff cuts. We therefore have to ensure that the review and possible reorganisation are carried out on a planned, analytical basis, not merely as a rearrangement of people or titles that provide no real benefits to the business.

■ Decentralisation and natural groups

The concepts and benefits of decentralisation are well established, but tend to be associated more with government departments and multinational corporations. The same concepts are equally applicable in small and medium-sized enterprises (SMEs) to provide improved communication, service levels and ownership of a single or limited product range. Based on the aims and principles of natural groups, decentralisation will frequently provide these and other benefits but it can have several outcomes: business units, production modules and single offices. They depend on the nature of the business, its resources and its products. These include:

Business units

Business units are created when a company decentralises its functions to form several smaller complete businesses, with all or most of the resources needed to give full responsibility for a product, a group of products or a range of services. Consider a single company formerly run as a single business with centralised support functions and separate production units making

- pumps
- valves
- control systems
- electric motors

After decentralisation, all support functions are deployed to four new business units, working as independent profit centres, with all necessary skills and facilities to work autonomously. This includes finance, administration, purchasing, planning, sales and production. Work between units is on a subcontract, costed basis.

Production modules and cells

Production modules and cells may be the further decentralisation within a business unit, or the decentralisation pattern for a complete company, where product type or complexity does not make separate business units viable, i.e. the end product requires services and/or products from all component modules and cells. For example, an aero-engine is a complex product with a large number of sub-assemblies and components. Modules and cells are most suitable, to decentralise several support functions and give production autonomy at the level of sub-assemblies and component families.

Single offices

A single office is a form of product- or service-based cell, describing the reorganisation of resources for all contributing functions into a single (natural) group and location, to provide all service and support needs of the product(s) being addressed. Single office is also a term used to describe this form of regrouping of resources.

■ Organisational aims: natural groupings

- Vertical integration
- Functional integration ⎫
- Reduce number of departments ⎬ Single office
- Reduce the amount of job fragmentation ⎭
- Reduce the amount of overhead services required

Commonly in the traditional organisational structure, work and information can flow backwards and forwards between various departments and functions, often involving travel between many different offices, wasting time and effort *instead of adding value* (Figure 6.2). No doubt there will be work done on the information or documents passed at each stage, but each activity has a lead time, a cost and a margin for error. This leads to

- long lead times for processing information

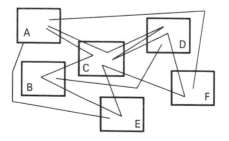

Figure 6.2 Information and document flow between offices.

- excessive paper for good communication
- lack of clarity and control, little ownership of an operation
- complexity and unreliability

Natural groupings

A proposed solution to these often ineffective methods is to redefine the essential processes into natural groupings of people and skills that can carry out and own one or more complete processing systems (Figure 6.3). Much as a manufacturing cell can be set up containing all the skills, resources and equipment to produce a product or product family, so we can create office cells or groups around information and documentation flow, often at a single location. Frequently the creation of these multiskilled groups, in open-plan offices, has a rapid and marked effect on improving response, accuracy and motivation. *Natural groupings can be formed as offices or cells of people and resources to provide all necessary services to a deliverable product or item.*

When analysing the activities and processing steps performed on information, this will often be drawn up as a flow chart of the stages involved. In a natural grouping, the multiskilled group would deal with all activities in

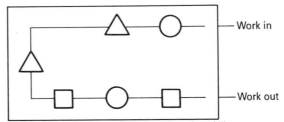

Natural group of office/information processing skills

(a)

Manufacturing Systems Engineering

Organisation

- Review the traditional multilayer fragmented/ compartmentalised organisation style

- Create a new integrated organisation style
  Eliminate restrictive segmentalism by a time phased recombination of over-specialised departments – get the organisation right – aim for T.Q.O.

(b)

Figure 6.3   (a) Natural grouping of office/information processing skills; (b) systems engineering approach to organisation.

such a flow chart, removing the need for interdepartmental travel and delay, and also removing any NVAs.

The concepts of natural and complete one-stop offices and groups of resources are equally applicable in both industrial and commercial settings. In industry, a work cell can be designed to provide all machining functions and skills for a family of parts, autonomously responsible for passing parts on to the next cell. In offices, different skills can be grouped together to deal with document processing in total, rather than sending paper or files between sections. Figure 6.4 illustrates the traditional and natural group concepts of floor layout.

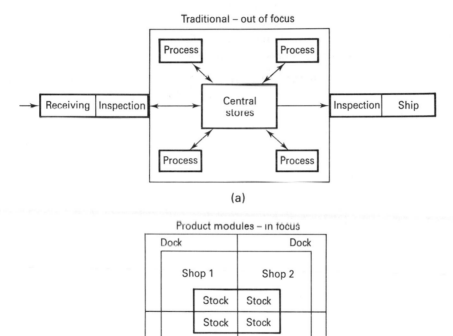

(a)

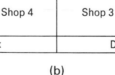

(b)

Figure 6.4 (a) Traditional centralised layout, unfocused; (b) natural grouping layout, focused.

■ Advantages of natural groups, cells and offices

By creating a unit that has total responsibility for a process, we produce an almost perfect communications environment, where talk and information can go directly across the table, with rapid and valuable feedback. Such communication becomes less complex and less reliant on computers, with fewer documents travelling shorter distances around the company. The group is also a natural area for thriving quality circles and continuous improvement groups.

The improvement group will have control over priorities, resources and performance for all group-based activities, giving clear responsibility, accountability and measures of group quality performance. As part of the internal customer/supplier relationship, each group will be a customer and supplier to other natural, single-office groups.

Natural grouping leads to reduced operating overhead, needing fewer supporting systems and resources once the NVAs are removed. For the group members, the multiskilled environment offers improved career opportunities, job satisfaction and flexibility. *Strive for ease of communication, rapid response and maximum flexibility by following these steps*:

- Review and redesign the structure; analyse department functions and purposes, using job design and departmental process analysis throughout the company.
- Move towards a business unit or single-office concept; eliminate fragmentation.
- Establish simple and flexible groups.
- Remove NVAs.
- Eliminate informality.
- Develop and document simple operational procedures.
- Provide performance measures reflecting customer/supplier needs.

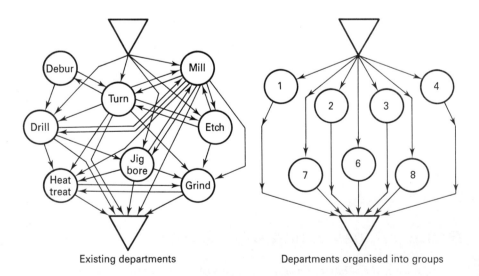

Existing departments   Departments organised into groups

Figure 6.5 Factory flow before and after natural grouping and flow simplification.

A major focus in the improvement programmes described in the text is the elimination of NVAs. Coupled with the improved communications, flexibility, control and speed of response offered by modular and group working, it makes a business reorganisation very worthwhile.

Distances travelled by parts and/or paperwork can normally be reduced and simplified in terms of the paths through an organisation (Figure 6.5).

■ Further advantages of natural groups, cells and offices

Systems engineering aims to give full autonomy to small business units – single offices that contains all necessary functions and services to plan, make and ship the product. The product can be tractors, package holidays or a repair service.

In many office and production situations, there is much to be gained from the reorganisation into cells or single offices, containing all necessary functions and expertise. But even companies organised around cells may not have effective operations. Their cells may be process based, not product based:

- A milling cell; a bending cell
- A scheduling function; a production engineering function
- A product cell due to a local initiative

Cells alone do not mean high levels of performance. Manufacturing systems engineering (MSE) is based on cellular concepts and addresses all areas and functions connected with product manufacturing.

Autonomy

- Define modules and cells with complete responsibility for manufacture or servicing of groups of products.
- Ensure autonomy so that all resources required to manufacture or service the product group are part of the cell.

Accountability

Conventional organisational structures are horizontal or functional in nature, with specialist departments reporting upward to each progressive layer (Figure 6.1). This does not give accountability for one product or product family, and results in low ownership of product quality and/or service level.

The vertical MSE-based/product-based structures effectively cut through the horizontal layers of organisation and poor communication, taking expertise and resource from each functional area and redeploying this into product-centred units with accountability for all aspects of one or more products (as in Figure 6.9). A combination may be required to give appropriate levels of control and autonomy

From functional accountability
Accountability to a central group responsible for a function, e.g. manufacturing engineering

To product accountability
Accountability to a group responsible for a particular group of products

■ Existing process layout or functional layout

Figure 6.6 is typical of a traditional machine-shop; it has similar plant and functions grouped together and is process based:

- High stock levels
- Long lead times
- Goal of high utilisation
- Lack of ownership
- Lack of accountability for the product

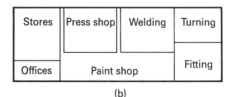

(b)

Figure 6.6 Process-oriented layout.

Flow lines or sequential shop

In flow lines (Figure 6.7) dissimilar items of plant are grouped (usually in a line) to allow a product to pass from one machine to the next with minimal travel and lost time, for example:

press – weld – paint – fitting – electrics – final assembly

And in the office context:

estimating – sales – BOM – production planning – materials – purchasing

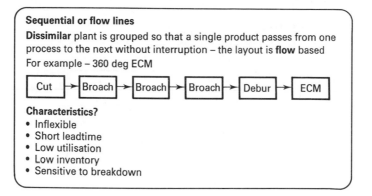

Figure 6.7 Sequential or flow lines.

These lines have the following characteristics:

- Short lead times
- Low inventory
- Low utilisation of overall plant
- Inflexibility
- Vulnerability to breakdowns

■ Manufacturing systems engineering: autonomous cells

Manufacturing cells (Figure 6.8) are now recognised as essential tools for removing waste and inefficiency from the manufacturing process. By moving parts and product in single or small batches through a grouped set of operations, we can eliminate waste in transportation, queues, storage and inspection. Originally of Japanese origin, these U-shaped cells consist mainly of simple, small equipment with operators feeding parts round the system using a simple pull control system. Each operator makes quality checks on the part before passing it on.

With MSE, the aim is to define and create cells that can have complete ownership and responsibility for a product or range of products of any form. We must provide the cell with all necessary facilities and resources to carry out the required tasks and jobs. This often requires the redeployment of service staff and resources into product-focused cells.

Figure 6.9 shows the traditional functions on the right, functions that were centralised to cover all products. Now, under a TQO, these functions may be decentralised down to the level of products or business units, appropriately staffed.

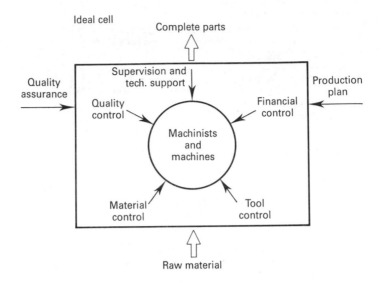

Figure 6.8   An ideal cell contains all the functions and skills it requires.

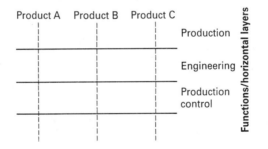

Figure 6.9   Horizontal versus vertical organisation.

We may have autonomous cells in

- Offices
- Manufacturing
- Design

In each and every case, we need to strive for a single office or business unit, aiming for simple, effective operation and accountability for both material and information flow. This often means moving away from traditional measures of efficiency, towards a new set of measures for effective performance and operation.

Plan the move from measures of efficiency towards effective operation

Operator performance	Production/job input and output for office and shop-floor functions
Department performance	Lead times for all tasks in a cell or office
Standard hours input and output	Inventory levels
Scrap and rework measurement	Quality conformance towards 100%

■ Measurements of performance for each area

Traditional systems measure direct labour efficiency and plant utilisation, poor indicators of actual business performance and ill suited to the new methods of flexible staff and flexible production. For all visible areas we must develop performance measures (Figure 6.10) related to local customer/supplier requirements and tied to overall business objectives.

Manufacturing Systems Engineering

Organisation

Weekly measure of performance – manufacture
- Stock turnover rate
- % achievement of planned product mix
- % quality achievement (aim for 100%)
- average cell lead times
  not standard hours or department or operator performance
Different measures for admin/service cells – related to business parameters

Figure 6.10   Some cell performance measures.

■ Systems engineering: effective versus efficient

Effective operation is fundamentally different from efficient operation. Many so-called efficient measures and practices lead directly to poor and uncompetitive ways of working. The following table shows how the two approaches compare:

Efficient	**Effective**
Conventional shop with functional layouts and poor flow	Flexible shop with group/cell layouts and good flow
Slow set-ups	Very quick set-ups
Large batches/long runs	Small batches/short runs
Maximises worker and machine utilisation, tends to overproduce	Avoids overproduction and redeploys to other quality/improvement duties
High inventory buffers	Minimised inventory
Long lead times/unresponsive	Short lead times/responsive
Makes to forecasts (inaccurate)	Makes to order
Complex production control	Simple production control
Difficult to manage and control	Easy to understand and manage
High overheads	Low overheads
High cost of sales and overheads	Low costs and overheads
Insensitive to customer needs, constrained	Responsive to customer needs, accommodates rapid change
Poor delivery performance	Meets all delivery requirements

Through the application of improvement methodologies and techniques, including cellular reorganisation, our objective is to become a much more effective operation, with the appropriate performance measures to gauge the level of effectiveness and continuous improvement. *Remember, this is applicable to virtually any form of business: manufacturing, service or commercial.*

■ Organisational analysis and redesign

Before continuing to describe the design and creation of more effective, often cellular businesses, it is necessary to consider the techniques and methods available to assist in analysing and indentifying potential cells, modules or other improved single-office structures.

We require a method to find out what a department does, how it does it and why it does it. Various methods are suitable, including IDEF (Chapter 3). Flow charting or I/O analysis is also used to investigate document flow and to refine the new basic structure for our business.

Whilst IDEF and other similar tools are very useful for mapping and information display, they do not, by themselves lead to restructuring proposals. Other techniques are available to provide better information and to determine the interactions between business functions. Departmental purpose analysis (Chapter 5) and job design are particularly good. Departmental analysis, as the name suggests, relates to the analysis of departmental functions, whereas job design refers to the analysis and respecification of individual jobs and people.

■ Cell and module definition

Referring back to the earlier section on natural groupings, we are defining a multiskilled multi-operational group or team that will own a complete group of tasks, operations or processes, and should therefore be located closely together as a cell, unit or office (as in Figure 6.11). The term natural group is often used to describe the office or support equivalent of a manufacturing cell, but in this text is used synonomously.

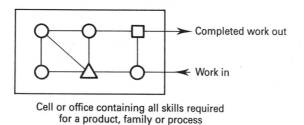

Cell or office containing all skills required
for a product, family or process

Figure 6.11 A single office owns all the processes it requires.

A cell is well defined when it is best suited to the market needs, with clear and obvious boundaries. A well-defined cell is simple to control with a high level of autonomy. When a cell is difficult to control, due to its size or range of activities, it may be sensible to create a module consisting of two or more cells that act together or individually as required (Figure 6.12).

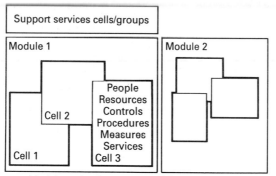

Module and cell structure within a business unit

Figure 6.12 Cells forming a business module.

■ Cell management and sizing

Each cell will normally be managed by a cell leader, who is responsible for all functions and resources within the cell. A module is managed by a module leader or business unit manager. This manager has responsibility for all component cells and any support office facilities that exist within the module.

Cell and module sizes vary according to the industry, product type and natural group sizes that are defined. For ease of operation, however, cells tend to have between five and twenty staff and each module contains no more than eight cells, equivalent to a production unit of between forty and two hundred people.

Cells are normally customers and suppliers to other cells in the same module, and also to other cells in other modules. Cell teams therefore need to be clear about the requirements of their customer cells, particularly regarding the performance measures for customer service. In turn, they will measure their supplier cell performances with agreed measures.

■ Cellular versus linear

Cellular manufacturing is not a panacea. Many industrial sectors, including the majority of the process and mass production industries, require specialised plant and equipment, frequently arranged in production lines. In fact, substantial benefits can be achieved from the application of JIT and kanban systems on small and large production lines. Where standard, repetitive products are required, the best solution is often linear.

Figures 6.13 and 6.14 show how an ineffective group of operators were

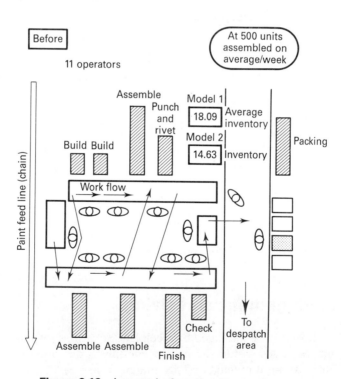

Figure 6.13 Layout before line improvement.

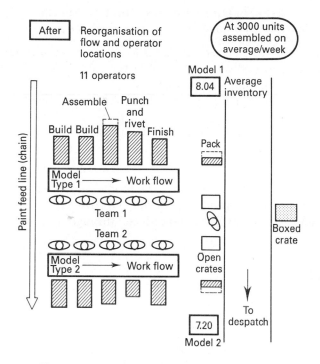

Figure 6.14   Layout after line improvement.

transformed into two product-specific lines or teams, greatly increasing throughput and lowering overall line inventory. This was largely achieved by reorganising the sequence of operations as shown. It only takes a slight reduction from these very high volumes for the benefits of cellular organisation to emerge.

■ Physical layout of manufacturing cells

The concepts of cells and their advantages have already been discussed: teamwork, ownership of a product, ease of communication, flexibility, short flow paths, visibility, etc. In order to realise these advantages, it is necessary to find the optimum physical layout of a cell. It is generally agreed that a U-shaped cell is ideal.

■ U-shaped cells

Where a group of machines and/or people are to work on a family of parts or products (gears, blades, motors, garments, furniture, specimens, etc), the U-shaped cell (Figure 6.15) is generally believed ideal because

■ The close proximity of operators leads to good communication.
■ Goods flowing in and out of the cell at the same end leads to effective movement and minimum travel distance
■ Movement around the U, from resource to resource, leads to ideal flow without backtracking.
■ There is high visibility across and around the cell.
■ By moving work from one station to the next, each operator has visual control.

This sequential grouping of resources commonly allows the elimination of interprocess WIP (stock), giving significant improvements (50 to 100%) in productivity and inventory levels, and reduced throughput time.

Example cell

Within a typical cell, there is flexibility between operators and machines, jobs and skills. This allows easy movement of staff between stations and provides coverage for absence. In many production cells, operators run more than one machine, or may move workpieces onto the next station once theirs is operating automatically (or in cycle). It has become commonplace for well-organised cells to have at least four machines run by a single operator, who loads and set then unloads the machines. Such cell have a very high throughput.

In this situation, the operator must be fully trained and experienced in each machine or processes, and the equipment tends to be smaller and simpler than for other cells. Small machines are relatively inexpensive to buy, simple to maintain and usually easy to operate.

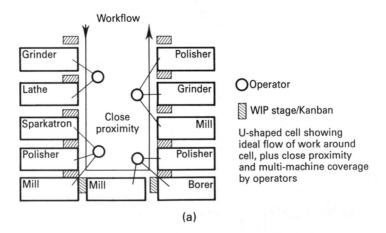

(a)

Figure 6.15 U-shaped precision machining cell.

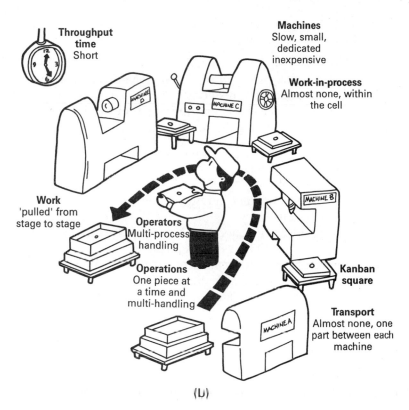

Throughput time Short

Machines Slow, small, dedicated inexpensive

Work-in-process Almost none, within the cell

Work 'pulled' from stage to stage

Operators Multi-process handling

Operations One piece at a time and multi-handling

Kanban square

Transport Almost none, one part between each machine

(b)

Figure 6.15 Continued.

■ Product-based cell definitions

We define a cell according to its

- physical size
- physical layout
- range of products to owned/ processed

To create viable cells or modules, we examine the features of products and families to identify common or similar areas. Typically check:

- Product/part type
- Manufacturing routing
- Process steps and types
- Process and overall lead time
- Market and/or customer
- Volume, range and frequency of parts
- Physical size and shape/geometry

- Requirements for precision
- Material
- Product life-cycle

In each area, we are looking for possible commonality and good fit between parts, products and families. The resulting cell may deal with part families that exhibit similar material, dimension and precision, or maybe another combination. The approach to forming the resulting production cells is based on

- Natural groupings of people, resources and products
- Manufacture in part families
- Simplicity, not complexity
- Small machines, not supermachines
- Flexibility, not rigidity
- Frequent and rapid changes, not large batches; drive to smaller batch sizes, reduce set up and lead times as far as possible
- A mobile and flexible workforce; eliminate demarcation
- Formal working practices in a quality environment
- Local planning, not remote planning
- Involvement of all workforce
- A united approach
- U-shaped lines, not linear lines
- Quality control by all workers, not by a centralised resource; build quality into the manufacturing process and methods

In designing and implementing new, cellular production, make best use of existing plant and equipment wherever possible, avoiding the need for yet more capital expenditure. Where new equipment is needed, employ sensible and cost-effective automation and technology.

If we identify a group of parts or products that all require the same resources for say 90% of the operations, we determine the machines and skills required to perform these operations on the expected part volumes. Thus one family and cell factor can knock-on to a second factor, and so on.

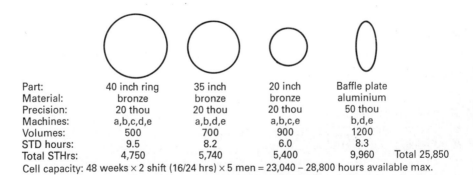

Part:	40 inch ring	35 inch	20 inch	Baffle plate	
Material:	bronze	bronze	bronze	aluminium	
Precision:	20 thou	20 thou	20 thou	50 thou	
Machines:	a,b,c,d,e	a,b,d,e	a,b,c,e	b,d,e	
Volumes:	500	700	900	1200	
STD hours:	9.5	8.2	6.0	8.3	
Total STHrs:	4,750	5,740	5,400	9,960	Total 25,850

Cell capacity: 48 weeks × 2 shift (16/24 hrs) × 5 men = 23,040 – 28,800 hours available max.

Figure 6.16 A family of ring parts.

Cell part families

In Figure 6.16, the parts are all suitable for making up a cell, based on the similar features of dimension, material/tooling, precision, machine and process routing. The volumes expected, with the standard labour required per part, indicate a cell of five machines and five workers.

■ Office improvements

Office systems and support for production can account for a substantial overhead and cost, and will normally benefit significantly from the application of improvement teams. In a cell or module clerical work and information passing can be reduced dramatically. In an aerospace multinational, the systems engineering team found that ten tons of paperwork were produced for each ton of engine! More significant than its quantity and cost was the labour time and cost penalty it represented. Through focused improvement teams, the labour and NVAs were reduced by over 6%, saving on cost, time *and* paper. Interesting and relevant though it may be, paper is only one way to measure the output of office systems. Even where there is much less paperwork or office automation, other improvements can still reduce wastage.

CASE STUDY 6.1

A single office: BOM, PC, material control, estimating

Background
Within a large manufacturer of engineered and finish-to-order commercial display systems, there were problems in order definition and production in time to meet customer due dates.

Quotes and orders were dealt with by the customer services function, reporting to sales. Mainly non-technical, these staff performed price estimating and outline job definition. From here, quotes or jobs requiring design or engineering action were passed to that function for progressing. Once design and/or engineering completed their work, their information was passed to the bills of material (technical definition) function, who then created a full specification and bill of material for the job. Once the BOM function had finished, all product information was passed to the planning function, consisting of material control and production scheduling. They would then process this information to order materials, and schedule parts and products to be made within the operations function. Frequently this resulted in late scheduling of work to the shop-floor, late receipt of purchased parts, and hence late delivery of product to the customers, who were very dissatisfied. Other information flows were also taking place, as shown in Figure 6.17(a), where various functions required information that was not provided by the job order, or was not available soon enough to meet customer deadlines.

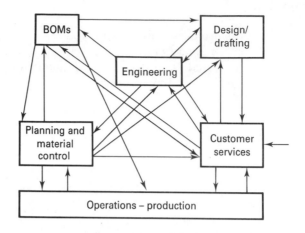

Figure 6.17   (a) Organisation and layout before reorganisation.

The effects of these problems, and the overcomplex paths for information passing, were resulting in the effective extension of the product lead time, causing customer dissatisfaction. From Chapter 1:

Total product lead time = estimate or quotation time
+ sales order processing time
+ design/modification time
+ full order definition time
+ time to source special components
+ scheduling time
+ production, assembly and test time
+ warehousing time
+ shipping/delivery time

The above organisation and operation was impacting on all areas from scheduling to shipping/delivery, which in turn affected total product lead time.

Problem analysis

During the review and analysis of the problems, effects and causes, it was quickly recognised there were several effects attributable to a main cause, showing it to be a more serious and widespread cause of disruption, ineffective working and low morale than was previously understood.

The root cause of the problems was the lack of technical expertise present in the customer services function. This was regularly producing incorrect job order specifications, causing disruption and wasted activity in most other functions. Additional corrective work was regularly needed, causing further delays in other work and adding to overall costs. The regularity and effects of these errors was producing interfunctional friction and lack of co-operation. Morale was low.

Problem solution

It was recognised there was a need to provide a more complete set of skills

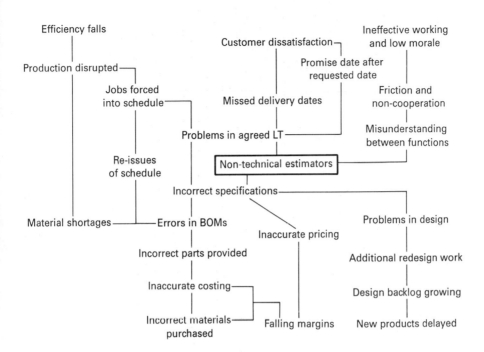

Figure 6.17 (b) Cause and effect chart of problem and implications.

(technical and sales-oriented) at this critical point of contact with both the customer and the production functions. At the same time, there was a potential to reduce the amount of duplicated effort and double-entry of information, if the related functions could be merged into a natural group, providing the majority of skill and resource needs within a single office.

This led to the reorganising of several functions, and the formation of a new unit responsible for order estimation, definition and production planning (Figure 6.18). In addition, a new computer-based order configuration system was developed to aid and speed up the ordering and interpretation process.

Outcome

Once implemented, this single office demonstrated significant improvements immediately. With departmental barriers removed, and team input to customer orders and specifications, accuracy grew quickly, eliminating or reducing most of the harmful effects described in Figure 6.17(b). Accurate information led directly to more accurate and achievable delivery dates, and paperwork was consolidated and reduced. Coupled with the new order configuration system, it reduced the overall lead time to develop and process full order information by over 50%. This in turn allowed staff cuts, further lowering overall operating costs.

Design and production functions were able to be planned and scheduled with greater accuracy and stability, improving their abilities to deliver on time, and directly improving customer service levels.

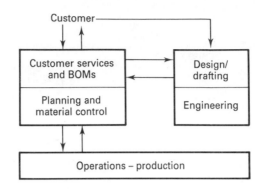

Figure 6.18 Organisation and layout before reorganisation into natural groups.

Interfunctional movement of information was greatly reduced, and lines of communication simplified, to provide more effective and direct information flow within the business. New group-based performance measures were established, including

- percentage complete orders
- Order details on time (for scheduling cut-off times)
- Full job definition accuracy
- Percentage designs and drawings on agreed
- Percentage delivery dates matching customer requested dates

CASE STUDY 6.2

Restructuring at ABC Engineering

ABC Engineering is a medium-sized UK mechanical engineering company whose main products are pumps, valves and filters. These products are mostly used in the oil, gas and chemical industries. The company has three manufacturing plants in the UK and a central sales and design office in London. The Scottish plant employs about 1,200 people and uses specialist materials to manufacture high integrity parts in four main groups: centrifugal pumps, axial pumps, filters and valves. Raw materials, mostly bars, forgings and castings, are bought in.

ABC has a basic functional structure. Specialist functions report to functional heads for

- manufacturing control
- quality
- purchasing
- production engineering
- factory services
- finance
- systems

Within each of these specialist areas there is further division into separate functions. Manufacturing is divided into

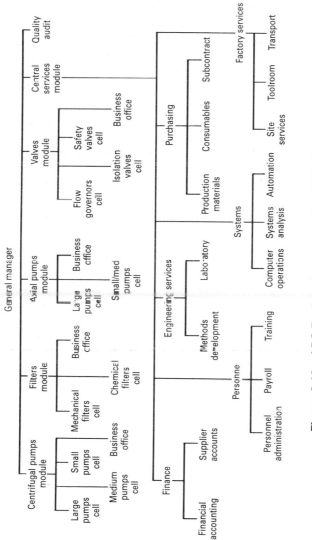

Figure 6.19 ABC Engineering, a possible MSE alternative.

- milling and drilling
- fabrication
- welding
- heat treatment
- assembly
- process
- inspection

This process-based structure forces parts and products to move between departments. The same is true for the associated paperwork and services, which are passed between several other departments, causing time and effort to be wasted as delays and queues are met at the various customer/supplier stages. Several drawbacks exist:

1. Lack of accountability for finished product because so many functions are involved
2. The sheer number of different functions and the associated communication problems
3. The number of interfaces and routes to be travelled by any product going through the system

Such a structure may also be top-heavy and overstaffed to cope with a fairly high degree of job duplication between departments.

A possible cellular alternative

Applying the concepts of single-office, cellular groupings, we can restructure this company to create product-based modules and cells (Figures 6.19 to 6.21). The objective in doing this is to group together all necessary services and resources required by a product or family of products. By doing this, and placing them under one module and module leader, it is quite possible to dramatically reduce the lead time for all aspects of part processing, including planning and production times. All module resources are controlled by and dedicated to the module delivery plan, avoiding any split loyalty or priority conflicts. Working as

Module
leader
|
Module business
planner
|
Business planners

Responsibilities:
    Management accounting
    Manufacturing planning
    Industrial relations
    Material procurement
    Sub-contract procurement

Figure 6.20 ABC Engineering, a module business office.

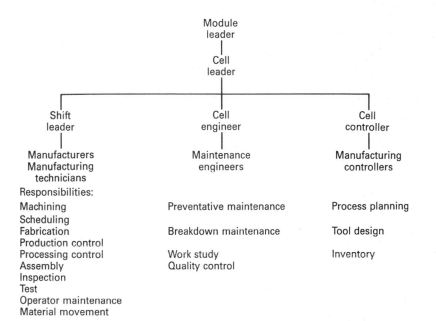

Figure 6.21   ABC Engineering: a typical cell structure.

part of a product- or service-oriented team, job satisfaction and overall service quality will also increase.

In practice some functions may be inappropriate or unsuitable for decentralisation into product-based modules. Typically, systems and personnel functions may remain centralised, depending on the size and nature of the company. Practicality and common sense are overriding in every case.

■ Job design in review and reorganisation

Key to defining the organisational needs of a firm is the identification of sensible responsibility groupings and to individual responsibilities at every level. When merging and reorganising functions, it is inevitable that job responsibilities of supervisors and operators will change. Job design defines each job in terms of information to be processed and tasks to perform. It should be carried out after the cellular structure has been fully designed. An example comparing a traditional functional structure to one based on cellular grouping is given below.

A systematic approach to job design

1.   Define tasks carried out by existing functions; use I/O charts.

2. Transfer tasks to proposed organisation structure: cells and single-office groups.
3. Develop job descriptions for each job within the proposed structure.
4. Consider implications of proposed structure and job designs, including any payment or compensation issues; amend to create new job specifications.
5. Develop structured plan for phased transition to new structure.
6. Negotiate with prospective candidates.
7. Implement with training and guidance throughout.

Job design: summary

Based on the same principles as natural grouping of functions, job design involves the application of these principles to grouping individual tasks and responsibilities at the lowest level, individual posts and people.

CASE STUDY 6.3

Job design in the single-office case study

1. Define tasks carried out by existing functions; use organisational structure charts followed by I/O Charts (Figure 6.22).
2. Transfer tasks to proposed organisation structure: cells and single-office groups (Figure 6.23). From analysis and rationalisation of the tasks, inputs and outputs for the existing jobs, new job descriptions were prepared as I/O diagrams for the new manager and team supervisor roles (Figure 6.24).

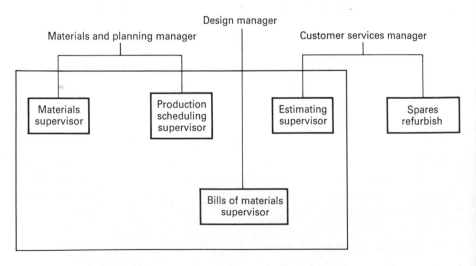

Figure 6.22 Previous job descriptions and structure.

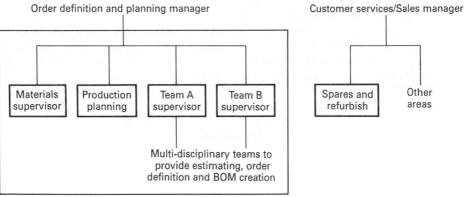

Order definition and planning manager   Customer services/Sales manager

| Materials supervisor | Production planning | Team A supervisor | Team B supervisor |

Spares and refurbish Other areas

Multi-disciplinary teams to provide estimating, order definition and BOM creation

Functions to be reorganised into a single office group

Figure 6.23 New job descriptions and structure.

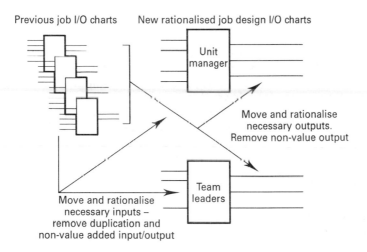

Previous job I/O charts New rationalised job design I/O charts

Unit manager

Move and rationalise necessary outputs. Remove non-value output

Team leaders

Move and rationalise necessary inputs – remove duplication and non-value added input/output

Figure 6.24 Unit and team functional responsibilities.

These contained the merged inputs and outputs from the previous job descriptions.
3. Develop job descriptions for each job within the proposed structure. The reorganised structure effectively merges the activities to form team leaders with responsibility for two groups that deal with all aspects of product support: order definition, BOM creation and maintenance, planning and scheduling to production. Separate service supervisors were retained for materials planning (which was a common resource but very awkward physically to separate) and for master production scheduling. Future reviews

of these functions were planned to check on scope for additional decentralisation of these functions.

4. Consider implications of proposed structure and job designs (including any payment or compensation issues); amend to create new job specifications. This included listing and review of all links and communications to the other functions within the company. Training for awareness and operation of amended procedures was given to all customer and supplier functions, to ensure adherence to the new operational procedures.
5. Develop structured plan for phased transition to new structure.
6. Negotiate with prospective candidates.
7. Implement with training and guidance throughout.

Stages 5, 6 and 7 are not described here; they would be agreed locally.

■ Problems and barriers

Resistance to change

Change to existing procedures and methods generates resistance from a large proportion of the workforce. However, when approached in a certain manner, change can be used to advantage, actively encouraging and motivating employees through consultation and involvement. The management of change can be a complex and far-reaching component in the route to world class performance. This subject is addressed in Chapter 8.

Headcount reduction problems

Problems will emerge more rapidly if there are staff cuts before the clerically and/or labour intensive jobs are simplified, computerised, or removed. This is a fact of life but companies should still plan strategically to overlap with incoming computer-based systems and simplified formal procedures. In the worst possible scenario, the staff is cut before formal procedures or effective computer systems are established. Here you can be faced with inadequate staff to carry out day-to-day tasks, let alone design and implement company-wide changes to procedures and systems.

Possibly the only positive aspect of being in such a poor position is that the workforce will be highly receptive to any improvement programme, and it may be easier to obtain the high levels of motivation and commitment to change than in a profitable organisation, provided morale has not fallen too far. Competent staff are needed to map out the path to improved performance and profitability.

■ Summary

A total quality organisation must be created to form a solid foundation for all other improvements. Reorganisation of a business to improve and streamline its performance is therefore one of the first stages in moving towards a total quality operation. The strengths and weaknesses of the organisation are addressed through a business review that produces recommendations for change. The structure and jobs required for the new organisation are defined through decentralisation, single offices and natural grouping supported by job design, DPA and other techniques.

The principles of systems engineering are applied to organisation and job design to reflect

- characteristics of specific products and services
- maximum flexibility
- complete accountability
- common organisation across the business
- independent requirements of site and product cultures
- maximum autonomy

A systematic approach should be taken throughout:

Elements of a systematic approach

- Treat everything as a process; processes have lead times and reliabilities
- List and eliminate the NVAs
- Look for natural groups of people and products; do not automate existing fragmented office structures
- Regroup service functions into an accountable unit
- Simplify paperwork and systems
- Measure the performance of everybody and set targets related to the business performance

From this, natural groups of functions are found and merged to form effective cells or modules, dealing with product-based families of work, and using cell- or group-based targets and performance measures. Detailed cell design and development is described in Chapter 7.

■ Further reading

Child, J. (1984) *Organisations: Problems and Practice*, PCP.

Kast, F. E. and Rosenzweig, J. E. (1985) *Organisation and Management*, McGraw-Hill.

Mintzberg, H. (1979) *The Structure of Organisations*, Prentice Hall.

Pritchard and Murlis (1992) *Jobs, Roles and People: the new world of job evaluation*, Brealey.

Skinner, W. (1974) The focused factory, *Harvard Business Review*.

7 Cellular Design and Operation

The modern approach must recognise contemporary demands for high variety, complex, high value products with short life-cycles, and pressures for continuous change and improvement . . .

Dr J Parnaby

■ Introduction

The previous chapter dealt with methods and approaches that are applicable across several business sectors. However, there are particular methods and techniques that apply to certain types of business. Cellular and modular product-based operating is widely applicable, but has specific and significant advantages for manufacturers. This chapter deals with the detailed design and application of cellular systems in product-based businesses.

■ Cell definition and design: a systems approach

The tasks of cell definition and design (Figure 7.1) need to be fairly rigorous and well tested, if the results are to produce workable, effective cellular operations. There is therefore a step-by-step process that is used in the analysis, definition and testing of cell groupings and capacities. This contains the following stages:

1. Preliminary analysis
2. Strategy definition
3. Cell grouping definition
4. Customer/supplier definition
5. Cell autonomy analysis
6. Static cell design: average sizing and resources
7. Dynamic cell design
8. Final cell definition

These stages provide a physical cell definition, its layout, its range of products and its operations. They also define two main aspects of how the unit will operate:

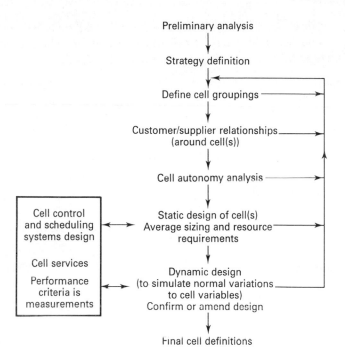

Figure 7.1 Flow chart for all stages of cell design.

■ Procedures and work instructions
■ Control and scheduling systems

These topics need to be dealt with as part of the overall cell design (for example, scheduling and control is part of both static and dynamic design), but are important enough to be dealt with as separate sections.

■ Preliminary analysis

Preliminary analysis is an overview of the company or area being considered. It may also identify which areas should be the initial focus for MSE improvements. This is equivalent to the review of operations and practices in Chapter 4. The task force collects information on

– Products and families
– Systems and processes
– Existing groupings and background reasons
– Departmental responsibilities

Using the basic concepts and rules of TQ and JIT, the analysis identifies what changes should occur in order of priority and benefit. The methods of cell analysis include MSE techniques described in Chapter 5:

- I/O analysis
- SWOT charts
- Cause and effect analysis
- Pareto analysis

Cell strategy definition

From the preliminary analysis, we determine a strategy for cell design in terms of

- materials used
- customers
- market demand
- functional route

We may find that a combination of approaches are required for different areas of the company.

■ Cell grouping definition

Using suitable tools and techniques, we analyse and group the activities and/ or process that each cell will perform. Several tools are available to assist in cell definition:

- Group technology principles
- Rank order clustering (ROC)
- Process/information flow analysis
- From/to analysis (Chapter 5)

Group technology principles

Group technology principles are well known and described by many texts. They create practical family groups of parts or products that share one or more important attribute, such as dimension, geometry, routing, material, etc. Various GT techniques and tools exist that can be usefully employed in cell definition.

Rank order clustering

Rank order clustering (ROC) is a method that uses part and product data – volumes, operation, times, processes, routes, – and calculates best-fit patterns of part grouping based on similarities in these attributes.

Frequently used for comparing hits on resources by each part or part family, ROC is ideally carried out by creating a matrix of part numbers versus resources. A hit is entered in the matrix where a resource is required to carry out one or more operations on a part. Treating the hits in each row or column as a binary number, ROC sorts each successive row in descending order, placing hits in the left-hand side of the top of the matrix. This has the result of displaying potential cell groups along the leading diagonal, as shown in Figure 7.2. The objective is to create blocks of hits, indicating suitable groupings of resources to service families of parts.

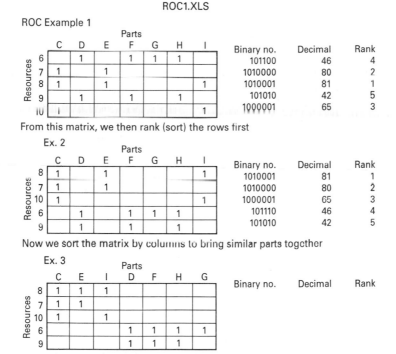

Figure 7.2 Rank order clustering using a PC spreadsheet.

179

Computer spreadsheets are a flexible medium well suited to perform ROC calculations, sorting and ranking of groups. ROC is not flow-oriented, however, and only gives an indication of part groupings, with no respect for flow or sequence. ROC tools will tend to produce a worthwhile shortlist of family groupings; detailed analysis normally has to be carried out with an expert eye combined with other flow techniques.

Two limitations can hinder ROC creation of cellular groups. Rogue parts, with operations that do not match in any way to the majority part/operation needs, can cause ROC to produce invalid clusters. The solution to this type of problem is to provide for part or work transfer outside the cell – to subcontract.

Where machines or resources are visited by a high percentage of group parts, a bottleneck can occur. This may happen when a first-choice resource is specified, and can often be solved by amending part routings to show alternative operations on another resource. For example, a CNC mill is specified for the majority of parts, with two other NC and manual mills available as second choices.

Process/information flow analysis

Process/information flow analysis forms cells based on existing process route documentation and data. It attempts to create an almost optimal cell structure that minimises inter- and intra-cell material transfer, and also groups similar or related resources and processes. Often used to design process-based facilities, process/production flow analysis has four stages:

1. **Group analysis** creates groups or cells around key machines. All available or relevant resources are classified into three groups:
   - key
   - intermediate
   - common/general

   Then, using the key machines group as a firm reference, the analysis continues, creating the other machine groupings by minimising movement of parts between the groups.
2. **Factory flow analysis** creates autonomous structures the size of a business unit, e.g. modules and focused factories.
3. **Line analysis** identifies and creates almost optimal layouts for processes and facilities by minimising material travel within the group or cell area.
4. **Tooling analysis** describes tooling groups or families to service the cells; it outputs
   - Defined lists of parts, families and/or products that will be the cell workload
   - A list of the resources and machines required to process the workload
   - A first-pass sequence layout of the resources to minimise backtracking of work within the resources

■ Customer/supplier definition

Using input/output analysis, we define and simplify the task groups required in each cell, to ensure maximum cell autonomy (Figure 7.3). By cell autonomy, we mean the percentage of work on the cell part families that can be done within the cell, against the percentage that has to be sent to another cell. For example, the initial cell grouping may define that ten machines are required to service all operations on all the parts. But one of these machines is only needed for two of the parts, which only represents a very small use of that machine. Part of the cell definition process is to create well-balanced cells, where resources are reasonably well utilised. If a cell resource is only to be used say 5% of the time then it may be more sensible to subcontract this operation set to another cell that uses the same resource to a higher loading (provided this additional load will not overburden it). If the resource is not required elsewhere, then it would be retained in the cell, ideally set up permanently or with rapid changeover tooling for the limited part set it has to process.

Invariably cells will have to subcontract some operations to other cells or modules, as it will not be economic to equip every cell with every machine and process. The same is true for office groups, where it can become a trade-off between hiring extra personnel to give each unit a skill and/or capacity, or maintaining a separate support services unit to serve a group of modules or cells.

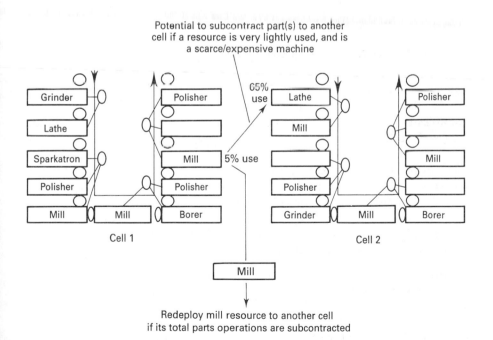

Figure 7.3 Cell autonomy and subcontracting.

Consider a CNC punch required by five cells, which together have a workload to justify only a single machine. Unless several machines are available, the single CNC resource should either be located in the cell with the greatest need, (with the others sending parts to that cell on subcontract), or located in a separate services cell. For maximum autonomy, additional resources may be purchased, or smaller, simpler machines used in place of the expensive CNC equipment.

■ Cell autonomy analysis

In turn, cell autonomy can be measured to provide a rating for comparison, indicating the degree of reliance on other cells or support functions. For example, cell 1 has 45 parts needing 500 separate operations. Of this 500, 50 have to be done outside the cell, giving it an autonomy of 90%. It may be prudent to compare the hours content of the subcontract: if the 500 operations represent 2,500 hours of work, and the subcontracted 50 operations represent 750 hours of work (more hours per operation), then the work content autonomy drops to 70%. The levels of autonomy required are heavily dependent on the individual business needs, but a general guideline is to aim for greater than 80% autonomy by either measure.

■ Static cell design: average sizing and resources

Having developed a cell in terms of outline layout, resource type and content, and products to be processed, we calculate the sizing and number of the resources that will be required to cope with the planned load. Static design defines the cell(s) with specific product and resource data, dealing with load, mix and other parameters.

This is an essential prelude to the dynamic design stage, where the cell design is exercised and tested under more realistic and variable conditions. Static design created the basic cell resources using quantitative data, which must be collected, verified and entered into the analysis tools. The data and the resulting output from static design is used as a rough cut at the basic cell configuration. Key issues in effective static design include

- Bottleneck machines or resources
- Batch sizing
- Set-up times
- Material/product flow
- Resource and labour needs
- Machine reliability

Recall some of the JIT goals:

- **Zero set-up time**: no delay between each activity, via continuous improvement
- **Zero inventories**: balance the system and avoid investment in wasted stock
- **Zero handling**: minimise the transport and handling of every item and reorganise to create multiskilled cellular units to avoid passing between departments
- **Zero lead time**: reduce the time taken for every task; eliminate NVAs to improve the response of the system; reorganise processes to minimise the number of activities

There is a high degree of interaction between several of these issues. For example, set-up time reduction will normally alter the batch sizing, and may influence which resources are bottlenecks, which in turn affects the resource needs. Set-up time reduction is a vital element and is described later in this chapter.

Resource requirements

Early in the static design stage we need to estimate the size and capacity of each machine or process resource that will be in the cell. This is done by calculating the actual demand/load in hours and comparing this with the number of hours available for each resource. The load should allow for

- operator performance factors
- cycle time
- set-up time
- scrap and rectification time

This should be based on actual and projected schedules for part loading, smoothed over a suitable period.

Although the cellular and JIT philosophy moves away from direct operator measurement, we will frequently be starting from a basis of existing work measures, and require to use these in this static sizing analysis.

From this comparison, it will quickly become apparent which machines or

Machine grinder 1: available 120 hours per week, 5,760 hours per year

Operation	Cycle time (hours)	Set-up (hours)	Scrap (hours)	Operator performance	Batch	Total time per batch (hours)
A	0.5	0.5	0.1	0.7	100	86.5
B	0.75	0.4	0.1	0.8	200	213
C	1.0	0.3	0.15	0.7	150	247
etc.						

Total hours required = 6,000
Load versus capacity = 6,000/5,760 = 1.04

resources will be bottlenecks, constraints in the cell, those with a load/capacity ratio of greater than 1. The identified bottlenecks will be the limiting factors on overall cell throughput, and must then (if proved to be real bottlenecks in dynamic design) be targeted for focused improvement, based on set-up reduction, reduced batch sizing (which requires set-up reduction), appropriate control and loading.

Labour/personnel requirements are also established in outline at this stage, based on the tasks and operations to be performed within the cell. However, later improvements to operating practices in the flexible cell structure are likely to alter this in the future, multimanning, multiskilling, etc. We are merely attempting to indicate the number of people required at this stage. Future detailed job design stages will fully address this aspect. Outputs from static design include

- Number and type of machines and resources
- Outline layout and position of resources
- Definition of machines/resources to be moved into and out of the cell(s)
- Number and skills of personnel required
- List of bottleneck resources for focused improvement
- First-pass batch-sizing proposals
- Detailed analysis and reports on cell capacity and loading data (including computer data discs)
- External support service requirements, including subcontract list of parts, volumes and load

■ Dynamic cell design

We now move from static design to dynamic design, investigating each area of cell operation and control under the variable conditions they are likely to encounter:

- Bottleneck resources
- Batch sizing and set-up reduction
- Material flow
- Detailed dynamic design and testing
- Scheduling systems
- Cell control systems

This will often be a lengthy and time-consuming stage of cell design, as can be seen from the detailed chart in Figure 7.4. Dynamic design (box 9 on Figure 7.4) is a core task in the complete design process, leading to other dependent tasks, such as control system design and service design.

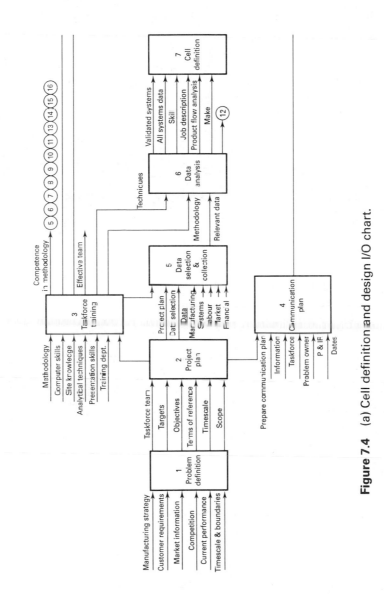

Figure 7.4 (a) Cell definition and design I/O chart.

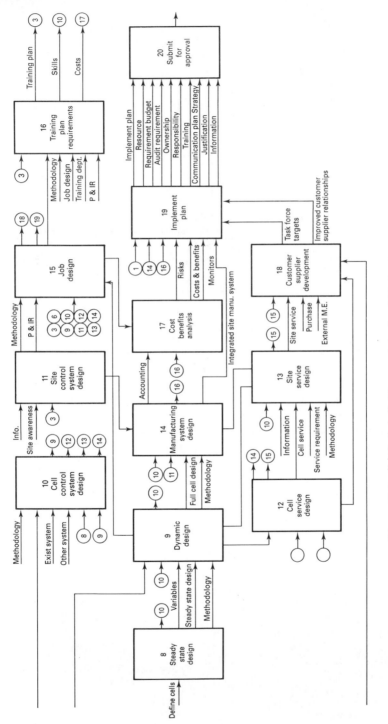

Figure 7.4 (b) Cell definition and design I/O chart.

Bottleneck resources and machines

In a bottleneck resource, the load is exceeding capacity, either constantly or periodically. There have always been particular machines and resources that cause hold-ups or bottlenecks in production of any items, from aircraft and aerospace products to loaves of bread. Bottlenecks are caused by several factors:

- True undercapacity
- Ineffective operation
- Inappropriate scheduling and control
- Set-up time
- Breakdowns
- Staffing

Many resources that at first appear to be capacity bottlenecks (in static design) can disappear when different scheduling and batch sizing are employed; the bottleneck often moves to another resource. With so many variables that can be altered, it can often be difficult to identify true bottlenecks, although the static and dynamic design stages will normally have addressed this.

The OPT approach

Optimised production technology (OPT) is a set of principles for identifying and correctly managing complex production systems containing several bottlenecks. OPT software manages bottlenecks in various business sectors. See Chapter 12 and *The Goal: A Process of Ongoing Improvement* by Goldratt and Cox. These rules concern the identification and control of bottleneck resources within a facility, tying production to the speed of these constraints. Fed by WIP buffers, bottlenecks are kept fully utilised; upstream resources are made to keep pace with the downstream bottleneck (Figure 7.5). Bottleneck rules are important factors in the scheduling and control of production cells:

1. **An hour lost at a bottleneck is an hour lost to the whole system or cell**
    Because the bottleneck is the limiting factor on total throughput, it is not possible to recover this lost time by increased output of other resources.
2. **An hour saved at a non bottleneck is a mirage**
    There is no point in running machines at a higher throughput than the true bottleneck. All it achieves is to pile up WIP behind the true bottleneck. Focus improvements on the true bottleneck.
3. **Bottlenecks control both the cell throughput and inventory in the cell**
    The amount of WIP and inventory buffer that is placed behind each bottleneck can be determined by analysing
    - variations in the supply of parts
    - variations in the schedule

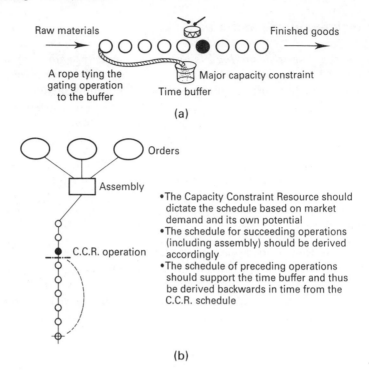

Figure 7.5   Bottleneck diagram and operating rules.

- reliability and uptime of machines
- labour performance variance
- levels of scrap and nonconformance

Buffer levels at a bottleneck can be calculated to keep the machine operating for a specific time period when and if the supply is cut off.

Targeting bottleneck resources

Engineering and improvement effort should be concentrated on maximising the effective capacity of bottleneck resources by

- batch sizing to suit the resource type, cycle times, and set-up times
- set-up reduction
- preventative maintenance
- manning agreements
- alternative resources

Batch sizing and set-up times

Batch sizing and set-up times are closely related. The set-up time for each part grows (relative to the total processing time) as the batch size is made smaller. Batch sizing has to be addressed in static and dynamic cell design and when defining the scheduling and control systems for the cell. In static design, the set-up time is the existing time; in dynamic design a *proportional reduction* in set-up time should be used in the testing and modification process.

> *Zero lot size: produce and pass on items as individuals or as small batches*
> *to save queuing time and to improve cycle time, response and inventory.*

The old economic batch quantity (EBQ) concepts were based on finding the least total cost of a batch, which was the low point of set-up cost versus inventory cost (Figure 7.6)

The JIT concept will be heavily resisted by shop-floor managers and supervisors, mainly because of the increased number of set-ups and increased set-up time, but also because the amount of material movement will be seen as increasing. This is where change to a JIT culture and world class performance is essential if the full benefits are to be realised.

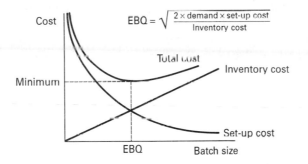

$$EBQ = \sqrt{\frac{2 \times \text{demand} \times \text{set-up cost}}{\text{Inventory cost}}}$$

Figure 7.6 EBQ method for calculating batch size.

CASE STUDY 7.1

Effect of batch size and set-up time on a bottleneck resource in an aerospace manufacturer

    set-up time   = 30 min
    process time   = 10 min
    old batch size = 50 (one week's requirement)

Total batch time = 30 min + 50 × 10 min = 530 min
Set-up time is 6% of the process time

To make ten batches of five units during the week (50 total) in a small batch scenario, with set-up not reduced:

Time for each batch = 30 min + 5 × 10 min = 80 min
Total time for ten batches = 10 × 80 min = 800 min (51% longer)
Set-up time is 60% of the process time

Outside the bottleneck this will not be a constraint, apart from the labour required for the set-ups, which could be better utilised. If we can reduce the set-up time by 50%, which is normally quite straightforward and easy to do, the equation becomes

Time for each batch = 15 min + 5 × 10 min = 65 min
Total time for ten batches = 10 × 65 min = 650 min (22% longer)
Set-up time is 30% of the process time

This 50% reduction in set-up time would not be the final goal, further improvement would be ongoing. Note that from a starting lead time of 530 minutes to process a part batch, this has been reduced to 65 minutes, an improvement of 88%.

Another is given in example Figure 7.7. A set of presentation slides covering the concepts and methods for set-up reduction is provided in Appendix H.

Batch size and set-up time are also dependent on market needs, and on the type of machines and resources to be used. For example, a large chamber or oven with a long cycle time (e.g. 24 hours) requires its capacity to be filled by a batch of parts if it is not to become an even greater bottleneck. Depending on the situation, it may be a longer-term aim to replace or supplement the large oven with several smaller units. Smaller batches lead *directly* to reduced WIP

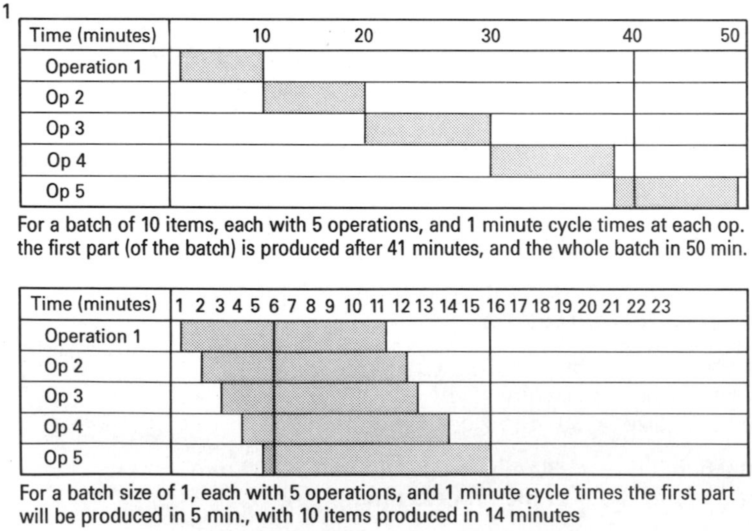

For a batch of 10 items, each with 5 operations, and 1 minute cycle times at each op. the first part (of the batch) is produced after 41 minutes, and the whole batch in 50 min.

For a batch size of 1, each with 5 operations, and 1 minute cycle times the first part will be produced in 5 min., with 10 items produced in 14 minutes

Figure 7.7   Effect of process and transfer batch size on lead time.

and inventory, which in turn increases cell response, flexibility and effectiveness. The reduced inventory is also of financial benefit to the business, due to the smaller investment at any instant in the partially made products that cannot yet be sold.

Varying batch size within the cell

Note that the process batch does not have to be the same as the transfer batch. This means that if we have a process batch of 20, we can start to move the completed parts onto the next machine as soon as they are finished, in batches of 1, 2, 4, 10 or whatever (Figure 7.8). This concept also reduces overall cell lead time. For example, process batch 50; transfer batch 5.

In general, the use of large batches should be confined to bottlenecks, with small transfer batches beyond the bottleneck station. The process batch should be minimised, taking into consideration:

- bottleneck capacities
- final product form
- raw material form
- material handling

Set-up reduction

Zero set-up time: no delay between each activity, via continuous improvement

The need to focus such set-up reduction on identified bottlenecks was described earlier in the chapter. Set-up reduction should focus almost exclusively on identified bottlenecks, only moving to other machines when the improvements on the bottlenecks have begun to relax the constraint (Figures 7.9 and 7.10).

If set-up times are not addressed effectively, batch sizes cannot be made small enough for maximum benefit. Therefore, the JIT philosophy views

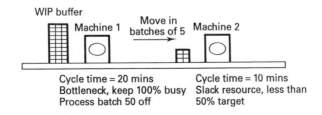

Figure 7.8 Varying batch sizes within a cell.

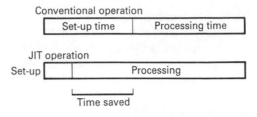

Figure 7.9   Set-up improvement.

set-up reduction as one of the key elements that lead to flow-based manufacturing. JIT provides a four-step methodology for reducing set-up times.

1. **Separately address internal and external set-up.** All set-ups can be divided into two components: internal and external set-up time. External set-up is activities that can be done while the machine is running; there is no need to stop or interrupt it. (e.g. preparing all tools and jigs). Internal set-up operations have to be done when the machine is not running.
2. Convert as many internal set-up operations as possible to external set-up. This reduces the time the machine is stopped.
3. Eliminate adjustment processes during internal set-up; adjustments normally account for a large proportion of internal set-up time.
4. Eliminate unnecessary operations during set-up and move towards simple, one-touch set-ups.
5. Strive for continuous improvement, as always.

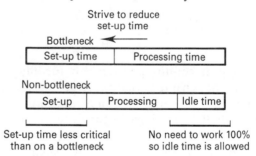

Figure 7.10   Different focus of set-up reduction and utilisation on bottleneck and other resources.

CASE STUDY 7.2

Set-up reduction in a Japanese steelworks

In a Japanese steelworks, a 100 ton forming press was a critical resource in the process. When first measured, the tool changeover time was averaging 4 hours with one worker carrying out the operation. This resulted in only 55% process time being available, which was the major concern. A large proportion of this set-up time was wasted in collecting and locating tools and material for the press, together with incorrect set-ups and damaged tool delays (Figure 7.11).

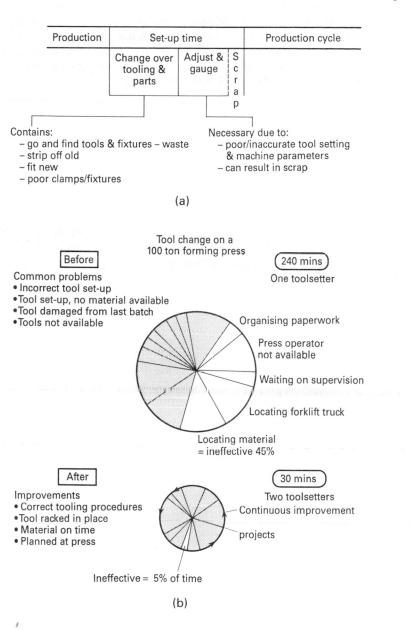

Production	Set-up time			Production cycle
	Change over tooling & parts	Adjust & gauge	S c r a p	

Contains:
 – go and find tools & fixtures – waste
 – strip off old
 – fit new
 – poor clamps/fixtures

Necessary due to:
 – poor/inaccurate tool setting
  & machine parameters
 – can result in scrap

(a)

Tool change on a
100 ton forming press

| Before | | 240 mins |

Common problems
• Incorrect tool set-up
•Tool set-up, no material available
•Tool damaged from last batch
•Tools not available

One toolsetter

Organising paperwork

Press operator
not available

Waiting on supervision

Locating forklift truck

Locating material
= ineffective 45%

| After | | 30 mins |

Improvements
• Correct tooling procedures
•Tool racked in place
• Material on time
• Planned at press

Two toolsetters
Continuous improvement

projects

Ineffective = 5% of time

(b)

Figure 7.11 Set-up reduction: (a) breakdown of set-up components; (b) before and after figures for a 100 ton press.

Following analysis of the existing procedures, action was taken by continuous improvement teams to solve the problems of tool availability and adjustment, material control, operator training and staffing requirements. This rapidly resulted in improved operational procedures, with double manning on the machine for tooling. The benefits were immediate. Tool changeover dropped from 240 minutes to only 30 minutes, due to the elimination of waste and waiting time. This resulted in the press being available for operation for over 90% of the available hours, dramatically increasing throughput.

Material flow

Zero handling: *minimise the transport and handling of every item; reorganise to create multiskilled cellular units to avoid passing between departments*

The optimisation and reduction of material flow between operations and resources is a fundamental aim of JIT and systems engineering, and needs to be addressed during cell layout definition (Figure 7.12). Tools used to assist in the detailed design of layouts include

- String diagrams

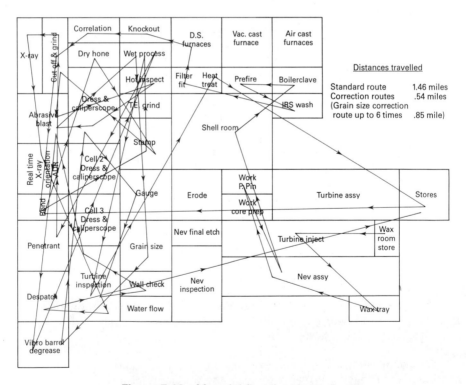

Figure 7.12 Material flow in a large factory.

- Travel charts or tables
- Process flow analysis
- Computer layout/distance-travelled optimisers
- Simulation packages to display and alter part routings/machine locations

When planning for optimal flow, several guidelines apply:

- Plan for all parts to have similar operation sequences
- Avoid backtracking within the cell or unit as this causes disruption
- Consider common tooling
- Where more than one machine of a certain type exists, consider dedicating machines to groups of parts

In most cases we are attempting to find a resource layout that reduces the travel distance for all parts around a cell and minimises backtracking of parts within the cell. Backtracking disrupts the material flow and ease of handling. As a general guideline, machines and workstations should be placed as close together as possible, within operational and safety limits. The aim of developing the simplest possible cell layout and material flow should also override any resistance or objection to the physical movement of machines

	NO IN	1 MAKES	mse	Part no	%STH 89	%STH 90		MIN	9	10	11
1				Cell 1,2 customer schedule						PERIOD	
2											
3	-----	-------	bat						----	----	
4	1	38	38	AX56321	.14	.06	STR	75h	38	0	0
5	8	24	48	AX60721	.7	.92	RPT	100	48	0	0
6	8	45	45	AX60987	11.7	6.35	RUN	85h	270	270	270
7	8	45	45	AX60988	11.66	6.35	RUN	85h	270	270	270
8	2	38	38	AX62627	.09	.23	STR	75h	38	0	0
9	8	15	45?	BR11337	.9	1	RPT	81h	90	45	0
10	18	68	68	B503230	1.09	5.84	RUN	145	68	68	0
11	14	195	195	EU11486	.91	.9	RUN	40h	195	195	195
12	8	80	80	EU24718	.7	.34	RPT	16h	80	80	0
13	8	80	80	EU24753	.02	.08	STR	16h	0	0	80
14	15	14	70?	EU32099	0		RUN	131	70	70	70
15	9	18	54?	EU35122	.09	.24	RUN	95h	54	0	54
16	3	36	36	EU35131	.02	.03	STR	53h	1	2	1
17	4	20	20	EU43710	.22	.28	RUN	24h	0	0	3
18		12	48?	JR21104	3.33		STR	130	0	0	0
19	15	14	70?	JR26086(EU32099)	.2	.2	RUN	+30	58	58	74
20	15	15	90?	JR29325(NDG1118)	.2	.2	STR	92h	0	15	0
21	1	29	29	JR29326(NPG1119)	.2	.2	ZERO	47h	0	0	0
22	15	15	90?	JR29327(NPG1120)	.2	.2	STR	136	0	15	0
									0	1	0
									0	15	0
									0	1	0
									40	83	55
									23	138	92

Number of piece parts made from incoming part

Proposed 'ideal' batch size

Total % standard hours by part in 1989

IN 190 Runner, repeater or stranger

Expected minimum time to process a batch of each part (assuming no cell congestion)

Schedule for production each month – used to define part as Runner (RUN) or Repeater (RPT) or Stranger (STR)

Figure 7.13 PC spreadsheet for static and dynamic analysis of a manufacturing cell.

195

Spreadsheet tools

The static design process again lends itself to the use of computer spreadsheet tools, which are ideal for the arithmetic and mathematical operations to be performed. For example, we can build up a part versus resource matrix in the spreadsheet, with labour resource time per operation by part type in the row cells. This can simply be calculated to give the total resource loads from different schedule and mix senarios (Figure 7.13).

■ Detailed dynamic design

Once the cell has progressed successfully through the static and early stages of dynamic design, it is necessary to more fully define its structure and operation, and to exercise the design thoroughly, by using data that varies with time. Dynamic changes will occur in reality, and our cell must be capable of dealing with them. These can include variations in load, resource and mix. Specific tools for this include computer simulation of the cell operations. Specially designed PC spreadsheets can also give effective results when used to compare different operating and loading scenarios (Figure 7.14).

Dynamic simulation of changes to variables

Providing the necessary cell data is available, computer simulation (and manual simulation) offers the ideal dynamic testing tool, allowing the team to load and run a very high number of test simulations in a short timescale, if required. Figure 7.15 illustrates a simulation model. In each set of runs, the cell parameters can be altered or fine-tuned, including layout, control system, batch size, process times and manning. This allows the effects of each change to be simulated and quantified then compared to other options. Simulation projects are fully described in Chapter 10.

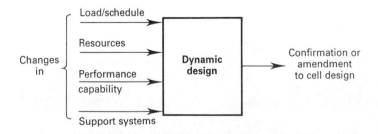

Figure 7.14　Dynamic design I/O chart.

Any significant deviations from expected performance may need closer scrutiny; previous design stages may need to be revisited. Important system variables should be fully explored to discover their effects and optimal arrangement under dynamic testing.

1. Batch size and set-up times
2. Cycle/operating time and operator performance
3. Machine breakdown and reliability
4. Scheduling

Batch size and set-up times

Earlier on, we stressed the interrelationship between batch size and set-up time. Their variation can markedly affect the performance of the whole cell or individual resources. Variation of set-up times, batch sizing and cycle times can be carried out through computer simulation, configuring each variable against the collected measurements, variations and any planned improvements (Figure 7.15).

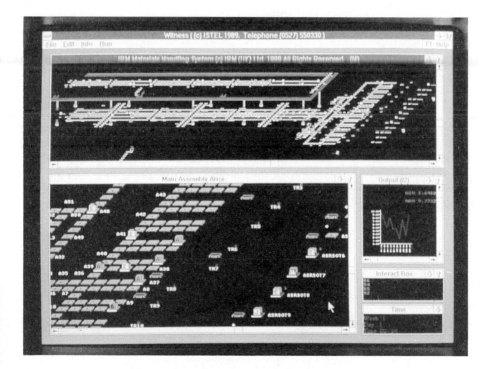

Figure 7.15 Witness simulation of IBM, Greenock, UK. (Courtesy AT&T Istel and © IBM UK Ltd 1989, all rights reserved.)

Cycle/operating time and operator performance

At the steady-state or static design phase, cycle times and operator performance were taken as fixed (normally average) values, with an adjustment for operator. In dynamic design exercises, this simplistic approach is no longer adequate, requiring more accurate treatment if we are to derive a realistic and trustworthy model for the cell. Typically, a variability factor will be built in to each potentially variable quantity, either as a maximum and minimum, or as a more sophisticated treatment using statistical distributions. This latter approach is best implemented using the powerful computer simulation packages mentioned elsewhere in the text.

For example, the standard time for a particular operation could be 5 minutes. But in practice this varies between 4 and 6.5 minutes. In a simulation model, this is commonly represented by a normal or triangular distribution, with mean of 5, minimum of 4 and maximum of 6.5 minutes. The simulation package will select the variable values from this distribution, in an intentionally random manner (Figure 7.16).

In addition to variation in cycle times alone, there may also be a requirement to include different operator performance values, which also will impact on the final cycle times. Virtually all simulation packages provide for the inclusion of operator performance ratings, either as a percentage or as a rating between 0 and 1.

Machine breakdown and reliability

Machine and resource reliability can play a significant part in overall cell performance. Low reliability, with numerous breakdowns and excessive downtime will cause any cell to perform poorly. In many businesses the equipment is highly reliable, so there may be no need to build a degree of downtime into the dynamic design model. But in other cases, reliability may be a long-standing problem. In practice very few firms keep detailed logs of machine downtime and failure, so analysis is difficult or, at best,

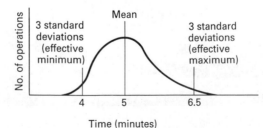

Figure 7.16 Normal (skewed) distribution of operation times.

approximate. Machine and equipment reliability should be analysed and normally expressed as

- mean time between failures (MTBF)
- mean time to repair (MTTR)

In all situations where fast response, cellular manufacturing is being considered, **a comprehensive preventative maintenance programme is essential**. Machines are critical to cell performance. A lengthy failure will kill cell output and let down customers, due to low inventory, small buffer stocks and small batch production. Statistical data on machine breakdown, line breakdown and repair performance is worth adding to the simulation, to accurately portray the effects of random machine failures on various cell operating scenarios (Figure 7.17).

Machine type: 1–6 (single station, batch, multistation, process, etc)
Machine name: Big Press
Reliability: MTBF (hours)   Distribution (select most appropriate)
          MTTR (hours)   Distribution (select most appropriate)

Cleaning/maintenance downtimes
   MTBMaintenance (hours)   Distribution (select most appropriate)
   MTTMaintain   (hours)   Distribution (select most appropriate)

Will machine be stopped for cleaning if processing? Y/N ☐

Additional conditions

Figure 7.17 Simulation resource reliability entry screens.

Scheduling

Scheduling is the planned loading of work onto resources. This may be a workload for several months or just for a few minutes, but the principles are identical. Taking into account the available resources, their capacity and the required date for product completion, schedules can be prepared either at broad or detailed levels.

The aim of scheduling is *not* to maximise machine utilisation and output, nor is it to keep operators busy. Scheduling aims to

- produce and deliver products to meet customer needs and deadlines
- produce products at or above the required quality conformance and at the lowest cost

To maximise the benefits from the simple and effective cellular and modular production units developed so far, it is necessary to organise and produce workloads for each module and cell that take full advantage of their potential capacities and strengths. These workloads must be correctly structured to match the cell characteristics, in terms of the volume and mix of work. This scheduling will be carried out on a rolling basis to match customer demand with cell performance and capacity.

■ Developing the optimal cell design

Once the cell development has progressed to a firm definition of groups of products, resource and machine requirements, staffing and physical layout, this configuration is explored in depth using the dynamic design tools described above. This uses the variable data for lead times, set-ups, batch sizes, operator performance and other factors within the dynamic model. Still to be defined and tested are the methods and systems for scheduling and controlling cell operations.

Scheduling and control systems

In our production environment, material control is required at three specific levels:

1. **Inputs from suppliers**: these are external to the business, such as bought out parts, raw materials and services
2. **Internal to the production system**: controlling flow of material to ensure correct processing and routing, to meet customer requirements and due dates, whilst achieving JIT goals of low inventory and effective operations with no waste
3. **Output to the customers**: delivery of product or services to the customer at the time and frequency required

Scheduling and control systems (Figure 7.18) are required to deal with the internal requirements and to service the supplier and customer requirements.

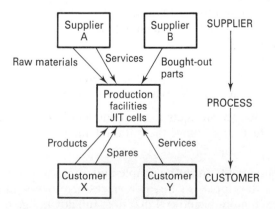

Figure 7.18 Material control and scheduling requirements.

After *application of* JIT concepts, there may be substantial changes in the way the company plans, orders and produces goods. This in turn may significantly alter the way planning and scheduling take place, changing the requirements for scheduling and control. They will probably be simpler because the overall planning and manufacturing processes are likely to be simplified and streamlined following the review. Some companies will still contain a relatively complex scheduling requirement, but it should be simpler than before the introduction of JIT.

Several JIT goals contribute *directly* to scheduling and control of production, e.g. kanban control, and all contribute *indirectly* by making production easier to plan and control. For example, by reducing scrap and rectifications, set-up times, unnecessary paperwork, and so forth.

Levelled schedules

Levelled scheduling is the term used to describe the process and means of achieving appropriate schedules. A levelled schedule is one where the required volume and mix of production has been smoothed out over several short periods of time. This is achieved by varying the mix of scheduled products to equalise and balance the loading on each resource in the cell. It has to be done inside the constraints of meeting customer due dates.

The cellular JIT system requires a master production schedule (MPS). This master schedule states the required production levels by product over a period. For levelled scheduling the JIT MPS will be created using two objectives:

Spreading and levelling product production evenly over each period
Spreading and levelling the quantities of each product over each period

Ideally, the schedule will plan for products to be made in a period equal to the shipping and selling period, whether this be daily, weekly or monthly. This is the JIT goal of 'sell daily make daily', 'sell monthly, make monthly'.

Where a cell is dealing with many types of parts, the need for varying weekly or periodic supply to the market can be well catered for by scheduling on a mix of the required parts as small, regular batches. So each day or week, a repeating pattern of small batch jobs is scheduled on to each cell, mixed model scheduling. This pattern has other significant benefits. People become used to the regularity. The problems of preparation, maintenance and part quality can be planned in advance. Where a regular scheduling pattern can be established, it allows the use of much simpler cell control systems.

Runners, repeaters and strangers

When parts or models are regularly demanded and scheduled every period, they can be termed runners. If scheduled regularly, but not every period, they are termed repeaters. Parts or models that only appear on the schedule occasionally are termed strangers. When scheduling, runners are placed first, then repeaters, to provide the basic high volume foundation for the schedule. Then the strangers are fitted in as appropriate, with final adjustments for levelling.

Depending on the final mix of runners, repeaters and strangers, defined as the load family for a cell, there will be differing requirements for scheduling and control. The more strangers and repeaters in the cell, the more uneven and complex the loading becomes, making control more difficult. Runners are the easiest to control and schedule.

Mixed model scheduling

Mixed model scheduling can best be shown by an example. If we have a requirement to produce three different parts 300 of A, 100 of B, and 200 of C during a month, this could be scheduled in various ways:

1. Standard sequential schedule, large batch: 300 A → 100 B → 200 C
2. Mixed model schedule, medium batch: 150 A → 50 B → 100 C → 150 A → 50 B → 100 C
3. Mixed model schedule, small batch: 30 A → 10 B → 20 C → 30 A → 10 B → 20 C, etc.

The benefits of mixed model scheduling are substantial, giving increased flexibility and response to changes in market demand. The best model mix is largely dependent on the industry, market and production facilities. There are no hard and fast rules; use what works best.

Figure 7.19 shows the differences between unlevelled and levelled schedules for the same part volumes. The old version has effectively three large batches based on the single set-up concept. The lower schedule has been levelled. During each period the load is more even and product is available earlier. This levelled schedule is further broken down into weekly schedules and smaller batches as the cell and JIT operations become tried and tested.

A safe harbour for holding schedules firm is required to let this form of scheduling work well. Where changes in customer delivery date and quantity are occurring regularly, the operational structure and facilities must be set up to ensure the process lead time is less than the customer lead time, otherwise the essential levelling and stability of the schedule will be compromised.

Exactly how the master production schedule is created is relatively unimportant. It may be carried out manually or using MRP. MRP is an ideal planning system, widely used both in the East and the West for planning and controlling manufacturing operations, even when they are run on a JIT basis.

Schedules to suppliers for component parts and raw materials should also be levelled wherever possible, since the supplier has exactly the same problems of trying to balance load and capacity as your own business. Again, deviations

Old, unlevelled schedule

Produce	Period	Volume	Period	Weeks					
				1	2	3	4	5	6
Part A (runner)	1,200		300	300	300	300			
Part B (repeater)		500						300	200
Part C (stranger)		50							50
Workload	1,750		300	300	300	300	300	250	
Cell capacity	1,800		300	300	300	300	300	300	

JIT levelled schedule

Produce	Period	Volume	Period	Weeks					
				1	2	3	4	5	6
Part A (runner)	1,200		200	200	200	200	200	200	
Part B (repeater)		500			100	100	100	100	100
Part C (stranger)		50		50					
Workload	1,750		250	300	300	300	300	300	
Cell capacity	1,800		300	300	300	300	300	300	

Figure 7.19 Unlevelled and levelled schedules.

from the agreed supplier schedules should be negotiated with the supplier rather than forced, and provided with as much advance warning as possible. This is part of developing customer/supplier relationships, discussed later in this chapter.

■ Control systems

Scheduling and control are not one and the same; they are separate activities that occur at different stages of any production cycle. Scheduling is carried out before any work is done in the cell, but schedulers need to know about the cell's control system. Scheduling is primarily concerned with providing a workload to the cell(s) that is achievable, cost-effective and meets the customer requirements. Control systems deal with cell operation, how and when items (such as tools and labour) are passed between workstations.

Figure 7.20 shows a theoretical control system. Here, any given process will have inputs, outputs and a feedback signal that can be used to determine whether the actual output is what was required. A control signal is used to alter the input to the process to bring the output back to the required level.

There are three primary requirements for control:

■   information on the current state: feedback or measurement
■   requirement for a desired future state: new output requirements

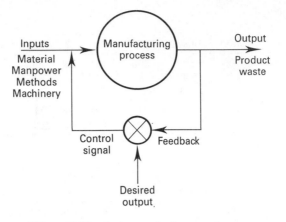

Figure 7.20   A theoretical control system.

■ an understanding and method of how to get from one state to the other: control system

An effective process control system will be able to maximise the controllable inputs and minimise the uncontrollable inputs, the disturbances. Where disturbances remain, the control system must deal with them in a suitable manner. Production disturbances include

■ Machine breakdown
■ Absenteeism
■ Material shortages
■ Scrap
■ Rework
■ Tooling non-availability
■ Industrial relations problems
■ Schedule changes with short lead times

Although disturbances are minimised wherever possible, the chosen control system must be able to cope with a certain amount.

Complexity of control

Most real production operations have several products, many machines and resources, and multiple flows. Cellular design reduces and groups these effectively, but will rarely succeed in reducing the cell to one flow and one product. Control system complexity is directly related to the number of flows, operations, resources and products (Figure 7.21).

	Control system	
	Pull	Push
Number of flows	low	high
Number of parts	low	high
Number of operations	low	high
Overall complexity	low	high

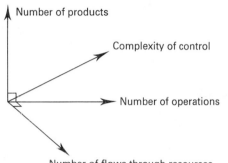

Figure 7.21   Complexity of control.

Approaches to production/cell control systems

There are two basic forms of production control: push and pull. As the names suggest, one system *pushes* material into the facility then *pushes* it from machine to machine until it emerges from the end of the production unit (Figure 7.22).

Push systems

These tend to load work onto production resources in successive stages of operation, with the launching and/or receiving of material being the trigger to commence work. MRP is a common example of a push system (Figure 7.23).

A push system starts by considering the operations to be carried out in a product routing, then loading the work onto each successive workstation from first to last, giving each station a starting and finishing schedule to work to. The finishing time of a batch from the first station becomes the start time for the second resource. If work with conflicting priority arrives at a workstation at the same time, priority rules are used to control the queue that develops:

- First in, first out (FIFO)
- Last in, first out (LIFO)
- Least time remaining (until delivery due)
- Shortest processing time

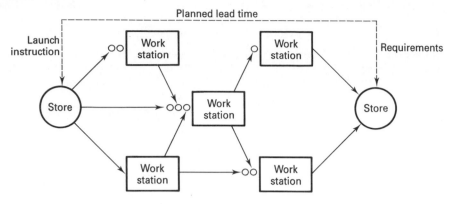

• System triggered by raw material launch instruction
• On completion batch moves to next operation
• Control by 'priority rule' if queue exists

(a)

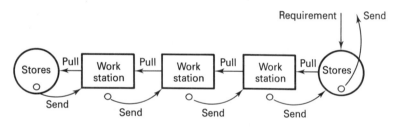

• System 'triggered' by demand on finished part store
• Each work station 'triggered' by 'pull' from next one
• On completion, batch remains at work station until pulled

(b)

Figure 7.22 Control systems: (a) push and (b) pull.

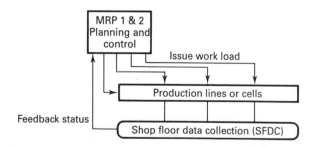

Figure 7.23 MRP push control and SFDC feedback.

Push control has these features

■ Works well with multiple parts

- Works well with complex flows
- Performance depends on good capacity management
- Performance depends on good, accurate data
- Once work is launched, push control adds value when capacity is available (unless deliberately stopped)
- If workstation stops, previous stations continue to work, producing a wasteful build-up of WIP
- Requires complex systems to operate and monitor (including MRP)
- Tends to have high (pushed) inventory and large WIP queues at each resource.

Push control for batch flow

One form of push control system that provides the benefits of levelled schedules and regular workload to a production cell is period batch/flow control. Based on launching or releasing sets of work at regular intervals, this repeating cycle schedules and controls the flow of mixed batches of material. The master schedule is based around a production period and its capacity is within that period. This determines how many of a particular product type, or mix of product types, will make up a production batch. The production has a mixed model format making up a period load.

For example, if we choose a period of 4 weeks, then each production batch or load will be scheduled 12 times each year in a 48 week year. Depending on the business sector, the batch for each period will be made up from both forecast and order information. Capacity within a period is the major governing factor. Once a period batch quantity has been reached, no additional orders are allowed to be added; then must be moved out to a future period. The processing time is then assumed to be a finite elapsed time from release of order, similar to letting a ball roll down the same slope every day, knowing it always takes the same time to reach the bottom (Figure 7.24).

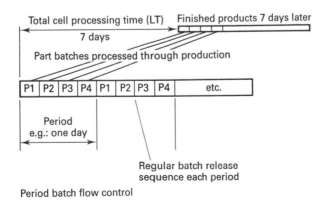

Period batch flow control

Figure 7.24 Period batch flow control with a seven-day cycle.

The benefits of period batch control are

■ Reduced inventory because the forward load is largely known, allowing JIT material delivery with low stock
■ Simple control because the processing pattern is regular and repetitive
■ Improved product quality because the regular production pattern leads to improved visibility and preparation for problems
■ Improved customer service because production and delivery are regular

Where this method is used in an environment of very high product mix, additional visual control can be used to identify batch priority within the cell, for example, marking by colour and/or number. In order to monitor batch and item progress through lines or cells with lengthy total process times, measurement points, or way points are often set up at certain positions in the path. However, period batch control cannot overcome the inherent disadvantages of push systems.

Pull systems

Pull systems draw work and material through the cell by placing a requirement on the last resource. This pulls product from the previous workstation, which pulls product from the *next* previous station, and so on all the way through the cell (Figure 7.25). If sufficient material or components are available at the final workstation to satisfy an order, this will be done. If insufficient product is available here, parts will be pulled from each previous workstation in turn, forming a pull chain from the last station right through to the first.

Pull systems are

■ ideal for simple flow
■ good with few similar products
■ good for steady demand
■ able to add value to satisfy demand as required
■ simple to understand, operate and monitor
■ unsuitable for frequent product changes, dissimilar flows and irregular demand
■ economical; when a workstation stops, all preceding workstations stop so there is no wasteful build-up of WIP
■ able to carry minimum inventory.

The control of production cells and lines that are operated along a JIT basis frequently operate on a pull basis. Push and pull systems would give the same excellent results if perfect information was always available and there were no uncontrollable, random effects. Rarely is that the case. It is therefore necessary to design or choose a control system that meets the specific type of control requirements and will cope with the variables within that production scenario. In many cases a conventional push or pull system may suffice; in others, a form of combined push/pull hybrid may be required. Many Japanese manufacturers launch a process using MRP1 then switch to kanban pull control downstream.

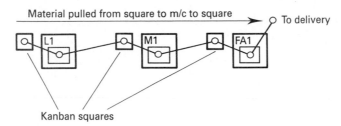

Figure 7.25 Visual kanban control on a production line.

Kanban control

Kanban is the Japanese word for card, so kanban control is often control using a card, sometimes called a kanban card. Kanban control of production facilities is now fairly well established in the UK, mainly in flow-line environments but also where production flows are simple or have low part counts. For more complex control needs, kanban is often considered inapplicable. This need not always be true.

Kanban is the system which implements many key aspects of JIT on the production floor. Like most forms of control, it is basically an information system that helps to monitor and control production activities, plus providing a level of management information.

JIT kanban pull control has several benefits that assist in making the cell more effective:

■ simplicity
■ visibility
■ discipline

In the kanban system, only the final stage of manufacturing or assembly receives or knows the actual delivery requirement for products. When this schedule is received, the station pulls material from the previous station, which then pulls from its previous station to replace the quantity pulled by the final station, and so on. This has the simple but valuable effect of synchronising all the production stations to a rate determined by the final station. It also can result in minimum work in process (WIP) inventory being produced and held in the cell.

Suppose there are multiple flow routes through production, or a high part count (with part variety), without a uniform customer schedule. What are the options? The common route is to use one or more scheduling and loading software packages, possibly an overcomplex solution to a complex problem! MRP2 may be part of this, together with additional bespoke or packaged software products. This can give a workable solution, but frequently leads to high inventory and WIP, whilst distancing operators and control personnel from the problems and solutions. MRP is frequently best left to the planning and control of material. It also generates the delivery schedule for these types of cell for 'pulling' products to meet demand.

Varieties of kanban signal

In addition to kanban squares and containers, there are many other signal elements that can be used in the control of material movement. These include cards, tokens and lights to indicate status. Kanban cards are very common and very flexible. They are described in the next section.

A kanban square may be painted or formed between workstations to hold a specified number of parts. Each square can hold one piece, a number of pieces or a container that holds a set number of pieces. More than one square can be painted where it is desired to hold more parts. Users frequently provide special containers that hold the pieces, aimed at keeping the quantities correct and visible.

In the example of Figure 7.25, if the final assembly station FA1 receives a demand for three finished items, it will finish the part in process, draw the next part from its kanban square and start to process it. The previous work centre M1 has seen the kanban square being emptied, so it processes its current part to fill the space. M1 then pulls a third part from the kanban square between itself and the previous machine L1, processes it and passes it to the next kanban square once it is empty. In this way, parts are rapidly pulled to FA1 to satisfy the customer order, but without excessive inventory or wasted effort in the cell. Normally operators will move to another station or other duties once the immediate production demand (pull) at a station is satisfied, i.e. the kanban square is full. The maximum cell inventory is the sum of the parts in progress at each station and the parts in each kanban square. In this case there is one part in progress at each of three stations and one part in each of three kanban squares, a total of six parts.

Kanban for bottleneck management

Where there is a need to ensure the key bottleneck machine is kept busy, or even to provide a higher degree of WIP buffering than single kanbans, a second or third kanban square can be added to the system. In many JIT factories, the line capacity would be balanced to remove the bottleneck by adding extra M1 resource, by subcontracting the operation or by set-up and cycle time reductions on M1 (Figure 7.26).

Kanban cards for multipart cells

Kanban cards can be used to control material flow in a similar manner to kanban squares. They give information about the part, its route and other

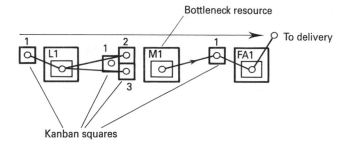

Figure 7.26 A bottleneck is kept busy using different numbers of kanban squares.

details, so they can be used in more complex control situations, such as multipart, multiroute cells (Figure 7.27).

The card normally tells:	Typical Kanban
• number of parts • part identification no. and description • source – where they came from • destination – where they are going	Part quantity 50 ME 1077 – canopy section Source – Welding Destination – painting

Figure 7.27 Kanban card details.

The control card is normally fixed to the container or box, which is moved from source to destination. The container itself is often used as the pull trigger, and may require a preset quantity to be made, filling the container for passing on to the next station. With this method of working, when several containers arrive at a station this will cause the production of enough parts to complete all kanban containers. If a container holds 50 parts and four containers arrive, this is signalling for 200 parts to be made, but each box of 50 will be pulled individually as the next resource is ready.

Control of the overall WIP and inventory in a cell or line can be further dealt with by controlling the physical number of containers in the system (Figure 7.28). More containers means more inventory; less containers means less inventory; This approach can be used when continuously improving cell performance. If one or more containers are removed from the cell, it reduces the inventory buffer.

There are two types of card signal systems in common use

■ Withdrawal signal kanban cards give instructions to withdraw or consume material
■ Production signal kanban cards give instructions to produce material

The withdrawal card system states the quantity of parts a station is to take or

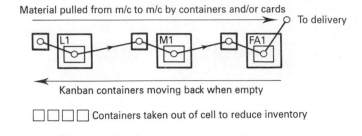

Material pulled from m/c to m/c by containers and/or cards

To delivery

Kanban containers moving back when empty

☐☐☐☐ Containers taken out of cell to reduce inventory

Figure 7.28   Kanban containers in operation.

withdraw from the previous workstation. Cards normally move between pairs of workstations, requesting and sending parts. Production kanban card systems state the quantity of parts to be produced by a supplier station, to replace parts removed for downstream stations. The two types are illustrated in Figure 7.29.

Kanban cards are also used extensively for other purposes, including subcontracting to other cells and suppliers, often in modified or colour-coded formats. High priority rework of an urgent batch can be shown a card colour or marking.

Withdrawal kanban card		Production kanban card	
User/home station ID	M1	Source station ID	L1
Source station ID	L1	Part ID and descr.	PE0097
Part ID and descr.	PE0097	Container ID	Box Y005
Container ID	Box Y005	Capacity	50
Buffer location ID	JA 22	Buffer location ID	JA 22
(a)		(b)	

Figure 7.29   Withdrawal kanban and production kanban.

An example of these systems in operation is as follows. Parts P1 and P2 are to be welded up then stored before call off into the assembly area. Assembly P3 can be created and further machined to make two different finished products, F1 and F2. The finishing machine receives a demand from the master schedule to produce a number of F1 and F2. From the store it withdraws the required numbers of P1 and P2 using a withdrawal kanban card.

Referring to Figure 7.30, we assume that ten production cards are required for P1 and ten for P2. And we assume an equal part mix. According to a JIT levelled schedule, production is planned as P1/P2/P1/P2/P1/ etc. The cards are passed to the welding area for action. As the units are manufactured, they are placed in containers, a production card is attached and the containers are taken to the storage area. When all the units are made, each of the twenty production

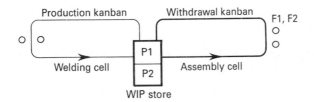

Figure 7.30 Kanban control by production and withdrawal signals.

cards is on one of the filled containers, and the line will stop until more production cards are received, or until some of the parts produced are taken further downstream.

Parts are taken downstream by the assembly unit. The assembly unit has been issued with withdrawal cards for parts P1 and P2. These cards are taken to the central store and used to withdraw the required number of containers of the two parts. Each time a container is taken from the storage area, the withdrawal card replaces the production card on the container. The production card goes back into the welding area to authorise the manufacture of more P1 or more P2. The withdrawal card stays with the container until the parts are removed for assembly. Then it goes back to the store to authorise issue of another container of parts. Together, the cards and containers form a disciplined and simple control system. No operation is performed without card authorisation. No container is authorised to hold parts unless a card is attached to it.

Modified kanban

The above programme of simplification, rationalisation and improvement of the manufacturing function should simplify and reduce the requirement for control, which may be addressed by methods other than powerful computer systems. We may not be able to achieve zero WIP inventory, but we can aim for significant reductions by designing a modified kanban control system, a hybrid.

Remembering that control system complexity increases with part and route variety, there are several different ways to adopt the basic kanban system to a more varied production environment. Using cards, containers or squares, it is possible to develop a custom version of pull control that can manage a production cell of at least fourteen machines. An example is given in the case study at the end of this chapter.

Pull system discipline

- simple
- versatile

- visible
- disciplined

The pull control system provides precise authorisation and control of activities in a cell or product unit, in the order required and with the required priority. However, the system will only operate correctly if all instructions are carried out fully and properly, with no one ignoring the system or trying some other way of processing work. This can be a difficulty that is often encountered when changing from a traditional high inventory and buffer stock environment. Operators may not be used to operating on a fraction of the inventory and with some machines idle for part of the time. Once these mindset problems are overcome, the pull system works simply and effectively, with little thought required.

Rules for scheduling and control

In certain cases it may be desirable to develop and use a number of simple rules for the scheduling and control of a cell, particularly when there are mixed models with different process routes and times. These rules are often required to assist in day-to-day decisions on batch sequencing and process order. The rules should aim to

- minimise interference at machines
- ensure no time is lost at bottlenecks
- minimise unnecessary set-ups
- effectively utilise staff

Rules must be closely tied into the cell schedule to ensure no conflict exists. A typical rule might be: *When a batch of MM2341 is pulled, do not release any MM2355 for 24 hours otherwise congestion will occur in areas of the cell.* Such rules will normally be developed from experience in running actual production and from simulations of the cell.

Training on the control system

Once a final definition of the control system has been fully tested and approved, it must be fully documented for future reference, for system operation and for training of cell operators. A comprehensive training exercise is a very desirable way of starting the cell personnel with its operation, and will often show up practical problems that can be resolved before going live. In these training sessions, it is normally useful to include a table-top model of the cell that can be used to manually simulate cell and control system operations. The computer simulation can also be a valuable visual aid.

■ Final cell definition

The final stage of cell definition includes the review of the developments and findings of the previous stages, bringing together all interested parties to attend the presentation and detailed discussion on how the cell has been evolved. This will include a full description and demonstration of how each aspect was analysed, developed and tested to reach its final state.

Demonstration of cell performance to back-up and prove the stated benefits is highly desirable, using the manual and computer simulation systems, where they exist. A proposed implementation plan and schedule, with all costs and resource requirements, should be prepared and discussed. As part of the costs and benefits proposal, it is common to include a study of any thoughts or plans for business improvement of any market area. This will often include make-or-buy analysis and recommendations.

Provided the review and cell design have been properly carried out and provided the results indicate potential benefits, move on to the next stage. Implement all aspects of the cell, including part rationalisation and the control system. This may be part of a larger multicell or module project. During implementation, planning and control become the new tasks of the project team. It may take several months to complete all aspects of the change. **To summarise the steps in control system development:**

Stage 1: First simplify
- Analyse products and processes
- Identify current bottlenecks
- Simplify where possible; plan for flow
- Target bottlenecks; reassess and simplify again, even where apparently impossible

Stage 2: Then develop the basic concepts
- Analyse and level the schedule
- Identify the real demand period; a false constraint may be causing schedule and loading disturbances
- Analyse and improve flow
- Consider product groupings; identify which parts and/or operations are best subcontracted
- Determine optimal batch sizes for the cell and/or cell sections
- Determine the control system; this will define the methods for limiting inventory and WIP

Stage 3: Develop and test the concepts to full detail
- Develop simulation models, manual and computer-based
- Use the manual model to develop control rules and educate personnel
- Use a computer model to test and fine-tune the control system and rules
- Develop a manual control board or a computer program for shop-floor use.

Ⓒ

CASE STUDY 7.3

Modified kanban cell control

A production cell in an aerospace component manufacturing company has been designed to be highly autonomous and effective, consisting of 14 mixed machines: grinders, lathes, mills, polishers, borers, etc. The cell is to handle a total of 46 parts, which are divided into 6 families. The parts are 60% runners, 35% repeaters and 5% strangers. Cell layout is U-shaped, with minimal backtracking occurring in processing the 46 parts. Through the design process and subsequent testing using spreadsheet and computer simulation, it was shown that the new cell design could dramatically reduce product lead time and inventory levels. It was also expected that the quality of product would rise sharply, due to the total quality and JIT philosophy for operator involvement and responsibility within the cell. Expected benefits:

Lead time to fall from 8 weeks to 4 days average
Inventory to fall from £500k to £50k within the cell

The products of this cell were very complex, high precision components, and actual machine time per part was between 50 and 80 hours. Corporate MRP-based delivery/production schedules were to be used, with reasonable (6 month) schedule notice. It was decided that local detailed MRP/SFDC control was not to be used, due to previous problems, and that a specific pull-type system was to be developed if possible.

Simple kanban square control was quickly discounted due to the more complex nature of the cell product mix. Developed instead was a version of kanban card and container control with a priority facility. This was a true hybrid of push and pull systems. The cell was to be given a push by way of orders being released to the cell. Then it was basically left to an internal pull system, which working effectively would produce the parts several days later, approximately on time for delivery. The hypothesis of the cell control system was therefore as follows:

1. Due to the mix, complexity and length of operations, a total pull system would not cope effectively.
2. Via manual and extensive computer simulation, it was shown that, once launched, part batches would reliably be pulled through the cell in approximately 1.3 lead times of the part (30% queue and blocked time).
3. A batch release, not a pushed launch, would make parts available for processing once the secondary pull system had space to hold them.
4. A form of modified kanban card and square control was to be used. The multipart/multiroute aspects were catered for by defining and displaying colour-coded squares and cards.
5. Dedicated kanban squares were provided for each route. The squares displayed all 'next' batch options to the downstream station.
6. Bottlenecks were identified and addressed by using alternative operations and set-up time reduction to minimise their effect. However, there was a degree of schedule-based movement of bottlenecks. Due to an expected and unavoidable variation of part mix at certain times, the bottleneck effect would occur on another station suggesting these were not true bottlenecks but schedule-based bottlenecks.

7. The kanban card system was to hold information on the fields shown in Figure 7.31.

User/home station ID	MI
Source station ID	L1
Part ID and descr.	PE0097
Family	Blue
Container ID	Box Y005
Capacity	50
Buffer location ID	JA 22
Release date (priority)	9/6/90

Figure 7.31 Withdrawal kanban card.

8. A set of basic batch release rules were developed. Simulation and testing showed they produced minimum congestion and cycle lead time in the cell. The rules were also to be built into the overall cell scheduling via MRP. Developed heuristically, they were based on the knowledge that specific parts and part mixes would cause heavy loading on specific cell resources. This suggested a need for localised schedule levelling. Typical rules were
   - If part P1234 has been released within the last 24 hours, do not release any batches of P4567 or P7890 otherwise heavy congestion of key stations will occur.
   - Release any available batches of P2345 when releasing parts P0001 and/ or P0002.
9. A large cell control board was to be used for showing machine and operator status, WIP positions and WIP availability. This board was to be located at the end of the cell and maintained by all operators. It would provide at-a-glance information on the complete cell status, assisting in operator awareness and control. The control rules described above would be listed on the control board for reference.
10. Sets of family-specific kanban containers were to be used, with facilities for holding the information card.
11. Varying process and transfer batch sizes were used at different sections of the cell to optimise flow.
12. An on-line cell simulation model was available at all times for cell supervisors to use and test schedule mix effects.
13. On examination of subcontracted parts, several items, once partially subcontracted due to undercapacity, were brought back into the new cell workload.

All machines were provided with kanban squares to act as an inventory control device; the actual number of squares depended on

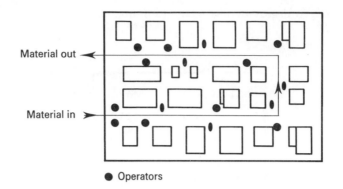

Material out ◄

Material in ►

● Operators

Figure 7.32 Hybrid card and batch release system, a modified kanban dealing with multiple parts and routes in a medium-sized machining cell.

whether the m/c was a bottleneck (there were two, depending on the product mix)
the number of feeder routes into a station (minimum one, maximum three)

Bottlenecks were given WIP squares to equal up to 2 days of production, as a buffer against previous machine breakdown. This was to be gradually reduced through time, as the system and cell became effective. Where three feeder routes existed, a dedicated square was defined for each route and no other route could place a container in these other squares. This provided a downstream station with potentially three containers of parts at any one time. A Kanban card system was to be used to provide normal source/destination and part count data, plus a priority (due date).

The part families were each colour-coded, as were the multifeeder squares. Those parts with an unusual routing were given a second symbol to add to the colour.

This cell was made fully operational in late 1991, and performed up to expectation within 3 months.

■ Summary

Using a systems engineering approach, production and business improvement is undertaken by analysing the current production operations then developing a cell/business unit design. Static and dynamic design activities are carried out to develop a detailed and tested cellular unit with defined control systems and work instructions.

Variations of this approach can address each area of the company, using and applying the methods described in this chapter and the rest of the text. Systems engineers evaluate, simplify and improve. Only then do they apply technology and only where it can produce measurable benefits in performance.

Once internal improvements have been implemented, we can develop relationships with suppliers and then with customers. This completes the chain of JIT business operation.

■ Further reading

Goldratt, E. and Cox, F. (1987) *The Goal: A Process of Ongoing Improvement*, Gower.

Hirano, H. (1989) *JIT Factory Revolution*, Productivity Press.

Imai, M. (1986) *Kaizen*, Random House.

Institute of Production Engineers (1989) *Lucas MSE Guide*, Institute of Production Engineers.

Lucas Engineering & Systems (1989) *Manufacturing Systems Engineers Handbook*, The Lucas Miniguides.

Parnaby, J. (1986) 'The design of competitive manufacturing systems', Int. Journal Tech. Management, 1.

Parnaby, J. (1988) 'A systems approach to the implementation of JIT in Lucas Industries', Int. Journal Prod. Res., 26.

Schonburger, R. J. (1982) *Japanese Manufacturing Techniques*, Free Press.

Schonburger, R. J. (1990) *Building a Chain of Customers*, Hutchinson Business Books.

Sekine, K. (1991) *One-Piece Flow*, Productivity Press.

Wantuk, K. (1989) *JIT for America*, The Forum.

8 | Management of Change

He that will not apply new remedies must expect new evils; for time is the greatest innovator.

Francis Bacon

Any major project or change in direction undertaken by a business will encounter a degree of resistance from a proportion of employees, due in part to a side of human nature that prefers stability and regularity. However, change must become a way of life if businesses are to remain competitive in a changing marketplace. Change projects and missions can cover all or part of a business, for example:

Enterprise-wide missions

■ World class manufacturing (WCM)   **Very high degree of change**
■ Total quality organisation (TQO)
■ Just-in-time (JIT)
■ Enterprise resource planning (ERP)
■ New business venture or diversification

Function-wide missions

■ Plant or facility rearrangement or relocation
■ New product development and introduction
■ Supplier integration and development
■ Computer applications
■ MRP1
■ Local performance improvement projects   **Low degree of change**
■ Capital equipment/machine specification
  and installation
■ Training programmes

Some of the function-wide missions may be adopted company-wide, e.g. new product development and introduction. These and other undertakings will all cause varying degrees of change. A rough guide to the degree of change is shown on the right of the above table. A very high degree of change means substantial disruption in all areas of the enterprise over several years. A low degree of change means only restricted, short-term disruption.

■ Internal resistance to change: a real obstacle

When a business has been operating for several years, people become used to the way it operates. Most people prefer familiar procedures and tend to dislike changes to their comfortable ways of working. In the majority of business, this understandable resistance to change will be encountered at all levels (Figure 8.1). This resistance should not be underestimated.

Facilitating forces	⟶ ⟵	Inhibiting forces
Poor performance	⟶ ⟵	Habit and complacency
Competition	⟶ ⟵	Low morale
Falling profits	⟶ ⟵	Apparent size of problem
Changing markets	⟶ ⟵	Cost
Customer needs and expectations	⟶ ⟵	Resource requirements
Known success stories	⟶ ⟵	Lack of knowledge of 'the way forward'
Internal and external expertise and ability	⟶ ⟵	Lack of expertise

Figure 8.1 Force field analysis of resistance to change.

Several aspects of resistance can be considered. Resistance to change will be encountered for several reasons:

> **Ignorance** of any problems or solutions may lead workers to believe they're 'too busy for that', frequently aligned with a fire-fighting management.
> **Inertia** makes it difficult to move in a new direction.
> **Fear** of the new, the unknown and the magnitude of the task.
> **Complacency** about the old ways; there is no need to change.
> **General resistance** to anything new or different, 'Not invented here!'

There are regularly those who exhibit the 'not invented here' syndrome. If I didn't think of it, I won't accept it. Sceptics are sure that whatever is being attempted will never work. In planning and implementing change, it is necessary to deal with these and other critics, at each stage of the process. Resistance is 100% a people problem, and is worthy of careful consideration.

■ Several common forms of internal resistance

Fire-fighting: a way of life for too many managers

The problems created by overall company inflexibility frequently result in the continual degradation of performance in the face of market demands for higher quality, lower cost and faster response. As the inadequate procedures and systems break down and fail under these pressures, more operational problems occur, and the situation will continue to deteriorate. Because of this, many managers spend a large proportion of their time dealing with crises in and around their area, fire-fighting. An effect of this is that very little time is available for planning and managing change, which therefore plays a much smaller role than is necessary. *Fire-fighting makes you feel good at the time, but you have only solved a small problem.*

Unfortunately, for many people these problems constitute a way of life, and moving them forward to a more effective way of operating will not be easy or rapid. Some managers even view their fire-fighting capability as their route to promotion and progress within a company, the hero syndrome. This is all part of an internal inertia, and requires careful translation to a new way of thinking. Senior and middle managers will need to become agents and facilitators of change, troubleshooting fewer problems and targeting more root causes.

Internal inertia

When a large, massive object moves in one direction, deflecting it even slightly will take enormous effort. A business is similar. As well as helping people to accept change, we need to engineer a marked *step* change in the company efficiency. This magnitude of change requires the total, unbroken commitment of all staff.

Inertia is the negative side; the positive side is momentum. Although it is difficult to begin a change, once under way, positive momentum *starts to* build up and actively helps to drive it onwards. This is partially due to the benefits of changes being perceived and experienced by employees, which leads to a desire for more improvement and improved quality of life. The key is understanding the need for change.

Regular change and improvement is already a way of life for a small percentage of businesses. These have either had an environment of change from early beginnings, or have undergone the transformation from old to new ways of thinking and operating. We must all strive to become masters of change, if we are to compete in the global market (Figure 8.2).

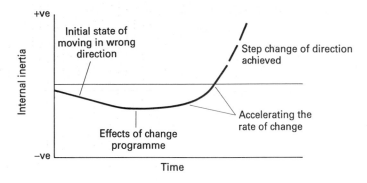

Figure 8.2   A change of direction in company performance.

■ Managing change

Once the directors of a firm have realised that significant change is required in their business, and possibly what these changes should be, they must then consider how to effect them. It is wrong to think that significant change can be brought about merely by an instruction or edict from the company executives. This may be possible in highly regimented companies, but most businesses are run on less dictatorial lines.

The aims of WCM and TQM can only be met and sustained if the whole culture and effort of the workforce is committed to improvement, which requires this internal resistance to change to be addressed and overcome. Once we understand that change is a factor that requires careful control, we can begin to plan and manage it for the benefit of the company.

Taking the resistance factors from above, we can apply suitable means to combat them:

Resistance type	Solution
Ignorance	Communication, education and involvement,
Fear	and demonstrate improvements and benefits
Inertia	,,
Complacency	,,
General resistance	,,

In practice, the solution to virtually all forms of internal resistance is the effective and sustained involvement of people, improved communications and a culture for change. Managing change must be an integral part of the complete move to world class, total quality operation, becoming a way of life for all employees. For this, it is necessary to consider all aspects and stages in the change process:

Awareness of the problems, the aims, and the route to achieving them.

Approach to the tasks ahead; planning the route and the methods to be used, including participation, education, communication; creating the cultural change.

Achievement of succeeding milestones on the route, measured and communicated to the staff.

Sustaining improvement and change within the company, through cultural change and continued involvement and motivation.

Reviewing progress and problems that may develop any time after the programme has begun and seeking continuous improvement in all functions.

When planning for change, a systematic approach has therefore proved to be valuable, assisting in the early identification and definition of likely problems and possible solutions. From this, plans, resources and timescales for change can be developed and prepared for (Figure 8.3).

Strategy for managing change

As with other major business objectives, managing change requires a strategic approach, with detailed plans, timescales and tasks being developed agreed and actioned. Figure 8.3 lists the major topics in managing change. Within the strategy for managing change, we must therefore create and include plans for several interrelated areas:

Resources
Project management
Education and training
Communications
Implementation of changes

As part of such a strategy, it is necessary to decide on the most suitable approach or vehicle to carry out the plans. The approach proposed in the text is to use teams or task forces.

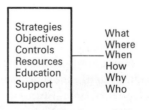

Figure 8.3 Managing and planning for change.

Resources for change: a team and task force approach

The ideal resource for this type of project is the Task Force or Project Team, as it directly ensures staff involvement, improved communication and motivation.

Why do we need task forces?

The projects or missions discussed in this text are major undertakings for any business and will require significant resources to stand a chance of success. There are many areas and business functions to cover in any improvement programme; inadequate resources will only produce an inadequate depth of treatment or excessively long duration. Consultants may be used to inject knowledge and experience but to any one project they should contribute no more than one or two experts or 5 to 10% of the resources. People involvement, motivation and accountability can only be realised if the employees are personally involved in all aspects of the project. It is little use in having the external consultants carry out most of the review, analysis and recommendations to the exclusion of staff who will be expected to live and work with these recommendations. In short, we have to put our own house in order. Teams and task forces have the following benefits:

- Most effective for company-wide review
- Build multirole, multidisciplinary teams
- Allow broad coverage with few people, a core of full-time members plus part-time members as required
- Part-time members come from all relevant functions
- Provide effective communications platforms
- Offer valuable labour and skills resources
- Provide project or mission credibility with the total workforce
- Help understanding and dissemination of information
- Lead to employee ownership and involvement
- Give accountability
- Assist in successful project implementation

As well as offering valuable labour and skills resources to a team or task force, part-time members also provide project or mission credibility with the total workforce, due to the effective involvement of every area of the business via the part-time member. This leads to employee ownership and involvement in the mission, either by being a task force member or through communication back to the staff via the team member. Understanding and dissemination of information is of great importance throughout, and this forms a reliable, easy communication channel, allowing employees to be kept informed of progress and objectives at each stage. A channel for feedback of staff concerns, questions and suggestions is also needed.

The part-time members also form the basis for spinning off and staffing other task forces and continuous improvement teams, once they have been trained and experienced in the concepts and methods of TQM and systems engineering.

Project organisation

Under the overall control of the executive team, the most effective organisational structure for this type of undertaking is shown in Figure 8.4. Here, a top-level steering group is set up comprising the chief executive, the directors or senior managers, and the project leader. There should also be a project champion at director level, who is responsible for the overall progress and control of the project.

The steering group is equivalent to a cross-functional quality council, meeting regularly every two to four weeks to review progress, project management and problems. The group is responsible for progressing management of change across the company, and needs to review progress at the business level, together with any spin-off problems and potential improvements from the project.

The champion or mission owner is ideally a senior manager, operating part-time. The champion is responsible for obtaining and allocating resources, monitoring progress and maintaining the plan's priority in the eyes of senior management.

Full-time leader

For major change programmes, the manager of the project or mission always needs to be full-time, devoted completely to all aspects of project management, planning and control. Part of this role, and the role of the champion, is

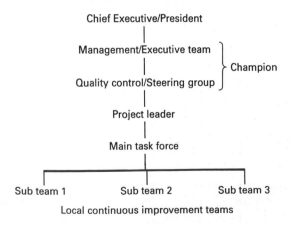

Figure 8.4 Resources and structure for change.

to prepare and smooth the way for task force members to carry out the review, analysis and implementation of change activities in each area of the business. Responsible for achieving the objectives, timescales and budgets of the mission, the project manager must be a stable, able and respected member of the workforce, able to work and communicate effectively at all levels of the organisation. Such a role enhances career prospects, as it provides wide and deep experience of most areas of the business.

Mission leader

Mission leaders need many of these attributes:

- **Dependability and stamina**: the mission may take up to several years to achieve worthwhile progress. Continuity is important, so an individual who is likely to leave the company during this period should not be chosen.
- **Communications**: The person selected must be able to communicate effectively with individuals at all levels of both line and staff departments. A person who is neither familiar nor comfortable with the shop-floor should not be chosen.
- **Understanding of functions**: embarking on a total quality mission also has support system information and control implications that the project leader must be capable of understanding in order to correctly guide the implementation. He or she may come from any of the functional areas of the company, such as engineering, materials, manufacturing, finance or management information.
- **Understanding of systems**: when dealing with the MIS/computer staff, the project leader will be representing all users, and needs to have enough knowledge of systems to communicate effectively with computer analysis. Where this is a weak area, education and training can provide adequate understanding.
- **Respect**: since the leader will be dealing with all the staff, mutual respect is required. This will assist in later stages of problem identification and solving. To be effective, the leader must command the respect and confidence of top management.
- **Sense of urgency and working to deadlines**: the leader should be the type of person who has a strong commitment to high standards of work and on-time completion.
- **Understanding of balance and importance**: not everyone will feel they get everything they want out of total quality; there will be many compromises. The project leader will be responsible for balancing the trade-offs in the system to maximise benefits to the whole company. Sound, unbiased judgement with a corporate viewpoint is essential.
- **Ability to use authority effectively**: top management must give the leader the authority to make decisions and to have those decisions carried out. He or she must be able to delegate and monitor well.

■ **High frustration threshold**: any job which requires dealing with a large number of people in different disciplines will encounter many frustrating situations. The mission leader must be tolerant, persistent and resilient.

Types of task force

The nature of team groups depends on the nature of the business and the project. Two main types are used:

Main project task force/project team
To address the major redesign of the facility: manufacturing, offices, sales, design, systems
– Local basis
– Cross-company remit
– Methodical approach
– Well briefed and trained
– Expert assistance provided

Small continuous improvement groups
Set up throughout the company to deal with specific targets and improvements in each function
– Developed from main task force
– Led by member of main task force
– Given basic training on systems engineering and TQM
– Run like quality circles
– Given authority for actioning improvements
– Can be given tasks by main task force

Members are drawn from each area of the business and educated in TQM and systems engineering. A mix of full- and part-time members is preferred. The size and number of teams depend on the size and complexity of the project. Very large projects will normally require several small teams, with the project manager controlling all the multidisciplinary teams through a group of team leaders. Large or medium-sized projects can be carried out by a single team, varying the number of full- and part-time members to suit the workload and area of activity. Use extra sales people to review sales, extra accounts people to review accounts. For maximum benefit, only proceed with creating and running continuous improvement groups (CIGs) once the main task force is well established.

Education and training

Education and training must be fully used at every stage of the mission. Starting with awareness education for senior management, a comprehensive and well-planned training programme must be developed and delivered. From chief executive to shop-floor worker, everyone should receive training on what is being done and why, dovetailing into an overall mission communications strategy that provides regular feedback to all employees.

Task force training is a very important section of the training programme, since it will equip these key groups with the knowledge and skills necessary to undertake the review and improvement of all areas of the company. For this reason, the highest standard of task force training should be obtained; no half measures will do. Task force members will pass on their knowledge and skills to other spin-off teams and improvement groups, so a 'train the trainer' approach should be used.

Best practice in project management and planning

Projects for change are often the largest undertakings of any business, so it is essential to maintain best practice for project management, planning and control. This applies from the steering committee right down to the CIGs. Project management embraces the provision of leadership, control of resources and tasks, creation of a suitable project environment, the use of project planning tools, activity tracking and defined review periods. Further reading on project management is given at the end of the chapter.

■ Stages in the mission

Typically, the structure of a change project can be considered as three stages

- ■ Preparation
- ■ Design and analysis
- ■ Implementation

■ Profile of mission stages

Preparation stage
1. Definition of aims, objectives and mission statement by management team
2. Selection of team leader and team members
3. Training of team leader and team members
4. Development of outline project plan(s), timescales and targets

Design stage
1. Task forces carry out review and analysis of the business and its operation

2. Regular and frequent monitoring meetings, with feedback, revised plans and additional or changed tasks and objectives, at each level of the change structure
3. Final review and presentation of findings and recommendations, including detailed plans, financial cost and benefits

Implementation stage
1. Execution of detailed improvement and restructuring plans, as and when agreed with steering group
2. Review and reshuffle of team members to provide best team coverage in each area of change
3. Establish more CIGs to drive change throughout the enterprise
4. Review progress at each major milestone during implementation, controlling and amending timescales and objective where changing requirements exist
5. Continue TQM council and CIGs with regular monitoring and communications

Project definition: terms of reference

Those running and managing the project should be made fully aware of why the business is undertaking the change, the perceived benefits and objectives that are to be achieved, and the level of commitment from senior management to its success. For this, the project needs to be clearly defined with specific objectives and terms of reference. This brief must clearly describe the scope of the project and the team.

Prepared and agreed by the management team and the steering committee, this brief should be in written form, and available for discussion with task force members. It should also be summarised as a one-page mission statement in communications with the workforce, to ensure they are fully aware of the project and its objectives.

Implementation planning

The implementation stage of the project commences once the findings and recommendations of the review have been accepted and signed off by the steering group and the management team. This is the phase where change begins to happen in the business, and where most resistance will be encountered. Management commitment will often be tested to its limits. Persistence, strength of purpose and endurance will be required by all those involved. This is a major undertaking across the total organisation, requiring adequate resources and support throughout its implementation.

On entering this stage, the task force will have reviewed and analysed the systems to be replaced and the new systems to be implemented.

Existing systems	**Incoming systems**
Strengths and weaknesses	Strengths and weaknesses
Organisational needs	Restructured organisational needs
Areas to be retained, resource movement	Links to existing good systems
Activities and jobs to be changed or removed	Effective replacements
Inappropriate measures of performance	New measures of performance
	Computer system requirements

Projects of this nature also contain risks of various kinds, which must be recognised in the plans, timescales and reviews throughout the project duration.

Detailed planning and control

The project is broken into manageable chunks, which are further broken into detailed tasks and activities. A common project breakdown is shown in Figure 8.5. An effective way to manage and control the team resources is a task-oriented approach, passing tasks on to subgroups and CIGs in a written, unambiguous format. Such a standard format of task allocation, recording and progress reporting is shown in Figure 8.6, and forms part of the overall documentation for the project.

- Break down project stages into sections and activities that can be monitored
- Create a project flow chart showing the sequence and dependency of these activities
- Prepare structured plans for each stage of the project
- Define start/finish dates for each activity, with review dates and milestones
- Plan education and team-building for the change period
- Plan user training on the new systems to take place before implementation
- Create and maintain a matrix and work plan of tasks and responsibilities for the team and its members
- Regularly review and adjust workloads to maintain project progress and team loading, down to staff-days or part-days
- Maintain project budget, using internal labour rate, external quotes and timed expenditure against plan, possibly over several years
- Expect and prepare for difficulties and/or changes to the project, including resources, resistance, difficult new systems and unforeseen problems during change; develop contingency plans wherever appropriate

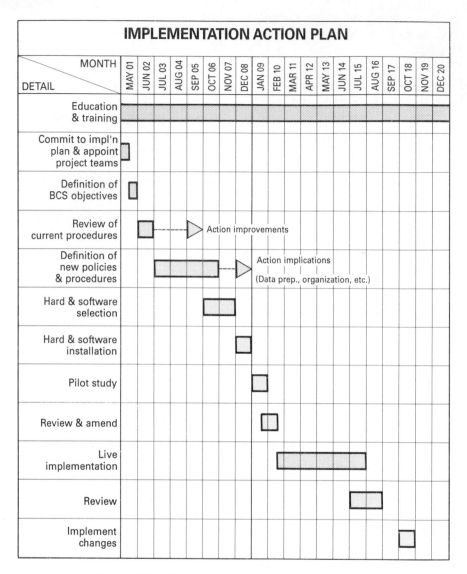

Figure 8.5 Mission stages for further expansion.

■ Follow the agreed plan for communications and feedback to other change groups and to the business workforce
■ Set up and use performance measures to determine the effectiveness of changes

Activity no. 41

Activity
Identify the content of the part master data files

Description
The information which the company wants to hold on assemblies, components, materials, etc. needs to be established. This forms the database known as part master data, a foundation of the MRP2 system

The manufacturing systems department will provide a list of what typically is held in this database to run an MRP2 system. This will be refined to suit the company requirements through liaison with the various departments.

Questions/considerations
Action
Manufacturing systems
Relevant departments

Figure 8.6 Standard task/activity sheet.

Project documentation

A large proportion of task force work is concerned with recording and processing information, which will be required for checking, reference and preparation of presentations. It is essential that accurate, well-organised documents are maintained for all activities. This includes diagrams, flow charts, audio visual material, computer software and data files. Personal computers can prove extremely valuable to a task force.

Use of personal computers in change programmes

Use of personal computers in change projects is widespread, offering many useful tools and packages for review and analysis of business information. Some of the more specialist systems are described in Part II, but the most commonly used systems include:

Type of package	Common examples
Word processing	Word, Wordstar, Amipro, Multimate
Desk-top publishing	Pagemaker, Pageplus
Spreadsheets	Excel, Lotus, Supercalc
Graphics and presentation	Harvard, Freelance, Concorde
Structured design/flow charting	Accelerator, Windows Chartlist
Simulation	Simfactory, Witness

Type of package	Common examples
Custom data analysis	Access, Visual Basic
Forecasting tools	Smartforecast, Spreadsheet modules
Expert systems shells	Crystal, VPExpert
Project planning	PC Planner, Project Manager, MS Project

Group software

Notes, introduced by Lotus in 1992/3, is an extremely useful mechanism for communicating and managing information and data within a group of people. Distantly related to electronic mail (email) systems, *Notes* group-based software provides powerful facilities for information updating, highlighting and prompting when and where activities occurred or should have occurred. Another example is Microsoft Exchange.

This form of software application can be of substantial value to project teams and cross-company groups. It is highly recommended. Security and storage of computer records need to be well organised and reliable, particularly when stand-alone PCs are used. Many of these applications are particularly useful for presenting information and proposals to the workforce and to other teams in the task force (Figures 8.7 and 8.8).

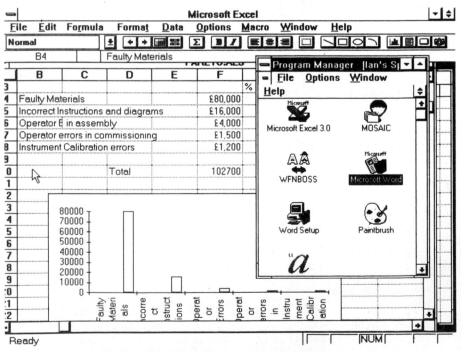

Figure 8.7 Windows-based PC applications.

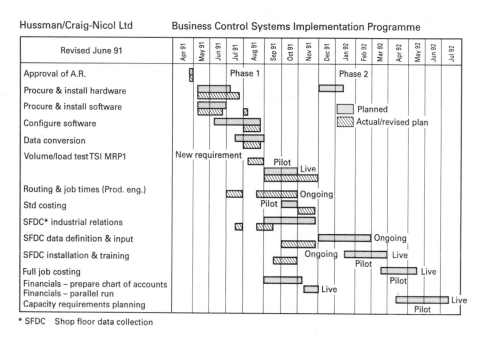

Hussman/Craig-Nicol Ltd   Business Control Systems Implementation Programme

Figure 8.8 Project planning chart from Microsoft Excel 3.

Monitoring and control

Regular monitoring and review of progress is essential between all levels of the change management structure, to ensure early identification of problems, findings and progress reporting against objectives. Monitoring intervals should be as follows:

Each day: task force and team leaders with task force members
Weekly or biweekly: task force and team leaders to core task force
Monthly: steering group/quality council with project manager(s)
Stage review: project manager, core task force to management team at middle and end of project

Effective project control needs to be exercised throughout the mission, using detailed plans and timetables for all activities, and the regular and accurate feedback of progress information through the change structure. Responsible monitoring and control by the project management will successfully achieve the aims and objectives. Ongoing project and mission success can be indicated and measured by several criteria:

1. Achieving the time schedules set for each successive task and/or stage
2. Adjusting and amending the change plans positively
3. Achieving the workforce communication and motivation objectives

4. Meeting the commercial and financial targets
5. Successfully implementing each change plan
6. Producing measurable improvements from each change project
7. Starting and sustaining CIGs to the plan, with regular, valuable output

Confidence indicators

At the strategic level, several important decisions and events can indicate the quality of the project management. Mistakes in or omissions from these early and important issues could jeopardise the whole project:

1. Selecting a strong, capable project leader
2. Selecting and organising the core team and task force
3. Training and educating the steering group, managers and project team
4. Establishing the review and monitoring structure
5. Selecting and practising the necessary improvement methodologies
6. Defining the communications strategy and plans
7. Defining measurable indicators for progress and quality standards during mission and project progress

Resistance once under way

Even once the project is approved and commenced, obstacles to change and eventual success must continue to be understood, expected and planned for, if they are to be overcome. The type and direction of resistance can also change, but can often be identified to a fairly detailed level. These additional threats include

- Little or no commitment from management
- Lack of a suitable champion
- Resistance from middle management
- Trade union concern and resistance
- Lack of adequate training
- Inadequate project management
- Lack of external expertise
- Compromise solutions

If any of these threats become evident in the project, they must be addressed and resolved quickly, before project progress or success is placed in jeopardy.

■ Areas and vehicles for change: culture, organisation and systems

In practice, the very aspects that need to be changed in a business can be used as weapons to combat and overcome resistance to change. Culture, organisation and systems improvements are cornerstones of TQM and world class manufacturing.

- **Culture** encourages, motivates and involves all our people
- **Organisation** forms natural, efficient groups of people and resources
- **Systems** help people to communicate and the business to operate; systems are based on defined, formal procedures accepted and carried out by all

A change to a new computer system will usually incur opposition from many of the users of the original system. But this can be largely avoided by approaching the need for a new system as a **catalyst for change**, involving the users and designers in defining a system that will be improved for both the user and the business perspective. Similarly, changes in organisational structure are perceived as threatening by most employees, but if many of them are involved in reviewing and defining the changes that should take place, the levels of understanding and motivation will be much higher.

Catalysts for change

Catalysts for change are applicable to virtually any project or major objective undertaken by the company and are normally very successful. Major business projects including MRP2 and BS 5750 are now commonly used as vehicles to start and drive change into an organisation. The difference is between doing the job and doing the job excellently. The rest of this chapter addresses the management of change and the cultural aspects of total quality. Organisational aims for improvement are briefly discussed but are covered in more detail in Chapter 6.

Existing organisation and structure

Independent of the size or nature of a struggling company, several common inflexibilities can usually be identified:

- separate, remote departments with poor interdepartmental co-operation and rivalry
- narrow, specialised jobs
- split responsibilities for complete processing of a product or service
- rigid production facilities

Traditional companies are structured like a pyramid, a hierarchy of departments that contain specialist staff and skill groups. This tends to result in a departmentalised attitude of isolation and separation, discouraging communication and creating barriers between people. The results are demotivation and inefficiency.

For those workers who are prepared to take on ownership and continuous improvement, this type of organisation often makes it extremely difficult to achieve change or to effectively apply the methods and techniques of TQM and systems engineering. Top-level commitment to the pursuit of change needs to be obvious to everyone. Establish all necessary resources and address the organisational structure required for effective change and improvement.

Organisational aims

The TQ, systems engineering approach to combat organisational weaknesses is as follows:

1. Break down functional boundaries and barriers
2. Establish simple/flexible groups – modules and cells – a single-office concept
3. Eliminate informality, fragmentation
4. Develop simple operational procedures
5. Provide performance measures reflecting customer/supplier needs.

In this approach, we are striving for ease of communication, rapid response and maximum flexibility. These changes to organisation and structure will also result in concern and hostility among employees, unless communications and change management is fully addressed. So, to realise the benefits of such restructuring, people's potential resistance and fear for job security must be understood and dealt with (Figure 8.9).

Systems and procedures

As soon as we have functional units responsible for a specific part of the business cycle, we require agreed operating rules and procedures to be carried out. (The term *systems* is used to describe both manual and computer-based procedures.) Life is straightforward when the company is small and

Figure 8.9 Resistance to change.

communications are good. But as the company grows, volumes become larger, more people are involved and communication often suffers.

Unless the existing methods and systems are perceived as particularly poor by the users, any proposed changes will be resisted, and will require user involvement in the definition of replacement systems, if the resistance is to be minimised. Again, the new procedure or system can be approached as a vehicle for change instead of a totally new practice. Examples of several manual procedures are contained in Appendix D, and a detailed treatment of implementing a major, company-wide, computer-based MRP2 system is contained in Appendix I.

Culture: participation and involvement

A total quality business is most easily distinguished by the attitude and actions of its people. The importance and value the people place on quality and excellence of every activity, together with their continued striving for improvement in all aspects of their work, are the foundations of a total quality, world class enterprise.

> *In a poorly motivated, non-TQM company: 'Quality of the product isn't my responsibility, I'm in Sales!' This can result in everyone abdicating responsibility, not checking their own work well or at all. In a TQM enterprise we would get: 'That product defect might be related to my late request for a design change. I'll check it out, and in future try harder to find out the exact customer needs earlier.'*

This total quality organisation will be totally committed to the following objectives:

- seeking quality before profit, as poor quality leads to reduced business
- communicating effectively throughout the business
- making decisions based on facts and accurate information
- continuously striving for improvement
- developing employees through training
- developing measurable accountability to the customer
- striving to delight the customer

Based on a willingness to change and adapt as necessary, these objectives will be achieved through universal participation by all employees, working both as individuals and as teams. Once achieved, this will result in improved customer satisfaction and goodwill, both for internal and external customers.

Resistance to change is greater when people have not been involved in the decision process leading to the changes. Make sure all those affected by the impending changes are fully involved at an early stage not only once the change has been decided by others. Early involvement leads to ownership, and this in turn leads to reduced resistance and easier adoption of the changes.

Participation is essential; work with people rather than controlling or confronting them:

- make people part of the process
- motivate people
- bring out the creativity in people
- provide personal development opportunities
- encourage people to try new solutions to old problems

Motivation

Essential to the success of company-wide change programmes is the creation and sustaining of employee motivation. Use various approaches:

- Teamwork
- Common working conditions
- Group rewards
- Opportunity for advancement
- Responsibility for complete product or service
- Regular formal and informal communications

There are many different theories and ideas about motivation. A large proportion of traditional managers believe motivation factors depend on a person's position in the company hierarchy. Though sometimes true, it is a simplistic argument and is often an *effect* of poor management, not a *cause*.

Perhaps UK management tends to work on a 'them and us' basis, even in the 1990s. Companies with this approach are unlikely to have any form of staff/worker participation and may well suffer from poor staff morale and industrial relations problems. These antiquated views are slowly dying out, and the modern, enlightened manager now correctly subscribes to staff involvement and empowerment. As you may realise, manufacturing systems engineering aligns with this management style. This can be summarised as follows:

- involvement and consultation
- challenge and responsibility
- achievement and recognition
- group working and respect

These aspects of working are key elements in motivation of people at any level of an organisation, and should become a way of life for all companies. Contributions from and capabilities of individuals will vary, but the basic rules still apply. Where there is a history of non-involvement and blame, it will take careful and constant attention to show people that this is a change in direction for the company. Don't underestimate this, and ensure regular communication and involvement always happens as planned.

Culture and motivation are heavily influenced by management style, philosophy and approach. Motivation tends to be highest where there is a caring, involved management style and lowest where there is a conflict, authoritarian style.

Mechanisms for motivation

- Keep everyone informed of plans, findings and progress; a team approach is a great help
- Start a company-wide suggestion scheme
- Develop local improvement groups with objectives and agreed authority
- Provide training and awareness education to all
- Involvement in the planning, design and implementation

The following sections offer general guidelines for good communication, applicable to individual and group situations. They allow people to feel at ease by treating them with honesty without confronting them.

Tools for motivations: communications skills

A key factor in the success of this type of initiative is communication between the practitioners and all employees. Where you will be involved with frequent contact with people from all functions of the company, good communications skills are essential. You may not be used to working with people, or only with people who share your specialist knowledge. A few people are naturally good communicators; you are probably a better communicator than you think.

As well as tools for motivation, effective communication will contribute to ensuring overall project success, by assisting in project control, information gathering and the take-up of ownership in the programme. This section covers several aspects of communication for change:

- Company-wide communications
- Running meetings
- PC tools for communication

Interpersonal skills and interviewing are described in Chapter 4.

Company-wide communications: keeping everyone informed

Senior management will be kept up to date through steering group meetings and reports, but these may not always filter down through the company. Several communications methods can be considered:

- newsletters to all staff (Figure 8.10)
- notice-boards and posters
- project presentations as part of regular company briefings
- development and running of general awareness seminars for each section of the workforce

HUSSMANN° / CRAIG-NICOL
A Whitman Company

MRP Project News
The way forward

May 1992

MRP - Improves Customer Service

Since the successful implementation of MRP1 at the end of last year, the development of MRP2 systems has made considerable progress. We are already benefiting from some of the advantages of the new systems.

In addition to the more obvious advantages of shortage reduction, better inventory control and faster processing of purchase orders, one of the most significant benefits is an improvement in the service we can give to our customers. Our customers require a fast response to their needs. In the past, the time taken to prepare quotations, issue Job Lines, place purchase orders and the issue detailed schedules to the factory has been a major bottleneck. This all adds to the overall lead time to ship cases to site.

The total lead time to ship an order is made up of all the functions in boxes A to F in the diagram below. By attacking the early stages of the overall process, significant reductions in overall lead time can and are now being achieved. MRP1 which was introduced at the end of last year has already greatly reduced the time taken to identify and schedule manufactured and purchased parts and the new Purchase Order Processing system has reduced the time taken to issue purchase orders to our suppliers. Last year it took

almost 10 working days to analyse inventory records, decide what to order, type and issue the purchase orders and update inventory records. Now this same process takes less than two days and is carried out at least twice per week.

The new features such as Sales Order Processing, Trial kitting and more automatic shop schedules, being developed for MRP2 will lead to further reductions in the time taken from receipt of order to the dispatch of cases to the customer.

Reductions in the preparation and support lead time have an additional advantage, as it allows orders and changes to orders to be processed faster. Our customers frequently change their requirements and our ability to process these changes quickly using these new features together with the other MRP1 features again improves our response time to our Customer's needs.

In this newsletter you will find details of some of these new developments. The effects of these changes are shown in the diagram below.

Total Lead Time Reduction - Attacking The Early Stages

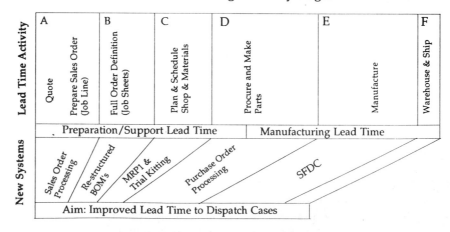

Figure 8.10 A newsletter extract.

The ideal combination of these methods will depend on your own organisational requirements, but should include as many sessions or events as possible. You can never have too much communication. To achieve successful progress on this journey towards total quality requires that communication be two-way. It is even more important to know what the staff are saying and thinking than to tell them what is happening. This feedback can be obtained through the project presentations, and via implementation meetings helds between task force members and their own sections.

Strategy for communications

It is valuable to prepare and work to a communications strategy for major change projects, detailing all aspects of project communications:

- What is to be communicated and how?
    - background and need for the change
    - effect on the business and the workforce
    - what will be done and in how many stages
    - methods, materials and resources that are needed
- When will communication be made and with whom?
    - before, during and at completion of stages
    - to whole workforce, sections, managers
    - length of time required
    - ways to obtain feedback from the workforce
- Will a project launch be used for plant-wide communication?
- How will the level of success of the communications activities be measured?

Communication exercises to try

- Quality circle exercise
- Writing a series of newsletters
- Running a meeting – role play

Running meetings

Most meetings are badly run. Frequently they are talking shops or platforms for individuals to state opinions or dictate to the rest of the group. Running an effective meeting is actually a very easy thing to do, provided you adopt a planned approach.

Preparation for a meeting

Firstly, define the purpose of the meeting. Is it a conventional meeting, a

243

brainstorming session, a seminar or what? Once you have defined the purpose of the meeting, this will point to the specific objectives and/or actions that may be required as outcomes. To give structure to this list of topics, we require a formal agenda (unless it is a brainstorming session) Because of the purpose and topics covered by the meeting, you will decide who should attend. They need to receive an invitation to the meeting, with an agenda and any other information required.

The meeting

If you are an organiser, you will have a specific role to play, maybe as chair or as a proposer. Make sure you understand your role fully and are well prepared to carry it out. As chair, you will have to control and guide the meeting to achieve its purpose by

- Ensuring the agenda is adhered to
- Firmly but politely ending discussions on side and unrelated issues
- Ensuring that appropriate actions and responsibilities are defined, agreed and noted, with timescales
- Keeping order and perspective
- Ensuring the minute taker issues accurate minutes
- Starting and finishing on time
- Following up on actions agreed in the minutes

Stress

Modern business creates personal stress, unhealthy when it includes continual goals and tasks that are extremely difficult or impossible to achieve. This is 'bad' stress, where the individual will be justifiably concerned about whether he or she can actually do the job, or may be dismissed. Psychologists divide stress into healthy and unhealthy. Healthy stress can assist us in performing at our best working well with just the right amount of pressure.

People tend to work better if there are known deadlines and timescales for a job, otherwise the job will frequently slip and/or be poor. Among the exceptions are highly motivated people, who often impose their own deadlines and high standards of work on each and every task they do. They create their own healthy stress level. In general, a reasonable amount of pressure is desirable and should be maintained for best personal and group performance.

A useful analogy is to compare stress with an elastic band. The band is serving no purpose when lying in a drawer, just as stress plays no useful part when undirected. When you take the rubber band and stretch it around a pack of letters or job cards, it becomes a useful aid. Likewise, healthy stress can become a useful aid to getting a job done. However, if the band is overstretched, wholly or in parts, it will snap or be damaged.

From this example, we can see that to be effective, stress should be gradual and fairly uniform all the time; overstressing will cause problems. Healthy stress extends staff capabilities and performances to safe and achievable limits. Healthy stress helps to derive the best business performance overall. Although we all may wish for a quiet day, forget it! Use any slack time for thinking or acting on improvements.

■ Summary

Management of change is wholly about people. It centres around communication, involvement and the culture within the organisation. When these critical factors are understood, planned for and effectively addressed, the success of any change programme is much more likely. An ideal approach to managing change is through the use of multidisciplined, cross-functional task forces and teams, which analyse, design and implement the changes and improvements required. Throughout the project, best practices of project management must be used to plan, control and review progress across each function and system being improved.

■ Further reading

Burbridge, R. N. G. (ed.) (1988) *Perspectives on Project Management*, Peter Peregrinus.

Jick, T. (1993) *Managing Change: Cases and Concepts*, Irwin.

Kanter, R. M. (1984) *The Changemasters*, Allen & Unwin.

Summary of Part I:
The Route to World
Class Performance

Figure S1 below shows the development of JIT-based internal cells for improved response. This is the first stage of a three-stage progression and to get there requires the following steps.

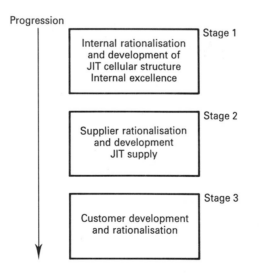

Figure S1   Full JIT operations at every interface.

1. From the identified needs of the market, **develop a business strategy and a product strategy** that define the resources necessary to service the market
    ■   machines and/or processes
    ■   control systems
    ■   people
2. **Simplify and restructure** both the office and shop-floor systems, to create

new flexible jobs and functions that have clear product or profit centre responsibility. Decentralise into local, focused structures that are measured by local performance ratios.

- throw out traditional concepts and beliefs of manufacturing methods
- break down barriers
- encourage fast communication
- speed up information flow by removing or shortening paths

3. Employ the small business concept to suitable production units; give them all the service facilities required to be autonomous.

4. Decentralise service functions to the small business units, to give them control over their own operation. Ensure both operations and development areas have the necessary service resources.

5. Strip out the traditional indirect labour overhead that was needed to support and operate the complex and/or ineffective systems. Simplify and restructure the remaining support functions using a systems engineering approach. Remove NVAs wherever possible.

6. Set up a team-based approach to continuous improvement, with multidisciplinary teams used to evaluate and improve each area of the business continuously. Establish performance measures for every area and continue to review and tighten the performance required from all sections and teams.

- think of how new methods can work, not why they can't
- don't accept excuses
- trace root causes by persistently asking, 'why?'
- perfection can wait; a 50% improvement now is just as valuable as a 90% improvement in six months
- correct mistakes immediately they are found; don't put them off

7. Combine visible performance measures, openness and hands-on management with clear leadership and involvement of the whole workforce.

8. Selectively introduce appropriate new technology – CNC, MRP, automation or whatever – then measure performance. Don't spend money on improvements unless absolutely necessary.

9. Aim for maximum flexibility of shop-floor staff and resources. Strive for set-up reduction and quick-change, mixed-product capability.

10. Plan to sell daily, make daily with low stock levels and rapid response to customer demand.

11. Create fast and robust personnel communication systems across the company, for example, a team brief for vertical communication, with improvement groups for horizontal information exchange. Make sure information can travel upwards as well as downwards.

12. Be sure to set up and deliver a core training programme across the company.

PART II
Selecting and Applying Appropriate Technology

Companies approach change in different ways – from slow, incremental improvement through to step-changes in the way the whole company functions. Most tend towards the former, due to reasons of low disruption, low cost and lower risk. Many shy away from major change unless it is forced upon them.

A company's attitude to change is also very relevant to the success or failure of advanced manufacturing technology (AMT) and/ or information technology. To successfully implement any IT-based system requires a committed management team that is not afraid of change. If they are faltering or failing in other improvement projects, there is a very high risk that an IT-based project will also fail. Those companies that have the vision, firmness and stamina for change will rarely fail in the application of AMT or IT.

In many cases, IT and AMT are facilitators or catalysts for change that might otherwise be impossible. In other situations, IT is absolutely critical for change to occur, for example, the use of MRP1 to transform manual inventory control from a massive, labour-intensive task into a streamlined, accurate process.

We must aim for seamless manufacturing, with information flowing to all those who require it, when and how they require it. The availability, currency and accuracy of information is vitally important to every business, although the type of information required varies with the industry or business sector.

■ Installation and implementation

With high technology systems, vendors are often attempting to provide a single solution to a complex (sometimes critical) organisational problem. Over the last twenty years, CAD, CNC, JIT and MRP2 have all been wrongly portrayed and perceived as panaceas. Most are well designed and function reliably but they are not panaceas. The problems and inadequacies mainly stem from the quality and completeness of their implementation.

Failure to distinguish between *installation* and *implementation* is a common mistake, and is a key contributor to system failures. *Installation* describes system fitting, mounting and operation. This in no way means it is working to benefit the business. Implementation means using and exploiting the system to the benefit of the business, including improved profitability, effectiveness and/or response. Proper, successful implementation results in people using the systems regularly, relying on them and having ownership of them.

Successful implementation frequently means the organisation and procedures have been restructured to suit and enhance the operation of the systems.

■ Technology toolkits

Piecemeal implementation of IT solutions has been commonplace, often with little view of the overall strategy (if it exists) and direction the business needs to take. This narrow and blinkered view can inhibit the integration of business processes, and slow or kill overall company improvement.

Selection of AMT and IT toolkits requires clear vision of the end result. First, determine the business strategy, then determine the IT strategy required to support it. The IT strategy will shape the specification and requirements of all IT and AMT systems, including the level of integration needed.

For integration between different systems and equipment we also require common or open standards, not vendor-specific proprietary systems that can only be linked to, or work with products from the same supplier. Open systems requirements therefore encompass:

- ■ physical communications
- ■ electrical standards
- ■ communications software
- ■ operating systems
- ■ platform-independent software packages

The following chapters describe the features and use of the most common and useful AMT systems, with particular emphasis on implementation success factors.

9 Enabling Technologies for Competitive Manufacture

When applying new methods and technology to a business, 70% of the benefits are normally derived from improved practices and procedures, and only 30% from the actual implementation of improved technology.

■ Do we need these technologies?

Many people equate improved competitiveness in manufacturing with the application of state-of-the-art technology, such as powerful computer systems, MRP2, CAD, robotics, leading to the greatly hyped computer integrated manufacturing (CIM) solutions proffered by numerous vendors.

Whilst it may be advantageous or necessary for a company to apply these and other enabling technologies in the pursuit of excellence, this should only be as part of an overall company strategy. If such technologies are applied to a company without cognisance of the needs or effects of other areas, at present or in the future, then it may be very difficult, if not impossible to integrate the patchwork of systems across the company.

So, do we need these technologies? Yes, appropriate advanced manufacturing and information technology is an essential to becoming and remaining a competitive player in the marketplace, and aspiring to world class. However, the technology is far less important than the achievement of simplified and effective organisations and operational procedures, described in Part I of the text. Without first achieving such an improved total quality approach and culture, any technology tools will only give minimal benefits.

Simplify before automation, and then only apply the appropriate technology

Failed systems litter most business sectors:

- New computer systems that fail to deliver the performance or functionality promised
- High technology plant and equipment that does not provide the expected increases in output, e.g. flexible manufacturing cells, MRP, CAD/CAM

These failures are frequently due to

- poor operational practices and ineffective use
- inadequate understanding of their potential

- lack of training on the systems (not merely how to use them but how to use them well, CAD is a prime example)
- the wrong solutions to the wrong problems

The challenge for each business is to undertake a mission to become a world class, total quality enterprise. This involves identifying and implementing appropriate philosophies, organisational changes *and technology to provide the levels of flexibility and response required by the market. For this, correctly identifying the specific needs of your business is vitally important.*

Technology does not guarantee success; its outcome depends heavily on the user. The rudiments may be easy to acquire but their judicious application is more difficult.

Costs and benefits

Unfortunately, many organisations continue to invest almost blindly in capital plant and equipment, with only token use of cost and benefit studies. 'We need a bigger/faster machine . . .' is an all too common state of mind.

Many such studies or justifications are often used solely to obtain approval of expenditure, rarely being subsequently monitored to see if (any) financial benefit is achieved in practice. Investigation of a large number of UK and European manufacturing companies shows that over 50% of investment programmes in advanced manufacturing (including FMS, robotics and CIM systems) fail to deliver the tangible benefits that were originally expected. (This also includes major plant reorganisations where advanced manufacturing equipment was included.)

Unfortunately, when faced with a major investment, many senior managers sign and forget. Signing the cheque or purchase order is the easy part, often too easy. Full and formal project evaluation and subsequent monitoring is essential. Anything less is unacceptable.

Precede capital plans with an appropriate review

It is now an acknowledged fact that well over 50% of any resulting benefits will be attributable to improved procedures and operational practices, rather than to the advanced manufacturing systems. It is therefore much more worthwhile to commence with a structured review programme examining the business and manufacturing areas, looking for potential improvements and opportunities in every area. The details for undertaking a company review were described in Chapter 4.

In many cases there are substantial benefits to the company that can be achieved more cheaply than advanced technology applications. Technology can play a major role in improving the company's manufacturing competitiveness, but only if applied in the correct form, at the right time and in the right manner.

Greenfield sites

Where there is a greenfield site it is often much more straightforward to specify, select and implement an integrated computer solution because there are no existing islands of automation or established methods to complicate the task (Figure 9.1). Human resources and industrial relations issues will normally be more straightforward, with virtually no problems of employee acceptance and use of this technology.

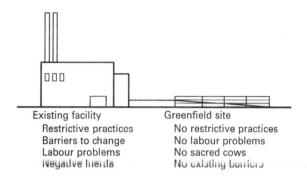

Existing facility	Greenfield site
Restrictive practices	No restrictive practices
Barriers to change	No labour problems
Labour problems	No sacred cows
Negative inertia	No existing barriers

Figure 9.1 Greenfield site approach to best practice.

This is the exception. The majority of plants will require a carefully phased implementation with incremental upgrading of existing systems, introduction of new systems and procedures, and the linking together of all areas necessary to facilitate the transfer of data across the factory. Human resources and industrial relations issues will have to be addressed early in the project. Employees may be concerned about job losses or skill levels.

■ Appropriate enabling technologies

The platform for virtually all the enabling technologies is the microprocessor-based computer system. Now established in all walks of life, the microprocessor provides an abundance of computing power at costs low enough to equip even the smallest machine or process with its own dedicated controller or communications system. At the other end of the scale, powerful minicomputers and interconnected microcomputers can provide company-wide information systems with tens or hundreds of user terminals. The optimal solution for the company will frequently involve a combination of company-wide and dedicated systems.

For the manufacturing and process industries, upgrading of equipment and systems will involve some or all of the following technologies:

- Computer aided design (CAD)
- Computer aided manufacture (CAM)
- Computer numerical control (CNC) equipment
- Flexible manufacturing systems (FMS)
- Process and control technologies
- Material requirements planning (MRP1)
- Manufacturing resource planning (MRP2)
- Enterprise resource planning (ERP)
- Scheduling systems
- Intelligent knowledge-based systems (IKBS)
- Simulation systems

Some or all of these systems may be implemented and integrated together on a single site, to achieve a form of integrated manufacture. The most important single result of this is the potential to create an integrated business and manufacturing database, which can then serve as common, core data for all manufacturing related operations.

To provide the necessary links between the selected systems, appropriate communications systems must be chosen. Figure 9.2 illustrates this concept. Communications systems are covered in Chapter 11. The above technologies are described in the remainder of this chapter.

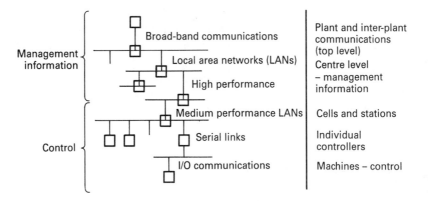

Figure 9.2 Communications between technology areas.

■ Computer aided design

Computer aided design (CAD) is one of the longest-established enabling technologies in industry, and is commonly the first experience a company has with the use of computers in the design/production process. Initially used as a virtual electronic drawing-board, CAD now has many more facilities, which are particularly important as part of an integrated manufacturing concept.

CAD can create an engineering design drawing onscreen then modify and display it as necessary. CAD can automatically create a product database of all design information, including dimensions, components and descriptions of key features. The product database is an organised structure that allows data to be accessed and manipulated in a variety of forms. This database is a key source and store of information in the engineering process and in other areas, such as bills of material. It can be considered as the hub of information flow between many areas (see Figure 9.6).

Types of cad system

Standalone systems

The initial aim of CAD systems was to improve the designer's productivity and quality of work; the computer system provided an electronic drawing-board, plus underlying facilities for calculation, modification, storage and retrieval of drawings and parts libraries. This can lead to a significant reduction in duplication of effort, through the ability to import existing designs and/or parts into a new drawing. Therefore in many cases, there is no need to design or draw anything from scratch. Typically, the productivity of a designer could improve by a factor of 3 by the correct application of a standalone CAD system, partly because drawings can be created and modified more quickly, but also because the designer can thus devote more time to design, rather than the drawing aspects. This often leads to improved job satisfaction.

CAD variants

CAD has evolved significantly from its origins as an electronic drawing-board, with specific variants and extensions available for different industry sectors. These are briefly described below.

Mechanical CAD

In the mechanical engineering sector, there are several other CAD software applications which go far beyond conventional draughting packages:

- **Design analysis** calculates properties, including volume, area, mass, centre of gravity.
- **Finite element analysis (FEA)** analyses an object by breaking it down into many finite elements then analysing each element's response and capabilities under external forces.

- **Three-dimensional draughting** creates and displays a *three-dimensional* model of an object, and manipulates the model in various ways. Three-dimensional CAD systems are now the norm, allowing wire frame or solid modelling of the design piece, and folding/unfolding for translation into sheet material. Drawing complete units and assemblies in *three dimensions* is often a faster design process than having to draw several *two-dimensional* views, again speeding up the drawing process. Typical areas of application for 3D systems are shown in Figure 9.3(a).

Electronics CAD

Several CAD packages have been developed purely for electronics design and manufacture:

- **Schematic drawing systems** use stored symbol libraries for circuit design and printed circuit board design.
- **Design analysis** packages for analogue and digital circuit design allow the testing of a proposed circuit or system before prototyping or manufacture. Simulation can also extend to analysing the heating effects of the chosen circuit configuration.
- **Computer aided test (CAT)** covers a wide variety of topics (including design analysis) but includes CAD generation of test software used to test the final product during and after manufacture.

Electronics CAD can also link to production machines, through post-processing software that produces the required part-programs. For example, a PCB is cut using a CNC laser then populated by a computer-controlled surface mount machine. The completed board is tested using programmed automatic test equipment (ATE).

Integrated CAD

Figure 9.3(b) shows that data from the CAD system can and should be used in several other areas of the business. The first facility provided by vendors was to allow the creation of a part-program for controlling an NC machine. The design data from the drawing is transformed, by a separate piece of software, into instructions for a particular machine tool that will machine the designed piece. This data can be punched onto paper or Mylar tape that is then loaded into the machine. Alternatively, the data may be downloaded directly via communications lines. This topic is covered in more detail in a later section.

CAD systems: AutoCAD

There are hundreds of CAD products and systems in the marketplace, running on a variety of computer hardware and operating systems. The majority of

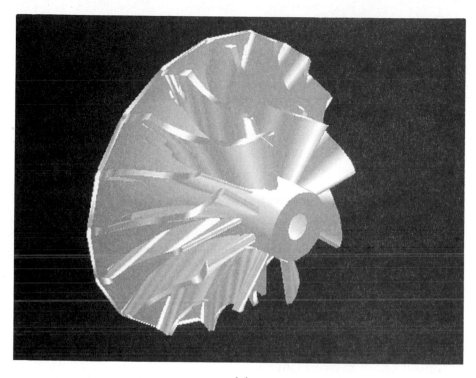

(a)

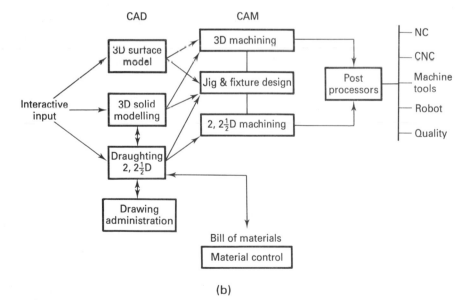

(b)

Figure 9.3 (a) Complex, solid, three-dimensional turbine rotor;
(b) schematic diagram of CAD to CAM.

CAD systems are now based on local networking of powerful computer workstations (see Chapter 11), which includes personal computer platforms. One of the most widely used CAD systems is AutoCAD from Autodesk, part of their integrated range of CAD/CAM systems. Fully supporting three-dimensional drafting and design, Autocad has evolved from a small PC-based system into something of a world-wide standard for CAD, with a market share of over 70%. Autocad runs on PC networks, Unix workstations and other operating systems. Further details of AutoCAD and Autodesk products are given in Appendix G.

Training: realise your maximum potential

Training is essential to derive the maximum benefit from your CAD system and the operators. This should not be limited to vendor training on package features and operation. Get an independent agency to provide regular training in CAD management and effective use. Failing to commit to this level of training can indicate that a company fails to understand the full potential of CAD and CAD/CAM; all they have is an electronic drawing board.

Potential benefits from CAD

CAD systems should also lead to reduced design lead time and more rapid turnaround of design amendments. The improved quality of output from the CAD system can lead to a reduction in the number of design changes, depending on the type of products. Benefits include

- Reductions in lead time
- Improved market response and customer satisfaction
- Improved design quality
- Improved drawing quality and reduced errors
- Increased standardisation
- Improved accuracy in both design and manufacture
- Improved integration of data

But without effective management, any benefits from this level of CAD can be eroded. Maximise benefits by carrying out regular procedural audits of the CAD department.

CAD audits

There is always scope for improvement in any area of business, and this is no exception. How do you determine the effectiveness or efficiency of your existing CAD system? By carrying out an audit of its use in the company. CAD audits examine

- utilisation
- operational policies and procedures
- service level to internal customers
- system administration
- training plans and requirements

The audit should also establish a set of benchmarks for CAD system performance:

- system utilisation, peak/off-peak usage profiles
- response time for various tasks
- number of jobs processed per station per operator
- average job size and time
- service level to customer departments
- backlog/lead time quoted

The CAD auditor

As for financial audits, the CAD audit is ideally carried out by an expert who is external to the department, allowing an independent and unbiased appraisal to take place. You may have a suitable auditor within the organisation, but in most cases the use of an external consultant is required.

Uses of CAD data across the company

Moving away from the shop-floor, data from the CAD system can be sent via the communications system to several other departments. Design data can be used in the process planning function, allowing planners to work out the operations and routing for the part during manufacture. Where a product is detailed in terms of the component parts, sub-assemblies, etc., the structure of the product can be represented as a bill of materials (BOM), key data for the manufacturing requirements planning (MRP2) system. This will produce the schedules and paperwork for buying or making all component parts, issuing them both internally and externally at the appropriate times. For estimating the costs of a product to a customer, the sales or estimating functions may be able to use the product database to derive lists of standard products and cost. The sales and marketing function will also be able to use the design data to create illustrations and diagrams for product sales literature and technical manuals.

Integrating CAD across the company

When CAD is integrated into a complete manufacturing information system, there is further potential to improve the Design-To-Manufacture interface. Due to a lack of communication, manufacture has often found it difficult and costly to make the items drawn by design. This is still true. With an integrated information environment, the ease of interfunctional communication facilitates much closer co-operation in cases such as this, with the manufacturing strategy and rationalisation setting a greater emphasis on design for manufacture and standardisation in design. For example, standardising on a limited range of materials and thicknesses, together with a limited range of bolt sizes, can lead directly to both simplified design/manufacture, and reduced inventory costs. The latter is achieved by having to carry a reduced range of materials, bolts, tools, drills, etc.

■ Computer aided manufacture

The dividing line between CAD and CAM can be hard to define, but in this text the term is used to encompass any activities that occur outside CAD as a design and draughting tool.

Sitting across the dividing line is a range of software that assists in the development of part programs, **computer aided part programming (CAPP)**, and applications that assist in creating and maintaining process routings, **computer aided process planning (CAPP)**. Unfortunately, the same acronym is used for both.

Computer aided part programming

In computer aided part programming (CAPP) a part is drawn on the computer screen in basic geometric and displacement terms. Different forms of CAPP are available, ranging from programming languages, such as APT and POLYAPT, to menu-driven packages, such as PEPS and PATHTRACE, that either stand alone or link into a CAD system.

When a CAD drawing or a CAPP program is to be sent down the wire to a CNC machine for actual production of a part, it has to be converted into a format for the particular target machine. No single format will cater for every type of CNC machine; the range is too large. The piece of software that converts a general design into an executable program for a specific machine, is termed a post-processor, meaning after-processor.

Various post-processors will be available for translating the toolpath (cutter location) program into a machine specific part program. The system can then simulate the toolpath movements necessary to produce the part, without having to cut metal. This obviously improves efficiency and reduces scrap. Despite the sophistication of the CAPP tools available, there still has to be human input in the translation from drawing to a workable part program.

Once the program is proven, it will be sent to the appropriate machine tool, either as tape or over communications lines.

CAPP is not normally necessary if a company has CAD with links to CAM. The decision whether to invest in CAD, CAPP or CAD/CAM may not be straightforward; factors of present and future workload, work nature, percentage time on design, programming and machining, suitability of machine tools for NC/CNC working will need investigation.

Potential benefits of CAD/CAM

- Reductions in lead time
- Improved market response and level of service
- Improved design quality and productivity
- Improved drawing quality
- Increased standardisation
- Improved accuracy
- Improved flexibility

Computer aided process planning

Process plans in manufacturing are core documents or databases holding detailed information on the routing and operations for a given part or product, and will normally include:

- path/route (s) through the facility
- machines and/or resources in the path, with alternatives
- operations and set-ups at each resource
- standard times for these operations
- tooling at each resource

Traditionally, process planning had a high clerical content, involving largely repetitive work in the use and duplication of process plans. Planning information was frequently held by a small number of planning engineers, making its wider use somewhat difficult. Computer aided process planning leads to several benefits:

- Facilitates company-wide use of standard routings
- Offers powerful edit and copy facilities for routing and plan generation
- Reduces planners' clerical workload and increases their productivity
- Allows the simple use of systems engineering computer tools and techniques for improving process flow and organisation; rationalisation is made easier
- Allows optimal templates to be created and used, best-practice process planning

■ Reduces lead times to prepare plans

CAPP is ideally used in parallel with CAD facilities, producing a related document set for each designed part. This data is a vital part in the building of a common manufacturing database.

The types of CAPP system available range from a simple word processor, which allows rapid generation of plans using manufacturing flow chart symbols, up to generative systems, which provide automatic creation of a process plan from previous data. These generative systems are now using expert systems in the selection and use of process algorithms and decision steps.

Common, integrated database

A highly desirable development from separate computer support systems is the creation of a fully integrated manufacturing database. Separate databases accessible only to parts of the company are unified into one database accessible to all (Figure 9.4). This solves the common problems of having several different sets of data for different company functions. For example, the need for different BOMs for design, manufacturing and sales, leading to problems of data integrity, maintenance and duplication, as well as wasted effort and resources. The advantages of the common, integrated database include:

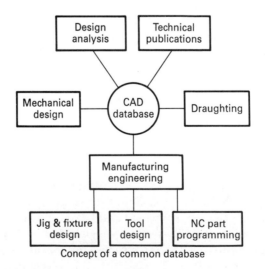

Concept of a common database

Figure 9.4 (a) CAD database as a building block to an integrated database.

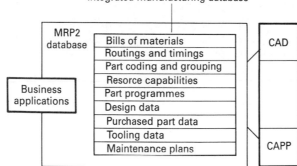

Figure 9.4 (b) Integrated product and manufacturing database.

- use of a single master data set
- accurate design and manufacturing work results from the accurate database
- reduced maintenance needs
- company-wide access and use of the single database
- simple interfacing to other software applications; no complex database hierarchy
- easy to measure performance and accuracy of the data

■ Numerical control machine tools

The development of microcomputer-based control systems for machine tools (NC machines) allowed them to be programmed to carry out a series of operations on a piece, with no operator intervention. Early NC machines were fed with punched paper tape which held the program data, or were manually programmed at a front panel, manual data input (MDI). These first-generation machines could improve performance over manual machines by up to a factor of 3, but were limited in memory, requiring a program to be entered for every part to be machined. There was little or no facility for automatic measurement or tool life monitoring. However, modern computer numerically controlled (CNC) systems provide greatly extended facilities, allowing the storage of many different programs, with sophisticated feedback measurement and tool control systems. The range of applications extends across virtually all types of industry, including electronics assembly, packaging, automatic test equipment, and conventional mechanical engineering.

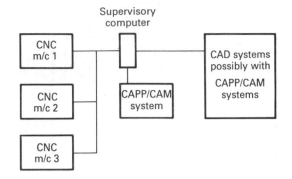

Figure 9.5 DNC of machines with a CAPP/CAM system in the design area or in production engineering.

Direct numerical control

Where there is a need for only a few NC or CNC machines on the shop-floor, it is normally adequate to use tape transfer or communications links from the CAD or CAPP system, if one exists.

However, where there is a greater number of programmable machine tools, and possibly a greater number of part-programs, it can be considerably more efficient to use a supervisory computer to handle all data transfers to and from the machines.

This is the basis of direct numerical control (DNC) where all NC machines are linked to the supervisory computer, which can store a large number of NC programs in its memory, allowing them to be downloaded to the correct machine at a specified time. Status information can also be uploaded to the supervisory computer on request. This would typically include information on the number of parts produced per machine, machine use breakdown, tool data, etc. This data can be presented directly as management reports, but for progress towards completely integrated manufacture, this data should be integrated into the manufacturing information system.

The DNC system should be linked with the prior systems that generate the part-programs, removing the need for tape programs and speeding up the process of program loading overall. Improvements in productivity of around 15 to 20% have been realised by the use of DNC against conventional NC systems (Figure 9.5).

The major weakness of DNC is its dependence on the supervisory computer; if it fails then the whole shop-floor may be halted. Larger organisations may have several supervisory computers to provide redundancy protection. The system is configured so at least one computer duplicates the work of another. Part-programs may be held at two different locations. And NC machines can be connected to operate independently if the need arises.

Sometimes a back-up DNC computer takes over if the prime machine fails. Back-up computers can be expensive unless they are usefully employed at other times. Nevertheless, they are fairly common where a company's production is wholly dependent on the DNC/NC systems. Flexible manufacturing systems (FMS) are examples of this.

Simple DNC links using a personal computer

Where the requirement is simply for passing an NC/CNC code file from a CAD system to a machine tool, it is possible to use a low cost personal computer to provide a basic link between the systems. At the most basic level, files can be passed from CAD station to CNC controller via floppy disk. When running on a network, PC communications software receives files from the CAD system in NC (G-code) format or via operator inputs using a standard word processor. The PC can then be used both for file storage and transfer to and from the CNC machine(s). An example of this is given in Figure 9.6.

Flexible manufacturing systems

Flexible manufacturing systems (FMS) have become synonymous with the factory of the future, highly automated and efficient with minimum set-up times, rejects and delays (Figure 9.7). Automatic tool changing and measurement can be included, together with automatic transportation systems. The benefits claimed include 40% reduction in lead times, 30% reduction in labour and 30% improvement in machine utilisation. FMS is applicable to many types of production shop but is mainly centred on mid-volume/medium part variety. For high volume/low part variety, dedicated transfer lines are superior in terms of throughput and cost, whereas groups of NC/CNC machines can best deal with high part variety/low volume.

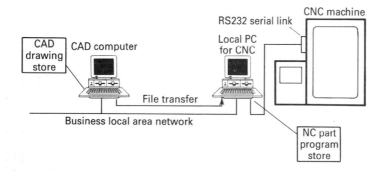

Figure 9.6 Using personal computers for file transfer to CNC equipment.

Figure 9.7   A flexible manufacturing system.

Supermachine panacea: a myth

Machine tool vendors are continually bringing out new and improved FMS, machining centres, and other supermachines. However, there are many situations when a company, considering the purchase of such a system, should carefully re-examine the requirements. The answer may NOT be a super-machine or an FMS; it may only require a rationalisation of existing machines, staff and procedures. The manufacturing review can solve many problems and save many thousands of pounds, often with no requirement for purchasing new major plant. Too many companies automatically assume that buying the big machine or system will solve all their problems. Usually it does the opposite; bigger is not always better (Figure 9.8).

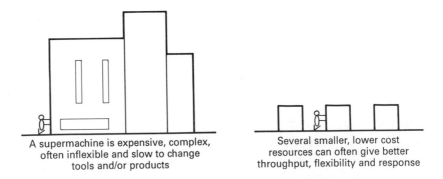

A supermachine is expensive, complex, often inflexible and slow to change tools and/or products

Several smaller, lower cost resources can often give better throughput, flexibility and response

Figure 9.8   Supermachine or simple machines?

■ Process control technologies

In this section we are concerned with the rest of the machines and processes that are part of production in any company, from heavy engineering to chemical processing. Where there are measurements to be made and data to be exchanged then suitable control equipment is required.

Control systems

Powerful low cost micro and mini computers are often used in both sequence and continuous control systems. Microprocessor-based control panels are small enough to locate at (or near) the point of final control, simplifying connection requirements.

In large processes it is now common for several microcontrollers to be used instead of a single mainframe or mini-computer, with resulting benefits in performance, cost and reliability. Each micro can provide optimal local control and can exchange data with other microcontrollers or a host supervisory computer (mini or micro). This is distributed control and it allows far greater sophistication than was possible with a single large computer, since the control function is divided between several dedicated processors (see Figure 11.16)

A distributed control hierarchy need not consist exclusively of mini and microcomputers, but can include other intelligent devices such as CNC machines, robots and programmable controllers.

Programmable controllers

The need for low cost, versatile and easily commissioned controllers has resulted in the extensive use of programmable control systems (PCs), also known as programmable logic controllers (PLCs), standard units based on a hardware central processing unit (CPU) and memory for the control of machines or processes. Originally designed as a replacement for the hard-wired relay and timer logic to be found in traditional control panels, PLCs provide ease and flexibility of control based on programming and executing simple logic instructions (often in the form of ladder diagrams). PLCs have a range of internal functions, including continuous control and communications making sophisticated control possible using even the smallest PLC. Although PLCs are similar to other computers in terms of hardware technology, they have specific features that are suited to industrial control:

- Ruggedness and high immunity from electrical 'noise'
- Modular plug-in construction allows easy replacement/addition of units, e.g. input/output modules with 10 to 1,000 I/O connections
- Standard input/output connections and signal levels

- Easily understood programming language, e.g. ladder diagram or function charts
- Easy to program and reprogram the PLC in-plant

These features make programmable logic controllers highly desirable in a wide variety of industrial plant and process control situations. PLCs operate at the lower levels of the control structure, connecting directly with shop-floor equipment and processes. PLCs can also act as supervisory units for data gathering, programme downloading to CNC/DNC machines, and for passing status data to higher levels within the company.

■ Electronic data interchange

From the previous sections on computer-based applications, it is obvious that the ability to easily transfer data between different equipment and computers is essential. When we consider the data transfer needs for business and commercial control systems, this requirement extends much further. Records and files are needed for sales and purchase order transactions, invoicing, stores and warehouse control, and so on. This is all additional to the data transfer needs of the manufacturing, design and engineering functions.

Electronic data interchange (EDI) describes the use of computers and telecommunications systems to transmit and receive electronic messages and data in place of their paper equivalents (Figure 9.9). Conventional fax machines are an example of partial EDI, where the paper messages are sent down telephone lines instead of by post. Since there is paper required at each end, this is not true EDI. Full EDI involves direct computer-to-computer data transfer between different companies or applications, normally using public or private telecommunications, such as the telephone system. But full EDI can also describe data transfer by disk or tape carried between systems.

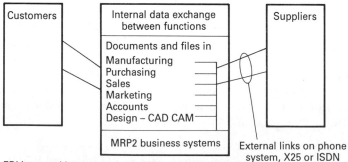

Figure 9.9 Electronic data interchange.

Communication systems are dealt with in greater depth in Chapter 11. Pursuing and achieving EDI with customers or suppliers can become part of an overall movement to closer working partnerships, following TQM and JIT principles.

Advantages of EDI

Electronic data interchange offers significant advantages of speed and cost compared to conventional methods, and many major corporations are now converted or converting to it. In a typical business, the general advantages could be as follows:

- Greatly reduces document generation time and cost; where paper documents were used, EDI creates computer files almost instantaneously.
- Eliminates interdepartment and/or intercompany transfer time for documents; replaces postal or manual couriers.
- Reduces or eliminates storage and transportation of documents.
- Improves response to customers and suppliers.
- Reduces errors and queries.

Individual sections may experience specific advantages:

Intercompany documents

Purchasing: MRP purchase orders and delivery schedules are sent by EDI into supplier MRP systems, reducing lead time to order, acknowledge and receive goods. This can reduce inventory. Supplier stock status and location data can be transmitted as required. Supplier catalogues can often be supplied on EDI disk, allowing customers to view and select goods on computer screen. It also reduces supplier costs for sales catalogues. Updates are by disk or file transfer.

Sales: estimates, quotes, price lists, catalogues, order invoices and acknowledgements can be sent by EDI to customers, saving on paperwork, lead time and labour. Marketing functions can link into remote marketing databases and information sources.

Accounts: EDI invoicing, statements and acknowledgements have similar advantages to those in sales. They can also lead to increased payment on time (payable turns) and improved cash flow.

EDI can therefore offer significant advantages, which will depend on the size and nature of the business. However, there should be a full appraisal of the costs and benefits. Due to the scope of EDI applications, it is recommended that the study is extensive in both inter- and intracompany areas. There are also legal and financial implications to using EDI, which must be investigated and understood as they may affect your company. For example, the placing of a purchase order with no paper copies being sent or held.

■ Business control and planning systems

To provide the levels of information accessibility and currency required by a modern business is a daunting task, requiring ever increasing degrees of data integration and accuracy. Manufacturing planning and control systems – manufacturing resource planning (MRP2), plus material requirements planning (MRP) offer solutions to these requirements.

It is totally wrong to consider any form of MRP as purely a technology or computer application. It must be pursued as a company and people system if it is to be successful. The implementation of MRP2 is a major undertaking, requiring a disciplined, team approach to developing procedures and policies that are prerequisites to its success. But MRP2 software is available virtually off-the-shelf from many hundreds of vendors as a product, running on a variety of hardware and operating systems. For these reasons it has been introduced here as one of the enabling technologies.

MRP2 evolved from material requirements planning (MRP1) and continues to use it within the overall MRP2 framework to determine net time-phased requirements for component parts to meet customer order due dates. MRP2 is much more (Figure 9.10). In its full sense, it can now be thought of as *the* company-wide information system. Now referred to as enterprise resource planning (ERP), it includes

- Sales order processing
- Long-term capacity planning
- Master scheduling for the facility
- Material requirements planning, including made and bought-in items
- Resource planning on a medium or short horizon
- Purchase order processing
- Shop-floor control, including works order generation and work in progress (WIP) tracking
- Inventory control
- Shop-floor data collection, which can streamline the input of WIP and time and attendance data.
- Job costing
- Financial ledgers and accounting
- Forecasting

These are the everyday functions of any modern manufacturer. There is no need to argue about the relevance of MRP2 to many different companies and many different structures. They may or may not need MRP1 but they are virtually certain to require a corporate business information system, a system to be found under MRP2 and ERP. ERP is now being adopted to describe planning and control systems across whole enterprises.

Many modular ERP systems offer integration between functions for companies ranging from 6 to 6,000 terminals. Whilst each system will have family similarities, each will also provide the above functions in a different manner. It is important to choose an ERP system that suits your company; therefore take considerable care over its appraisal and selection.

MRP and MRP2 are certainly not new, but are steadily increasing in functionality and application areas. They are, or should be a key part of the business information systems strategy. Because they form such an important strategic and operational tool, Chapter 13 of this text is devoted to their introduction.

Implementing and using MRP2 needs to be founded on the total quality policies and procedures described in the previous chapters. Effective project management and management of change are also essential. These topics are discussed in detail in Chapters 12 and 13.

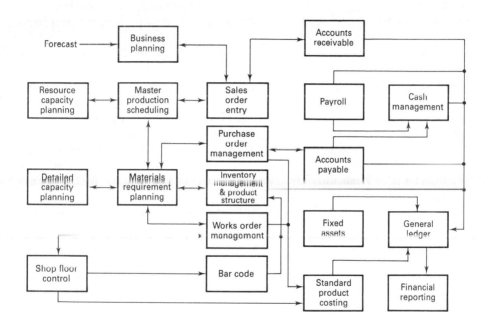

Figure 9.10 MRP2 functional diagram.

Knowledge-based systems

Expert systems are rule-based computer systems that can contain the knowledge and reasoning ability of a human expert or experts. When applied to a particular business area, expert systems can be constructed to contain knowledge and rules relating to particular choices and their resulting effects.

Background

Intelligent knowledge-based systems (IKBS or KBS) or expert systems are a part of artificial intelligence (AI), a branch of computer science concerned with the development of software that exhibits intelligent behaviour. Until the early 1980s, much of the work done in AI was academic research; great claims were made but few real applications were developed. During the 1980s, *soft* AI was developed, a more practical side concerned with the development of specific applications. These included expert systems or KBS designed to solve real problems.

Following many successes and failures, individuals and organisations are now claiming that KBS are the ideal AI development, and companies world-wide are embracing the technology in many different applications. The 1990s are seeing KBS technology become much more widely used as integral parts of company information systems.

Advantages of KBS

Expert systems (Figure 9.11) can be loaded with the knowledge and experience of one or more individual experts. Users exploit this knowledge without having to acquire it themselves. They need not rely on a human expert; they can tap into a computer instead. Among the users of expert systems are company directors and clerical staff.

Once operational, an expert system can provide information faster and more accurately than hitherto possible. Consider a set of batch sizes for

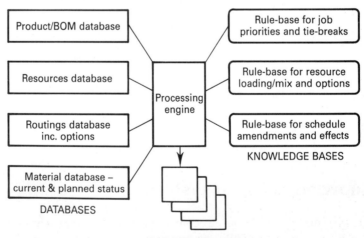

Figure 9.11 Elements of an example expert system operation.

274

different shop-floor schedules and part mixes. The expert system would analyse current and planned production loading then apply selected rules and strategies. The results of this simulation could be implemented by MRP2 or used to simulate a different strategy.

Learning and updating

The expert system can also be configured to learn from such operations, storing results which can later be defined as solutions to future scenarios. This ability is particularly important because expert systems tend to investigate areas that are difficult and/or time-consuming to maintain using conventional methods.

Penetration into business sectors

In 1991 the Department of Trade and Industry (DTI) surveyed KBS in over two hundred UK companies. According to its results, usage had extended from large companies and KBS had been extensively taken up by small and medium-sized businesses. Applications in manufacturing and finance were predominant but many other areas were also using KBS. Of the surveyed companies, 45% identified time savings and 36% identified cost savings. The widespread use of MPR2 in manufacturing companies creates many opportunities for expert systems to be placed within or alongside, providing even better business analysis and control facilities. The DTI survey contains a wide range of useful information, including best practice advice on KBS and several case studies.

Expert system languages and tools

The concept of creating a computer programme that can incorporate rules and knowledge is not particularly novel. All early examples were designed using conventional programming languages, such as Pascal, C and Fortran; some applications are still created this way. However, rule-based languages were especially developed for writing expert systems. Two of the most well known are Prolog and LISP.

These specialist programming tools are not readily understood except by computer scientists and programmers. In order to provide more user-friendly access to expert system fundamentals, several expert shells were developed using Prolog or other languages. Largely based on the use of English language commands and syntax, expert system shells have been widely used in academia, industry and commerce to experiment with expert systems. Popular shells include

- Crystal (intelligent environments)
- VP Expert
- Expert Ease

Mainly PC-based, these shells are low-cost tools that can be integrated to other business control or automation systems, to control or offer advice on data provided by the user or a linked database.

A shell is commonly used to develop application ideas and a working prototype, and may remain the platform to develop the final system. Alternatively, the ideas and concepts derived may then be used to create the final application in a different AI language, or even in a conventional programming language.

Areas of application

Applications exist in business and industry wherever knowledge and rules are difficult to maintain, use or duplicate in other people. Manufacturing applications include

- Planning and scheduling
- Design
- Product configuration
- Sales and field services
- Machine control
- Diagnosis and troubleshooting

For many years companies have evolved extensive database management systems (DBMS) and high levels of knowledge from human participants and operators. Until the advent of expert systems, it was difficult or impossible to capture this knowledge base within the company information system.

This is especially true in areas such as *product configuration* in a high variety or make-to-order business, where deep knowledge and experience is held in the heads of a few experts, such as engineers and estimators, who evolve and apply the rules of configuration for a product.

Possibly the most important contribution expert systems have to make to manufacturing is in providing the ability for business decisions to be made not purely on the basis of shared common data, open to varied interpretation, but also on shared knowledge. The traditional functional divisions of a manufacturing organisation can also result in divided knowledge of the product and manufacturing process, creating compartmentalism – one of the major stumbling blocks in the achievement of manufacturing integration and excellence. The main potential of expert systems is therefore to facilitate the recording, use, and accessing of, organisational knowledge, allowing better decisions to be taken. Many expert system products, such as Crystal, use standard database storage and query principles (SQL), allowing easy interrogation. Further details of Crystal are in Appendix G.

Promoting expert systems

To promote expert systems in UK industry, the DTI has established an expert systems group, which can provide information on the range and scope of applications and products. Over the last few years extensive work has been carried out by users and suppliers in linking expert systems into business control MRP2 systems. Two very different applications are now briefly described.

CASE STUDY 9.1

Product configuration and estimation using Digital's XCON

One of the earliest and best-known applications of expert systems is Digital's XCON product configuration system. Digital had great difficulty in selecting and configuring the many hundreds of units to suit each customer's computer equipment requirements for Vax and PDP11 computers, due to the massive range of items they produced, and the amount of possible permutations. The task was also very time-consuming and prone to errors. Conventional bills of material were inadequate for determining the option configurations.

Digital therefore designed and created an expert system called XCON. It contained data on all parts and units made plus a full set of rules on what to select in a wide range of situations. Configuration knowledge was in the form of standard if-then-else programming rules, and the most difficult part was knowledge elicitation, defining all the rules, options and knock-on implications necessary to produce a correct configuration.

XCON if-then-else rules (for power supply (PSU) configuration)

```
IF: THE MOST CURRENT ACTIVE TASK IS ASSIGNING A PSU
    AND A UNIBUS ADAPTER HAS BEEN PUT IN THE CABINET
    AND THE POSITION IT OCCUPIES IN THE CABINET IS KNOWN
    AND THERE IS SPACE AVAILABLE IN THE CABINET FOR A PSU
    AND THERE IS AN AVAILABLE PSU THAT MEETS VOLTAGE & FREQUENCY
    AND THERE IS AN H7101 REGULATOR AVAILABLE
THEN: PUT THE PSU AND H7101 REGULATOR IN THE CABINET IN THE
      AVAILABLE SPACE
```

This example is part of the large rule base that deals with physical layout and positioning of components inside the equipment cabinets.

XCON is also capable of resolving conflicting rules and situations via several conflict resolution strategies.

XCON is a continuously developing product; by 1987 it had over 12,000 rules in the rule base. The XCON system was very successful, consistently outperforming human experts in configuration task comparisons. It has served as a model for many other applications and competitors in industries with high product variety. Using Prolog, Nixdorf developed the CONAD (CONfiguration ADvisor) expert system for its own products. XCON now forms part of a suite of networked programs used in Digital's Knowledge Network.

Make-to-order configuration

The previous example was for a standard product assembled to order.

Companies in the make-to-order sectors supply individually designed products to a particular customer's specification. For these companies, the task of configuration is of paramount importance. Several rules apply:

1. The product must satisfy the customer's specification but not conflict with any constraints imposed by design limits.
2. Costing of the product must provide adequate profit margin but remain competitive.
3. Producing the product (and promising its delivery) must not violate planning and scheduling time fences, otherwise manufacturing and purchased parts may not deliver on time.

Additionally, the configuration task is important for those products that have already been designed but may require modification or additions to the finished design.

In a finish or make to-order environment, the conventional MRP2 bill of material system is again unlikely to offer adequate flexibility to satisfy the user. Even the *features and options* modules of otherwise excellent packages can fall well short of requirements. Major customising of the package may provide a solution, but is both expensive and restrictive. Further examples of expert systems in configuration roles are given in Chapter 12.

■ Simulation systems

Simulation of manufacturing or any form of discrete item processing can be an extremely cost-effective way of testing and improving new or existing systems. By allowing the user to adjust the elements within the simulation model and observe the effects, it is possible to create and tune layouts, operating methods and control systems before physically implementing them.

The use of computer-based simulation tools has greatly expanded in the last five years. Several powerful PC-based packages and languages are available at relatively low cost. This has made simulation available to all sizes of company. Chapter 10 is devoted to simulation systems and their use. Figures 10.2 to 10.7 illustrate the use of simulation in manufacturing and other event-based applications.

■ Summary

This chapter has briefly described a selection of products to make a business more effective. None of them is particularly difficult to use, but all are difficult to use well and to their full potential. Rarely does technology guarantee success; that task lies almost totally with the user. Careful evaluation is essential before purchasing any technology. Once obtained, the proper installation, training and continued use of the product is also essential. Too many systems have been installed with little or no payback advantage to the company. To ensure competitive business performance, not only must the best IT solutions be selected, they must be properly used throughout their lifetime.

■ Further reading

Computervision (1980) CADCAM Handbook.

Crumpton, P. (1992) *Introducing CIM for the Smaller Business*, NCC/Blackwell.

Mair, G. (1988) *Industrial Robotics*, Prentice Hall.

Rooney, J. and Steadman, P. (eds) (1993) *Principles of Computer Aided Design*, Pitman/Open University.

Warnock, I. (1988) *Programmable Controllers: Operation and Application*, Prentice Hall.

10 Simulation of Business and Manufacturing Systems

Simulation can prove the worth or inability of new equipment and/or proposed changes to production systems without risking major investment

Because of the growing importance of extracting the best performance from all types of production and process resource, the ability to test and try before committing both capital and time to a major project is extremely attractive. This is essentially what computer-based simulation offers, and this chapter describes the potential benefits and the use of simulation systems.

■ The role of simulation in business improvement

When examining an existing or planned production system, we ideally want to use a tool that allows us to experiment with the layout and/or operation of the system against desired objectives, aiming to improve the throughput or efficiency of that system. This could be a manufacturing floor, an airline baggage-handling system, or the flow of paperwork through several offices.

When a company is considering investment in additional or rearranged plant, to increase throughput in an area, or to reduce lead time, the investment is often made on little more than a gut feeling. In many cases this has led to inappropriate equipment and/or layouts being used and little or no practical benefits being realised.

Computer simulation tools are available that can quickly create a representative model of your system in question, allowing the effects of different plant, operation or flow to be compared in a short timescale. Such tools should not be ignored by the managers responsible for designing or operating any discrete part system.

■ Options

When attempting to analyse the operation of any system containing discrete items that move between two or more points, there are basically three options:

1. Observation of the existing plant or system
2. Precise mathematical analysis
3. Simulation

Observation is possible if the plant or system is available for study but is limited in trying out different load, mix and schedule patterns. They need to be physically tested, possibly disrupting production flow and output. Each test will require considerable time and effort.

Precise mathematical analysis is inappropriate for systems that contain several random factors, such as plant breakdown, varying operation and process times.

Simulation is most appropriate and offers significant advantages in the study and optimisation of simple or complex systems.

■ Computer simulation

Using a suitable simulation package on a computer (Figure 10.1), a model of the system under investigation is created by programming and configuration of the package or language. Once the simulation model has been created for an application, the computer model can be used to simulate and evaluate its operation under a wide variety of situations and scenarios. For example, the operation of a manufacturing system can be simulated to determine the effects of change to:

- Material flow
- Plant layout
- Equipment/resources
- Staffing
- Scheduling

Computer simulation has several benefits:

- Very fast speeds depend on model complexity, e.g. one year simulated in one hour real-time.

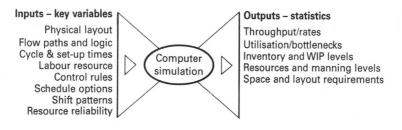

Inputs – key variables
Physical layout
Flow paths and logic
Cycle & set-up times
Labour resource
Control rules
Schedule options
Shift patterns
Resource reliability

Computer simulation

Outputs – statistics
Throughput/rates
Utilisation/bottlenecks
Inventory and WIP levels
Resources and manning levels
Space and layout requirements

Figure 10.1 Benefits of computer simulation.

- Realistic and conclusive models can be developed in relatively short timescales, days or weeks.
- Different strategies and rules can be effectively evaluated over long and repeated runs.
- Questions are quickly processed and amended. For example, what if we move machine A to the other line? If we speed up this process, will another bottleneck be created?
- Data is automatically collated and analysed. Several packages provide facilities to store input and result data after each simulation run.
- Easily understood graphical outputs are available. Mimic diagrams of the simulated system are the norm, plus animated display of moving items, machine status and processing statistics.

Correctly used, simulation can reveal

- Potential problems in a proposed system
- Under- or overcapacity
- Bottlenecks
- Queue sizes and/or WIP amounts
- Flow problems

Some cynics will argue against dynamic computer simulations, suggesting a spreadsheet will suffice. This is a simplistic view of the complex and interrelated variables that govern most operations. Full understanding and analysis of these dynamics requires a computer simulation.

■ Validity of simulation

Computer simulation of discrete events uses logic and statistics to ensure that models, correctly developed, are valid, accurate and reliable. Real-life operations contain random events such as

- machine breakdowns
- variations in cycle times
- arrival times and intervals
- operator performance

Random events are often represented by probability distributions, where statistical distributions are believed similar to the real operation or performance of an element. Some of the probability distributions are

- normal
- triangular
- exponential
- binomial
- alpha, beta

■ Poisson

Machine cycle time may vary according to a normal distribution with a mean of 5 minutes and a standard deviation of 0.5 minutes

Standard deviation (SD) is a standard way of describing the spread of results or events around a mean or average value. An actual cycle time may vary up and down the distribution, with 98% of the cycles within three standard deviations, or 3×0.5 minutes = 1.5 minutes. Basic normal distribution theory is that approximately 98% of all events occur within three standard deviations. More details are contained in Shore's *Quantitative Methods for Business Decisions*.

Each simulation package will have straightforward and appropriate ways to describe the characteristics of a particular resource, machine or plant. Typically, various standard distributions and user-defined series can be used to provide a realistic operating profile.

Random number streams

If we then wish to model this machine by simulation, a method is required to decide where and when events and actions will occur, events and actions relevant to each and every resource on the created model. In each time segment, the simulation selects a different random point on the chosen distribution to provide the appropriate probability.

Several different random number streams are provided in most simulation packages. This allows models to use different number streams for each element and for each iteration. In order to deliver high levels of accuracy, we must ensure that certain key criteria are followed:

1. Run the simulation for a long time to improve the accuracy of the model and to obtain representative results.
2. Use random numbers to represent probabilistic events.
3. Change the random number stream before each repeat run; average the results of all the runs.

■ Simulation packages

Simulation software packages are predominately based on PCs, using the powerful graphics and processing capabilities of 486 and Pentium technology. Packages are available to cover a wide range of user needs, from the relatively simple, low cost products, through to extremely powerful and flexible systems.

More simulation packages have evolved from virtual programming languages into more user-friendly, non-specialist application packages. This evolution has been extremely beneficial, as it allows the application expert (manufacturer, MSE team member or business analyst) to define and create the necessary models, instead of needing a programmer intermediary. Not only does this streamline the process of model creation, it also keeps ownership of the models within the team. Examples of popular and effective simulation packages are

- Simfactory (CACI)
- Witness (AT&T Istel)
- Siman/Cinema (Logica)
- Simple 1

Simulation systems have comprehensive user interfaces for modelling manufacturing processes, and in their own right are now popular tools for analysing and scheduling manufacturing plants. As described in Chapter 12, there are now integrated simulation/scheduling packages available and in use, allowing the testing of various schedules before selection and activation of the best fit (e.g. PROVISA, by AT&T Istel).

Simfactory 2.5 simulation package

Simfactory from CACI is one of the easiest packages to use, being almost entirely menu-driven. Having been significantly upgraded in recent years, Simfactory 2.5 provides a fully mouse-driven interface, where all options and simulation elements are selected by mouse.

Model creation is straightforward, with most field options being displayed via pop-up windows

Sophisticated screen graphics are a feature of Simfactory, with user-definable icons and screens that provide excellent animated displays. One disadvantage of the menu/graphics front end is the processing overhead it can appear to cause. Improved since the previous version, this is not evident with small and medium-sized models, but as model size and complexity grows, the run time of each iteration also grows. However, this is minimised by using a fast 486 or Pentium PC, and execution speed is unlikely to be an issue for most applications.

Comprehensive facilities are provided for linking Simfactory to external database files for data upload and/or download, very worthwhile for loading process planning data (routing, cycle timings) into a model. Version 2.5 now includes certain user-defined options, such as statistical distributions. The basic range of configuration options on each type of resource or machine is adequate, allowing selection of resource or machine type, manning, operation range and times, and breakdown pattern. Statistical variation can be applied to items such as cycle and set-up time, and breakdowns.

Simfactory 2.5 remains largely a package solution, with only limited tailoring possible outside the provided options, in both model and element operation. However, it is straightforward to create large and complex models,

Figure 10.2 Witness display showing the layout of a
discrete parts manufacturing plant.

provided the user understands Simfactory's way of working and applies them
accordingly.

Witness simulation package

Witness from AT&T Istel operates differently to Simfactory, and its displays
look different (Figures 10.2 and 10.3). Arguably more powerful, but less user-
friendly, Witness offers several additional facilities:

- more definable parameters for workcentres and processes
- user-defined logic and rules for loading and sequencing
- automatic storage of data per run; programmed reruns
- ability to program many elements and operations
- selectable run speeds for very fast execution
- user-defined graphs and histograms

These features make Witness an ideal choice for larger, more complex
simulation models, but they do require a greater level of knowledge and
expertise (Figures 10.4 and 10.5).

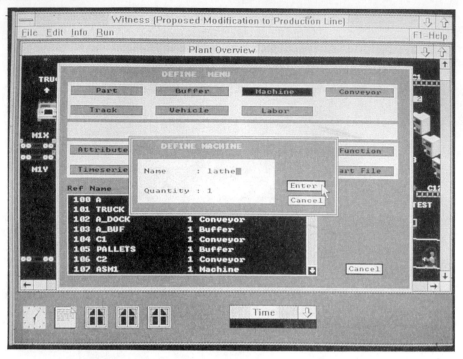

(a)

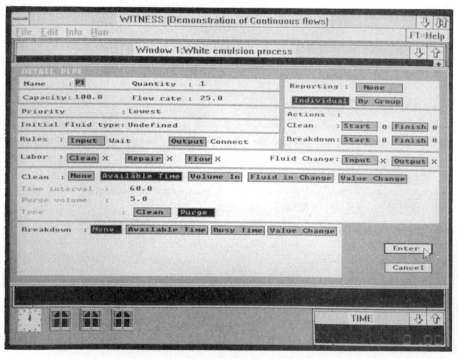

(b)

Figure 10.3 Witness displays: (a) machine configuration options when building a model; (b) menu screen used to enter or alter rules and parameters.

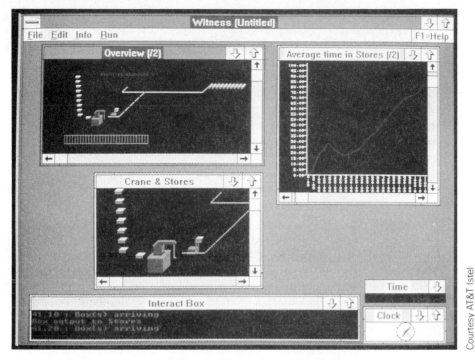

Figure 10.4 Combined Witness display of model processing graphics and graphed data.

MACHINE STATISTICS

Name	Number of Ops.	%Idle	%Busy	%Blocked	%Setup	%Down	%Waiting	
TRUCK	80	70.01	21.81	0.00	7.63	0.00	Setup :	0.55
							Cycle :	0.00
							Repair :	0.00
ASM1	46	2.39	62.96	34.65	0.00	0.00	Setup :	0.00
							Cycle :	0.00
							Repair :	0.00
M1	69	46.98	28.22	0.00	0.00	0.00	Setup :	0.00
							Cycle :	24.81
							Repair :	0.00
ASM2	34	63.42	36.58	0.00	0.00	0.00	Setup :	0.00
							Cycle :	0.00
							Repair :	0.00
M2	35	2.45	37.87	59.68	0.00	0.00	Setup :	0.00
							Cycle :	0.00
							Repair :	0.00
ASM3	32	74.50	25.50	0.00	0.00	0.00	Setup :	0.00

Figure 10.5 Witness report screen showing data collected on resources during or after a simulation.

Following a one-week training course on Witness, a reasonable level of expertise will, on average, take two to three weeks of modelling to develop. By comparison, Simfactory 2.5 competence will take approximately half this time to acquire, but Simfactory has less scope for complex models. However, the release of a Windows-based version of Witness has made the product more user-friendly, now offering power, flexibility and ease of use. Further reading on simulation is given at the end of the Chapter.

■ Carrying out a simulation

Selection of an appropriate simulation package is only a small, although important part of a complete simulation project. The software is only a mechanism for holding and processing the model data. Ensuring the accuracy and validity of the model is the highest priority.

With this objective, a simulation study must be organised and run using a formal project management approach. The study consists of seven identifiable stages.

1. Definition of objectives
2. Design and validation of a conceptual model
3. Collection and organisation of subject data
4. Model creation using appropriate software tool
5. Verification of the simulation model against the real system
6. Running of 'what if' scenarios
7. Analysis of results and presentation of findings

1. Definition of the objectives

A prerequisite for any study is the definition of the objectives. These should be precise and specific, fully understood from the outset. If not, the project will undoubtedly end in failure or dissatisfaction. *This task should be completed by senior management, not left to the simulation team*. The objectives should be written in the project brief and may be as simple as: determine the effect of providing a duplicate resource in the PCB line. However, this should be fleshed out to define what measures are to be used and under what range of operating conditions, otherwise the direction taken in the model may be inappropriate. For example, this objective could be further defined: 'Determine the effect of providing a duplicate resource in the PCB line with

1. The current level of manning
2. Additional manning
3. The current maximum product flow into the area
4. Flow increased 25% and 50%

5. Track queue/buffer sizes under all scenarios

This illustrates how a vague aim can become a specific objective, with unambiguous constraints and variables. This allows the project to focus correctly right from the start.

Team approach

A simulation project always crosses several functional boundaries in a company: manufacturing, plant/maintenance, scheduling, etc. MIS may also be involved for acquiring data from other systems. The project should therefore be run using a team drawn from the appropriate areas, to ensure adequate input from each function. The team will carry out each of the six stages of the project, obtaining additional resources as required.

2. Design and validation of a conceptual model

Before even considering computer software models, it is necessary to create a conceptual model (on paper) of what you intend to analyse and how (Figure 10.6). This stage commences with consideration of the study objectives, relating them to the operational function or area concerned. From this, we develop the outline concept model, which will also start to define the data requirements. A concept model will normally describe and define the following:

■ scope and boundaries of the model
■ constituent elements: resources, labour, parts, products
■ variety of elements and operational scenarios
■ operational rules
■ criteria for handling contention for resources

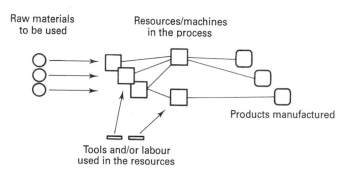

Figure 10.6 Conceptual model of a facility to be simulated.

289

- data required for each element, range and measure of data, e.g. machine reliability as mean time between failures (MTBF), mean time to repair (MTTR) or other.
- layout and flow options

The key to good simulation modelling is to build in enough detail and data. Too much data will overburden and complicate the model, possibly making it difficult to understand or analyse. A simulation consultant can assist in the early stages of model definition, concept design and validation of a concept model against the real or planned system.

Validation

The idea of validation is to see how well a model fits reality, and hence to provide a measure of confidence. Validation is essential to allow key users and executives to accept the direction of the study. This reduces the possibility they will later reject the model findings and conclusions.

3. Collection and organisation of subject data

When attempting to analyse any production or service function where improved performance is the goal, we must be able to identify and quantify the component elements. Only if the data for these elements is properly obtained will it form a reliable basis for creating a model of the system.

What data?

The data required, and its precision, depends on the objectives of the study. The previous stage will have defined exactly what is to be modelled in the study and how the model is to operate. This will in turn identify the areas in which data is required, its type and quantity. Deciding what data is necessary is an important step in this stage, as it relates directly to the complexity of the model. Keeping a model free from excessive or inappropriate features and data requires careful consideration at the design stage.

In a manufacturing situation, where repetitive items are processed, the data required would usually include

- part routings and alternative routings
- operation times
- machine cycle and set-up times
- operator efficiencies and manning patterns
- shift patterns
- schedule patterns (if any)
- part/product mix
- machine/resource downtime and repair data
- processing logic, e.g. LIFO, FIFO, priority.

Several of these data areas may be inappropriate to a given model, whereas less obvious data may be required. Whether any or all of the necessary data is available depends on the type of company and how it is being run. Typically there will be some data collection tasks to be undertaken:

- focused work study and measurement
- collation and analysis of existing data
- downloading of computer database information to a suitable format for the selected simulation package.

In many cases the data will be directly keyed into the package entry screens or downloaded using file transfer utilities. For bulk written data, two stages may be more efficient. Fast keyboard operators create a simple database then these files are transferred to the simulation package.

4. Model creation using appropriate software tool

The differences between two popular simulation packages (Simfactory and Witness) were briefly described above. All packages have strengths and weaknesses, so package selection should be based on the requirements of the model(s) to be created, together with due consideration of any in-house expertise.

Model building

Like any project using computer software, the majority of the time should be spent on preparation and design, not on actually coding or creating a model. Poor simulation models are often those where inadequate time and effort have been spent on the prerequisites, and model development has started too early. Mistakes and inappropriate model features will result, leading to a poor base model.

Model building should begin only after appropriate preparation and validation. Choice of package alone will rarely influence the success or failure of a simulation project. Of much greater significance are goal definition, model validation, verification, data accuracy, appropriate simulation runs and results analysis. A formal project management approach throughout is required.

Model testing

Testing on the model should be carried out at each stage of development, not left until all elements and features have been created. Incremental testing will cause bugs and misoperations to be discovered and cured early, because model complexity will be relatively low. Testing must include full documentation, noting

- date and tester

- stage of development, version number
- test(s) being carried out
- results and problems
- diagnosis of problems and solutions
- retest information

5. Verification of the simulation model against the real system

Once the computer model is developed from the conceptual model framework, we require to test or verify this against the simulated system, if it exists. Verification is essentially a confidence check of the model, to ensure it produces expected (realistic) results that compare closely with the target system running under similar conditions. Several different verification tests may be required for complex models.

'Back to back testing'

A manufacturing simulation model may be tested by running an actual production schedule (or schedules) through several iterations. This is a known product mix that will have established operating and throughput times, which can each be checked against the simulation results Over several runs, one would expect to see comparative results. A typical schedule might be

- 100 units loaded, processed in 2.4 hours
- Resource A1: 70% utilised
- Resource A2: 85% utilised
- WIP queue 1: maximum 33 units, minimum 2 units

Resource and labour parameters (efficiency, uptime, cycle times) will be set to mean levels initially. On satisfactory testing of this configuration, the appropriate statistical distribution parameters can be enabled for all elements, and further testing and adjustment carried out. Where testing yields poor or inconclusive results, modification to the model may be necessary, followed by further testing. This should continue until verification test results are satisfactory.

Greenfield applications

Where the simulation is for a proposed facility, or for substantial modifications to an existing system, relevant local control comparisons may be impossible. Partial testing of sections of the model may be done where there are similar existing facilities, using the output results to feed the proposed sections.

Another option may be to seek out similar equipment and/or layout configurations in other companies. In many cases the suppliers of plant and equipment can assist with the location of suitable sites and provision of access.

6. Running of 'what if' scenarios

Once we achieve a verified model that has met any agreed test criteria, the study moves onto the next stage, loading and executing different simulation runs.

Exactly which variables will be altered and investigated depends on the aims and objectives of the study. For example, investigating the staffing levels on a cell or line of machines:

1. Create a set of representative schedules, SCHED.
2. Prepare a set of manning options, MAN.
3. Generate a set of random number streams.
4. For each (SCHED, MAN) pair carry out the simulation using the first random number stream, then using the second random number stream, and so on until you have performed the simulation using each random number stream.

Four SCHEDuling options A–D
Three MANNing options X–Z
Decide to do four simulations for each schedule, using different random streams S1–S4

Pattern of runs

```
SCHED A      B      C      D
S1
S2                                    MANNing option X
S3                                    16 RUNS
S4


SCHED A      B      C      D
S1
S2                                    MANNing option Y
S3                                    16 RUNS
S4


SCHED A      B      C      D
S1
S2                                    MANNing option Z
S3                                    16 RUNS
S4
```

In this manner, all combinations of interaction are covered, and the same randomness is used in each MANNing comparison. Schedule A, stream S1 is used for MANNing options X to Z, then S2, S3, etc.

In this case, 48 different runs had to be made. Depending on the application, it may be required to run more (or less) simulations. If we had used different random streams at each test, items affected by the random numbers would operate in slightly different ways, possibly obscuring the effects of the varied manning patterns.

The set of stream results for each (SCHED, MANN) pair can be averaged to obtain an indication of performance. This can be compared against the averaged results of the other patterns. Model entities or operating variables often need to be revised to fine-tune the simulation. This tends to be a gradual, iterative process. In summary:

- Run the simulation for a long time to improve the accuracy of the model.
- Use random numbers to represent probabilistic events.
- Change the random number stream before each repeat run; average the results of all the runs.

7. Analysis of results and presentation of findings

Remember that the senior managers, who will take decisions based on the simulation study findings, will require a formal report with a summary of the findings and recommendations. Simulation runs will have produced statistical information that will have to be

- compared with other simulation run results and against practical expectations
- presented in a clear, straightforward and unbiased manner

During the project, it is prudent to hold regular progress meetings with the managers responsible, keeping them up to date with progress and any significant developments. In doing this, care should be taken to avoid misleading staff, possibly by drawing attention to simulation results that are not complete. These intermediate reviews should focus mainly on model validity, number and type of simulation runs carried out, and comparative results. Where necessary, the reviews should authorise (or direct) the team to pursue more promising directions of study.

In the final presentation, it is useful to show the layout and operation of the computer model(s), allowing the audience to see first-hand operation of resources, queues and throughput of product. Graphical representation of the simulation data is normally the most appropriate medium for presentations, with summary statistics and explanations of the findings and recommendations.

■ Manual simulation in scheduling

Before computers, simulation was developed in a readily available and easily used form, simulation exercises were sometimes carried out manually (Figure 10.7). This involved a team of people around a table, each controlling one or more resource stations. Each person would manually simulate the processing of parts through the station. Part arrival and movement was normally governed by a random means, such as dice throws or random number tables; timing was by stop-watch, to record the passage of processing time at each stage.

Manual simulation carries out similar functions to computer simulation, but requires several people to enact and log each discrete event, which can often take as long as the process being studied. Nevertheless, it remains a useful technique. Study teams find manual simulation worthwhile because it provides them with a close-up of the subject and increases their involvement. Sometimes it clarifies complex or rapid event sequences that computer simulation has identified as unusual or critical. Many MSE teams therefore use the technique at least once during a project. Manual simulation is

- simple to understand and operate
- good for demonstrating/practising system operation
- event based
- slow to operate, e.g. 10 × real time
- rarely effective at incorporating random variables
- time-consuming for 'what if' questions
- unable to collate and analyse data automatically

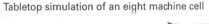

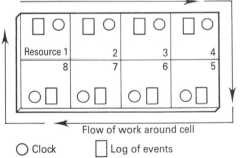

Figure 10.7 Manual simulation using clocks and dice.

Scheduling and simulation

Many packaged scheduling systems for manufacturing and process operations now use simulation as their 'what if' engine, running the range of scheduling options through the simulation system then comparing the results. Although simulations are different from expert systems, the two approaches are related.

Simulation-based scheduling uses simulated runs (iterations) of one or more plans, logging the effects and (simulated) performance of each run for later comparison and selection. **Expert (KBS) scheduling analyses** the production status and rule base in a more logical fashion, calculating a result purely from rules and data entered at the start.

■ Simulation as a scheduling decision support tool

- ■ simulation can be used as an on-line support tool for shop-floor scheduling; different attributes are required.
- ■ up-to-date shop-floor data from SFDC systems
- ■ tight distributions for probabilistic input
- ■ accurate output for short term
- ■ simple entry and evaluation of 'what if' questions

Simulation systems have comprehensive user interfaces for modelling manufacturing processes, and in their own right are now popular tools for both analysing and scheduling manufacturing plants. Istel's Witness simulation package has an integrated scheduling co-product, PROVISA, allowing the suite to be used as a simulation-based scheduler. Interfaces are provided to read and write data to and from MRP2 systems, allowing integration of the scheduling system. A case study of a PROVISA application is included in Chapter 12.

CASE STUDY 10.1

*The Witness simulation in aerospace manufacture**

Expensive, high performance materials, and sophisticated manufacturing processes meant that inventory was one of the company's largest single costs. So it was an obvious target for attention.

Initially, task forces were assigned to set up cost-saving initiatives of which inventory control was just one. But due to the complexity of the manufacturing processes, their analyses were often long and tedious. For instance, production rates, methods, tooling requirements and labour availability all varied depending upon individual projects. It was not ideal and the task force teams themselves were not cheap. Clearly, an alternative method for evaluating any changes that have been designed to improve efficiency, minimise work in process (WIP) and reduce inventory was needed.

Enhancing the existing computerised control system was considered but soon dismissed. Costs and timescales were both prohibitive. The result was that we found ourselves at the centre of a plan to use simulation techniques to improve inventory and logistics.

---

* Courtesy of AT&T Istel, from an article by J. Cruise of British Aerospace.

By modelling processes within the factory, simulation would allow the testing of alternative manufacturing methods before committing significant time, labour and capital to trial.

The overriding criteria for the selection of a simulation package was that we did not need a simulation specialist to use it. That would only reduce any potential savings. After all, it was British Aerospace (BAe) engineers who understood the problems and who would be committed to making any recommended changes work. Explaining manufacturing problems and options to a third party would not only be time-consuming but it would raise the risk of misunderstanding. Speed of model generation and the ability to make real-time changes to model parameters were also essential functions.

After reviewing three simulation package options, we chose Witness, developed by AT&T Istel Visual Interactive Systems, because it met the majority of our selection criteria in terms of functionality and ease of use.

The interactive nature of Witness allows simulation non-specialists to build a computer-based model of the factory operations using animated colour graphics. Large and complex models can be built and tested in small steps, and parameters can be changed and rerun at any time. We made use of our existing IBM PS2 personal computers to run Witness. Although it was important for us that we could do this, it did not inhibit our choice of package. Most offered a choice of running media. What was important was that our engineers would have access to the system as part of their normal work tools. Despite being PC-based, Witness can model large production facilities, a zoom facility allows you to look at local areas identified as bottlenecks.

Manufacturing wings for the European Airbus is one of BAe's major contracts. Demand for the Airbus is putting increasing pressure on the company to deliver more from its manufacturing facilities. By the end of 1990 BAe had delivered over 746 sets of wings, and had firm orders for several hundred more aircraft. So any improvements, even small ones, could make significant contributions to the company's profitability.

One of the first applications of Witness was at the Chadderton factory near Manchester, where a Cincinnati long-bed CNC mill was installed to machine wing spars for the new A330 and A340 variants of the Airbus. The 70 foot bed (for machine logic known as two 35 foot tables) is traversed by a single three-spindle gantry. Three spars can be machined simultaneously on either table.

The simulation package was used to examine potential efficiency improvements on the machine through testing a variety of billet loading procedures. The objectives of the project were twofold: to maximise machine utilisation and minimise WIP. Each billet of raw material cost £35,000.

Various machine loading options were explored (see Figure 10.8). For instance, at the expense of machine utilisation, WIP could be reduced by machining less than the maximum three spars; or by various combinations of port and starboard spars – three port and three starboard (a three-set) or six of one type followed by six of the other (a six-set). Pendulum loading is the unloading of completed spars from one end of the table while machining takes place on the other. It, too, reduces WIP.

Whichever option was selected, however, it could not be viewed in isolation. The implication downstream in the manufacturing process had to be taken into account. After rough machining, the spars are heat treated as matched pairs, port and starboard bolted together to balance distortion forces. The oven can only take one pair at a time. After heat treatment the spars return to the mill for further machining to their base. Spars are then turned over onto the other side of the bed for finish machining.

After we had taken our engineers through an awareness and basic model-building course, development engineers at Chadderton spent some time experimenting with simulation.

For the spar milling project it took less than a week to create the model. It was built by taking each side of the table as a separate machine and using standard Witness commands to ensure spars were marshalled into matched pairs. In addition to the animated graphics, which showed the status of spars and machine tables, a running commentary was added alongside each machine. This explained the current status of parts during their complex movements between locations. With the commentary added, we found the model was clear enough, not only to derive useful results but also for use in familiarising personnel on plant operation.

The model produced a realistic projection of the outcome of various production strategies. For example, net spar output using a three-set cycle turnaround could not match the requirements of a rising programme, even though the average time to make a spar was lower. While machine utilisation was significantly better with the six-set cycle, work in progress levels were higher. However, machine utilisation rose from 77.5% for the three-set cycle to 98.9% for the six-set cycle with pendulum loading. Without simulation it may not have been possible to meet the forward plan for wings from the existing machine.

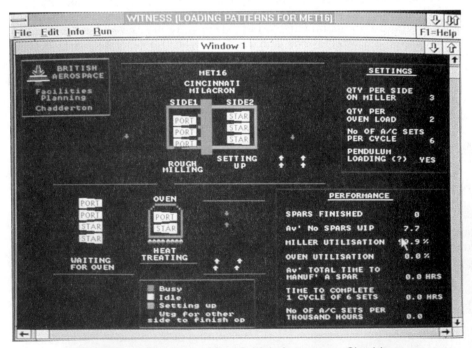

Figure 10.8 Witness model of British Aerospace, Chadderton.

As BAe has found, models can be created quickly and refined easily. New machines can be added, old ones deleted and loading and movement logic updated. It is just as easy to start from scratch when creating a model for a new location.

Witness is now used at six BAe sites for capital equipment justifications, streamlining the paper-intensive operations of the design print room at Hatfield, and the materials handling logic for Airbus subassemblies at Chester.

At Chester, the model actually proved what the shop supervision had known but was unable to quantify: the shop efficiency was being impaired by bottlenecks created by lack of crane capacity needed to move skin panels. After simulation analysis it was discovered that crane utilisation was on average 30% but it peaked at well over 100%. A second crane has now been installed.

CASE STUDY 10.2

Simulation application in the manufacture of chemicals and dyes*

The problems of success obliged Huddersfield based James Robinson Ltd., manufacturers of dyes and speciality chemicals for the hair colour and photographic industries, to rethink the issues of capacity management. Interactive visual simulation was used to identify plant bottlenecks, establish possible courses of action, and evaluate and quantify the results of each option.

The company's customers were forecasting market growth for a particular hair dye of 15% in 1991 and a higher rate in 1992, driven partly by energetic marketing and partly through the opening up of the Central European Market. Concern was expressed at the ability of Robinson to meet demand, prompting the company to make a detailed examination of its production processes and capacity management.

At this point James Robinson looked at simulation as a means of evaluating the various production options, having decided that the interactions of its plant were too complex to model via spreadsheet methods. Neil Grazier, a graduate chemist with James Robinson, was subsequently given the task to implement a simulation using AT&T Istel's 'Witness' package.

Hair dye production is a batch process carried out in a series of reactor vessels. A batch involving several stages can spend more than one week in the process cycle, placing a premium on accurate timing for each stage. Some reactions are exothermic so there is a limit on the amount that throughput can be increased by simply raising the pumping rate.

The plant has seven reactors of types A, B, C and D (Figure 10.9). Feedstock is introduced into either of the A reactors and is passed to reactor B. Following reaction in B, the fluid is filtered and pumped to one of the C vessels to yield ultimately either higher grade or standard grade material. The product is finally precipitated out of solution in either of the D vessels and centrifuged to separate solid product from liquid waste.

Process simulation

A series of simulations was carried out using Witness on an IBM PS2 machine.

----

* Courtesy of AT&T Istel.

The objective was first to identify production bottlenecks and then to experiment onscreen with potential solutions. Finally, the effect on throughput of each solution, or combination of solutions, was quantified.

Initial studies showed that throughput was ultimately controlled by the availability of labour on the centrifuge and total capacity on the first two stages, which were close to 100% utilisation. Times for the various operations on different reaction vessels varied widely. The number of factors involved in the process meant that a number of separate and combined options needed to be examined. Initially these were

■　Increasing the feed rate to both type A reactors with increased cooling to control the exothermic reaction.

■　Investment in a parking vessel for the filtration stage following type B reactor, thus allowing the other stages to run at full rate without having to take into account holdings in type C vessels 4 and 5.

■　Investment in a second type B reactor prior to the filtration stage.

■　Adoption of an evening shift to overcome the labour bottleneck at the centrifuge.

Various combinations of these options were then modelled and the improvement in throughput quantified.

The results indicated that the most effective increase in production would be obtained by increasing the feed rate to the type A reactors, investing in a new type B reactor and adopting an evening shift. Knowledge of the capital and operating costs involved in each element allowed Grazier to prepare a hierarchical plan of action by which the company was able to meet different levels of demand cost effectively.

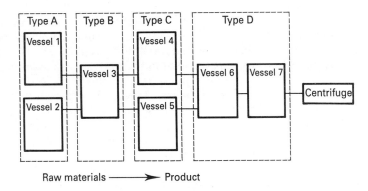

Figure 10.9　A schematic of the reactor layout at James Robinson Ltd.

■ Summary

A successful simulation project requires:

■　definition of the achievable goal of the project

- the correct mix of essential skills (a project team)
- user participation
- correct choice of simulation tools
- model verification and validation
- full analysis of the results
- formal presentation of results and conclusions

Correctly used and interpreted, computer simulation offers powerful and valuable facilities for selection, configuration and operation of many forms of production and discrete item processing. Before implementing changes, test them using a simulation, to prove or disprove them, and to fine-tune them until the desired benefits can be achieved.

■ Further reading

Carrie, A. (1988) *Simulation of Manufacturing Systems*, Wiley.
Shore, B. (1978) *Quantitative Methods for Business Decisions*, McGraw-Hill.

Computers and Communication Systems: The Platform for Manufacturing Systems

We've got a computer room the size of their whole factory and can't get the systems or performance they've got! What are we doing wrong?

Production vice-president of a Fortune 500
US manufacturing company, 1992

■ Introduction

In modern business, IT and computer support systems are vital to effective operation and continue to grow in importance. Nowadays most businesses rely on computers to keep their labour costs down and their profits up. Managers need to maintain a firm grasp of basic computer principles. In industry, the sheer demands for speed and volume of data processing, retrieval and reporting demand computer support. But the fields of hardware and software are both wide and deep, criss-crossed by hundreds of paths and strewn with many choices. There is therefore a heavy responsibility for specifying, selecting and implementing systems solutions that reflect

- The business needs, both now and in the future.
- The computer systems available, their strengths and weaknesses.
- Trends in the IT and computer industry, and their likely impact.

It is necessary to ensure the suitability and cost-effectiveness of IT systems and solutions in the company, with awareness of the strengths and limitations of the various hardware and software appropriate to the business. This knowledge must also be kept up to date.

■ Movement in the IT industry and its effect on MIS support – 'downsizing'

IT progression is particularly necessary in today's rapidly changing environment, where hardware platforms and software solutions are largely turning away from traditional mini and mainframe systems for corporate data processing. There is a widespread move to downsized and distributed systems with better performance and cost, with subsequent changes in data processing (DP) and management information systems (MIS) to service them.

This chapter does not attempt to describe the nature or operation of computers; many other texts exist that cover these topics exhaustively (see further reading list). Instead, it outlines some trends in hardware and software that are particularly relevant to business and manufacturing applications. It also covers the allied communications systems for company-wide data integration.

Planning and realisation of computer and communications systems should be provided from the company IT or MIS strategy, as described in Chapter 3. Failure to work in this structured manner invariably leads to MIS/DP facilities that suit the computer people but not the company, a very expensive and time-consuming mistake.

■ Integrated data across the company

The importance of intermachine communications has greatly increased over the last few years. This trend will continue as more industrial companies strive for improved business control and efficiency through linked information and automation systems.

This has far-reaching implications for plant design and operation, where discrete elements, such as CNC machines and component transfer systems, are being integrated into complete manufacturing systems. Existing flexible and dedicated manufacturing systems are also being linked to planning and control systems, as part of factory-wide information systems and integrated business and manufacturing control systems.

This objective can only be met through plant-wide communications systems that link all constituent areas, allowing information to be passed to and from relevant subunits.

Integrated systems will allow supervisory and management levels to examine shop-floor production data, status, etc., purely by accessing the appropriate database. Management information is presented using various handling packages, including spreadsheets, graphical analysis and database management systems (DBMS), as illustrated in Figure 11.1

The use of these decision support aids allows data to be filtered, summarised and presented in a form that promotes the recognition of production trends and areas for concern/action as well as simple report generation. The integration of data into a factory database from all areas of a manufacturing plant allows production and resource planning based on current levels of demand, sales, stock and work in progress.

Individual information systems may sit on their own computer platform,

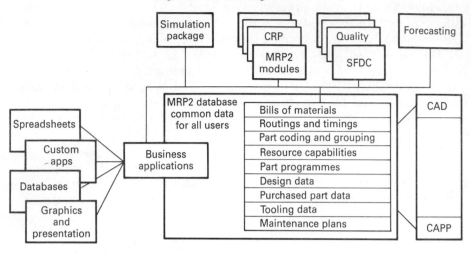

Figure 11.1 A company-wide information system.

be it micro or mini. The MRP and financial systems may run on one or more systems, with a small or large number of terminals connected to the computer via separate serial ports. However, when it becomes desirable to expand the system, or to connect computers and databases, this dedicated interconnection can prove inadequate and restrictive.

Communications are an integral part of any company information system, and are considered in detail later in this chapter. The current revolution in computers continues to affect the supporting communications systems. Computer downsizing cannot be ignored.

■ Mainframes, minis and micros

Rethinking the computer: downsizing

The power and flexibility of personal computers is threatening and now displacing traditional mini and mainframe systems. Poor performance, high maintenance and upgrade costs, vendor lock-in and limitations of packaged software have led many companies to downsize their computers. Moving from mainframe to mini, or from mini to micro offers several benefits:

- ■ lower initial and future capital outlay
- ■ low maintenance costs
- ■ fewer support staff
- ■ similar or better performance

Suppliers are also rethinking their approach and product range. NCR are moving away from mainframes and developing a range of systems using Intel microprocessors. IBM are now focusing strongly on AS400 minicomputers and the RS6000 UNIX machine, both capable of performing as network manager/ data storage centre in a local area network (LAN).

Information systems requirements are commonly organised in a multi-level hierarchy, which indicates the approximate type and size of computer hardware for each level. As a consequence of downsizing, networks and micros have reached the enterprise level (see Figure 11.14). The divisions are now disappearing but the hierarchy is still applicable. The majority of functions are now provided by a common distributed system.

Relative performance

Among the performance measures of a computer system are processing speed, disk access and terminal response time. As an example of processing speed, the NCR 486-based low-end system outperforms an IBM System 390 several times over (source: *Management Computing*, December 1990) and a Pentium 90 PC is more powerful than all but the top-end minicomputers and workstations. For further comparisons of micros and minis, see the time data for MRP explosions later in this chapter.

Microcomputers: IBM compatibility

Until recently, personal computers have been regarded by many as just that, suitable only for small, individual applications, with no comparable performance to minicomputers. This was true until the mid-1980s. Then the IBM PC and its DOS operating system emerged as *de facto* standards in personal and desktop computing. Rapid development brought their performance up to that of many minicomputers, and mounted on a network they could service the same number of users.

The new fileserver and workstation:
a 486 PC at over 100 MHz

The initial IBM PC used an Intel 8088 microprocessor with an 8 MHz (eight million cycles/second) clock and an 8-bit databus. Since then, microprocessors have advanced considerably: from 80286 to 386, 486 and Pentium. The 486 runs at over 100 MHz and uses a full 32-bit databus, a leap as great as from push-bike to racing motorcycle. If the original PC has a performance index of 1, the 486 has an index of 150+ (Figure 11.2) and a Pentium 90 of 400. Further comparisons are given in Figure 11.3.

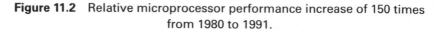

Figure 11.2 Relative microprocessor performance increase of 150 times from 1980 to 1991.

Another trend is the continued reduction in price, largely due to increased imports of Pacific Rim clones. IBM retaliated by moving away from the now established PC architecture onto Microchannel PS2 machines, designed to make cloning difficult, if not impossible. Hard disks have also become much cheaper; 10 Mbytes a decade ago cost as much as 500 Mbytes today.

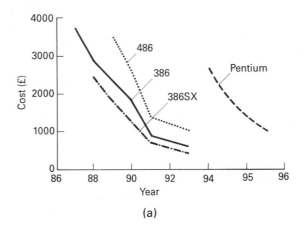

(a)

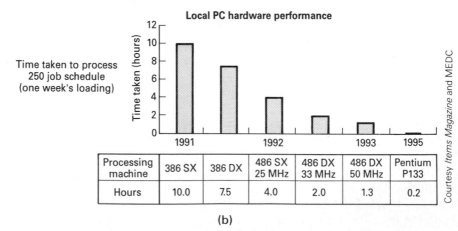

Local PC hardware performance

Time taken to process 250 job schedule (one week's loading)

Processing machine	386 SX	386 DX	486 SX 25 MHz	486 DX 33 MHz	486 DX 50 MHz	Pentium P133
Hours	10.0	7.5	4.0	2.0	1.3	0.2

(b)

Courtesy *Items Magazine* and MEDC

Figure 11.3 (a) Cost/performance of PC and disk systems; (b) execution times for various PC systems executing a large MRP run.

RISC systems

A non-PC development in computer hardware is the reduced instruction set computer (RISC). These are ultrafast microcomputer systems, forming another generation of power computing at relatively low cost. IBM launched the RS6000 UNIX range in 1990, believing it would serve much of the future mid-range market. Using the UNIX operating system, these RISC-based computers can run several times faster than conventional computers, with obvious performance benefits. Figure 12.9 shows relative performance figures for RISC UNIX systems against other PCs and minicomputers.

■ Operating systems and applications software

New computers come with an operating system to run the internal functions and any applications software installed by users. Operating systems vary and can often be specific to a single type of computer. For example, the operating system of the IBM AS400 operates on no other systems, not even the IBM System 36 or the IBM RS6000. Nor will it operate on any other make of computer. Applications software also tends to be specific. A single application may be written in several versions, one for each operating system. But some popular applications will run on more than one operating system; this increases its potential market. Until the late 1980s, most operating systems remained proprietary, so it was difficult and expensive to integrate hardware and software from different suppliers (Figure 11.4).

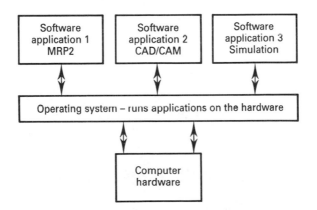

Figure 11.4 Operating systems, software applications and computer hardware.

Personal computer operating systems

Since the early 1980s PCs have standardised on a single disk operating system (DOS), allowing almost all the massive development of applications that have worldwide markets. Whilst DOS gave a level of compatability and use not previously seen in the computer world, it was a fairly restricted operating system. A major limiting factor to using PCs in larger applications has been the limitations of the DOS operating system, possessing no multi-user capabilities and several other essential facilities. Now, several other viable and attractive options for microcomputer operating systems are available:

■ replacement network operating systems such as Novell Netware
■ the UNIX operating system on 386/486 and Pentium machines

The combination of current microcomputer hardware, multi-user operating systems and communications networks allow the creation of systems with well over one hundred attached user workstations (nodes) for use as company-wide computing solutions, at low cost and with easy upgrade paths. A large business system using a PC LAN is described in the case study at the end of this chapter.

UNIX, the emerging standard for operating systems?

For many years the UNIX operating system has been put forward as the future standard operating system for multi-user computing (PC DOS is single-user). Emerging from a turbulent history of several non-standard versions and mainly academic use, this is now becoming a reality, with major OEMs such as IBM placing UNIX up alongside their traditional proprietary operating systems on minicomputers.

Most major original equipment manufacturers (OEMs) market comparable UNIX-based computer systems, offering the user a wide choice of hardware. Software applications under UNIX are now plentiful, including MRP2 and accounting systems.

Paths to UNIX

From certain PC systems, there can be fairly straightforward routes available for transferring to UNIX, should this be required. This is discussed further on.

■ Modern software development: 4GLs

The tremendous advances made in hardware speed and power have been complemented by the development of much more powerful software development tools and languages. Fourth generation languages (4GLs) provide significant advantages in development time and functionality, thus allowing the design and development of large and complex systems in very short timescales.

Many of the recently developed MRP2 systems have been created using 4GLs, and support new and useful features not found in traditional systems, such as pop-up windows, report generators, user-defined calculations and screens. The 4GL development facilities can also make the creation of custom software much more cost-effective, reopening the way to company-specific software without the long lead times and prohibitive costs of traditional bespoke development.

Software engineering

Having managed several major industrial software projects, the author has found the 4GL route to be much more reliable at delivering the specified system on time. Best practice requires regular reviews of progress and problems. Software development should be carried out using software engineering principles:

- Formal specification
- Full and regular user involvement
- Review of specifications, prototypes and test versions of the software
- Formal documentation and testing of the system
- User training and acceptance

Resistance to change

The rise of micros and small minis has been seen as a major threat by many traditional DP departments. Many IT departments resisted downsizing because they believed their power was threatened. They frequently opposed or ignored the potential savings and advantages offered, to the detriment of the business. Mini/micro solutions tend to require much smaller budgets and fewer support staff. Large departments may be cut; workers may fear enforced unemployment. During the restructuring, managers need to retrain their staff to run the new systems.

Control over PC-based systems

Concerns over security and misuse are valid; proliferation of PCs can lead to lack of control. Problems can be averted by firm IT management from the outset, with appropriate system/node configuration and user education before and after the hardware change.

■ Communication systems

Of equal or even greater importance is the interconnection of computer-based equipment across the range of activities performed by a modern business. A large proportion of the advantages of advanced manufacturing and control are dependent on communications systems, from simple machine-to-machine links through to local and wide area networking where tens or hundreds of intelligent machines communicate over a common data highway. The rest of this chapter looks at business and industrial communications, considering their outline operation, application and potential for the future.

Serial communications

Computers and plant equipment are normally provided with serial communications facilities for transmission of information, for example, sending logged data to a printer or display screen (VDU) over a **standard serial link**, where the data bytes are transmitted point-to-point, one after the other. There are standards for data transmission, but differences exist in the way manufacturers operate them. The most common standards are RS232 and RS422/423, a later derivative. These are serial communication standards issued by the US Electronics Industry Association. When several computers exchange data with a single computer, signals can be merged using a concentrator unit (Figure 11.5).

However, when it becomes desirable or necessary to interconnect a large number of terminals or intelligent machines, a **communications network** is

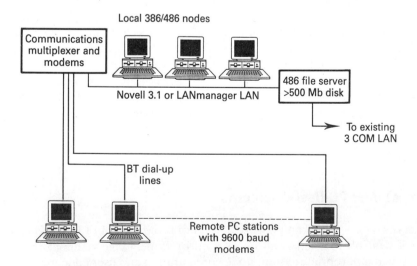

Figure 11.5 Linking several PCs for data logging into a single supervisor.

often used. A **local area network (LAN)** provides a physical link between all the computers as well as an overall data exchange management (protocol), ensuring that each device can talk to the others (Figure 11.6).

Local area networks

Local area networks are a relatively recent innovation in microcomputer development. In the 1970s the use of versatile and inexpensive microcomputers became firmly established as people moved away from the constraints imposed by large mainframe installations. LANs offered a solution to this problem, providing a data transmission system linking computers and associated devices within a radius of 10 km, although 1 km is more typical. Most networks use a file server to store shared files or software and may also provide access to shared peripherals such as printers and hard disks. Among the reasons for installing a network instead of point-to-point links are

- All devices can access and share data and programs.
- Dispersion of equipment makes cabling for point-to-point impractical and prohibitively expensive.
- A network provides a flexible base for communications architectures.
- There is an easy and low cost upgrade by adding or replacing nodes with more powerful nodes and/or file servers.

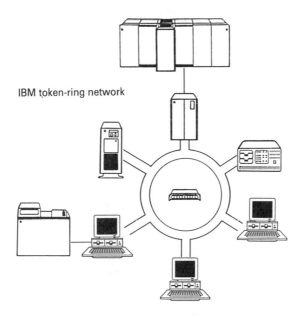

IBM token-ring network

Figure 11.6 Local area networks.

■ Compared with minicomputers, the node cost is much lower for the same computing power

In many companies, LANs have become the information backbone, crucial to accommodating a variety of hardware and operating systems. A network is a compromise between long distance and high-speed data flow. LANs provide the common, high speed data bus within the local area, often running at up to 100 Mbps (megabits per second).

Business applications

LANs have commonly been used in business applications to allow several users to share costly software packages, peripherals such as printers and hard disks, and company-wide information systems, including electronic mail and messaging. A wide range of options has developed. Many different networks are available: DECnet, Ethernet, Arcnet, IBM token ring, Ten Base +, etc. Each has its own physical, electrical and data protocol specifications. One family of micros may be compatible with just one network, preventing future expansion to include different computers or intelligent devices. For this reason, international standards were drawn up and published for all to use.

Network standards: ISO and IEEE

Several manufacturers have followed the International Standards Organisation (ISO) model for open systems interconnection (OSI). This model was defined in 1979 to assist with communications between dissimilar computer systems, and is now the standard that computer and equipment manufacturers have adopted or will provide as an option to their in-house specification.

A communications link between items of digital equipment is defined in terms of physical, electrical, protocol and user standards. The OSI model attempts to fully define all aspects of communications from user to user with a seven-layer breakdown. It provides a framework for network communications standards, but not an actual specification for communications protocols. However, specifications are available from both the ISO and the Institute for Electrical and Electronic Engineers (IEEE) for users to incorporate into their own OSI system in whatever combination they require. There are at present three international standards that define layer 1, which essentially covers the physical cable, and how data is transmitted and the access method.

IS 8802.3 CSMA/CD (used in Ethernet)
IS 8802.4 Token bus (used in MAP)
IS 8802.5 Token ring (used by IBM token ring)

Transmission media

Wires or cables, the physical media that carry the business data, are an important part of the overall system. The operating environment of the network also affects the choice of medium. Electrical noise has always been a problem in manufacturing industry, where electrical plant, welding and cutting machines produce electromagnetic radiation. It may be difficult to obtain reliable, high speed data transmission when communications cables pass close to noise sources.

Twisted pair cabling is commonly used on the factory floor, but may have to be routed through grounded steel conduit to obtain satisfactory communications. (Recent twisted pair systems have overcome this, offering high speed and noise immunity.) **Coaxial cable** can often operate at higher data rates than twisted pair and does not require additional shielding. Transmissions may be simple **baseband**, without the use of a carrier, and with only one channel defined in the system. Alternatively a **broadband** system may be used, having several channels multiplexed (separated) in frequency across the wide bandwidth of the coaxial cable (Figure 11.7). Broadband systems are relatively unaffected by noise, ideal for the factory environment, but they are many times more expensive than baseband systems mainly due to their frequency modulation/demodulation at each node.

Twisted pair and coaxial cabling is being superseded by fibre optic lines, which have much greater bandwidth and noise immunity; they are also smaller and more flexible. Fibre optics are particularly useful for electrically noisy shop-floors that may have heavy electrical machinery, arc welders, etc. Most new cabling installed is now fibre optic or twisted pair.

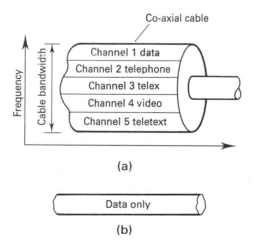

(a)

(b)

Figure 11.7 Schematics of (a) broadband and (b) baseband communications.

Network configurations

Although there are several hundred different network systems on the business and industrial markets, they all possess certain common features. Each device on a network, referred to as a station or node, has to have a suitable interface. All stations are linked into the system by lengths of cable to carry data from one network station to another. The network requires controlling software to correctly handle all file transfers within the system, dealing with station access, data validity, etc.

The topology of a network is the physical arrangement of the stations and their interconnections (Figure 11.8). There are three main patterns in use: (1) a bus topology consisting of a central cable with all stations connected to it by spurs, (2) a star configuration with stations clustered around a single, central device that acts as a file server and (3) a ring topology consisting of stations connected together in a complete circle or loop.

Each topology has its own strengths and weaknesses, with the bus and ring layouts vulnerable to breaks in the LAN cable which may disable the whole network. This risk is often reduced by dividing a network into several segments fed by different cables. However, a network is not fully defined by its topology; each configuration can operate in several different ways.

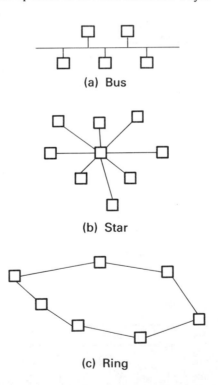

(a) Bus

(b) Star

(c) Ring

Figure 11.8 Bus, star and ring topologies.

Channel access and control

With several stations on a network, there must be a mechanism for deciding which station gains access to the common channel to transmit or receive information. Under heavy traffic, more than one station at a time will be trying to access the network, causing response times to deteriorate (often dramatically). It is therefore important to control the traffic on the network, allowing smooth, efficient operation and reducing the chance of data corruption caused by the collision of two data streams.

Protocols

Congestion can be eased by communications protocols. The two main ones are CSMA/CD, used by Ethernet, and token passing, used by token ring and MAP.

CSMA/CD stands for carrier sense, multiple access/collision detection. Stations gain access to the network on a first come, first served basis. Carrier sensing means that a station listens on the network to check for other traffic. If no carrier signal is present (no traffic) the station will access the network. If a carrier is detected, the station will wait for a certain time before trying again. Looking before crossing the road reduces the risk of data collisions but does not eliminate them. When data collisions are detected, data is retransmitted, which slows down the network. **Multiple access** means that any station may transmit data as soon as it senses the channel is free. CSMA/CD techniques offer fast response at low traffic rates, but as the load increases so does the waiting time. In practice this is unlikely to be detectable or perceived by any users until very high traffic levels are present.

Token passing

Token passing uses a special token or data packet that passes control from one station to another. Any station wishing to transmit information must wait until it has been passed the token. Having completed transmission, it passes the token to the next node. Token passing is used in ring and bus topologies, providing a relatively slow response at low traffic rates compared with CSMA/CD, but with little deterioration of response time as the load increases. In theory, token systems should perform better with more user nodes and/or more traffic on the system.

What is an open system?

Many network manufacturers and vendors purport to have 'open systems'. Perhaps this suggests that networks from different sources may have compatible architectures and protocols, but this is rarely true. Open systems are simply networks with a specification that is open and available to all. Openness allows users to engineer their own interfaces; it does not imply compatibility. It does not mean that other open system products from various vendors will operate correctly when connected to a given OS network.

This is a step in the right direction, however, as the open specification encourages second sourcing for products to connect to a specific open system, with potential customer benefits in cost and elsewhere. This will only happen where there is a large enough demand for that particular open system to make second sourcing worthwhile. This shows the attraction of an open system that is internationally accepted and supported:

1. There is a large market base
2. The open specification is available

Ethernet is a long-standing example of such a system, although it is only partially defined in the specification. Many companies adopted the Ethernet specification but ultimately their products were different, so very few Ethernet systems can be directly connected to exchange intelligible data.

Ethernet: leader on popularity and practicality

Many network systems are designed on a layer principle, including Ethernet, IBM's SNA (Standard Network Architecture), and Digital Equipment Corporation's DECNET (based on Ethernet). Ethernet is mentioned here firstly because it was the first major network product offered using non-proprietary protocols and interfaces, and secondly because it is arguably the most practical and popular LAN for business and industrial use. Ethernet was created jointly by Xerox, Digital and Intel, who defined products based on published communications standards. Ethernet systems are widely used in both industrial and office environments, with compatible products available from a wide range of suppliers. Unfortunately, the original Ethernet specification did not cover all aspects of communications protocol, and equipment from one vendor is unlikely to operate alongside another's. The following are a few of the companies that use Ethernet as the base for their communications:

■ Digital Equipment Corporation
■ Gould Electronics
■ Motorola Information Systems 4000 series LAN
■ Novell
■ ISOLAN/Bic
■ Taurus

In the author's experience, Ethernet has proved to be an acceptable LAN platform for both business and manufacturing use. With the current breadth and depth of Ethernet usage in industry, there is arguably no great need for any other standard LAN to be adopted (such as MAP, covered later in this chapter).

Ethernet: options and links

Links into other Ethernet-based systems are possible through the use of LAN bridges and routers, devices that translate communications protocols between two dissimilar systems. Bridges (Figure 11.9) also allow the interconnection of very different LAN systems, such as Ethernet and MAP. In this way, it is possible to interlink existing and incoming systems to provide integrated data availability. Ethernet also comes in thick or thin coaxial cable, fibre optics and low cost twisted pair cable. This gives the user options on each area of application.

IBM compatible LAN standards

In the mid-1980s, following emergence of the IBM PC and DOS operating system as *de facto* standards, many major LAN vendors redeveloped or redefined their products to be fully PC compatible as part of a world-wide strategic movement. Several software houses adapted their LAN/Ethernet software for PC nodes and workstations, viewing this as the way forward. Novell and Taurus were two of the main players, and for the first time IBM experimented with the token ring configuration.

Ethernet/Novell: the business standard?

Novell built upon its early entry to the PC LAN market by developing Netware, software that offered reliability and performance to a rapidly growing market. Several package software vendors, e.g. Lotus, Pegasus and Ashton Tate, adopted Novell Netware as their LAN software. With the rapidly growing LAN market, several MRP2 products are now available under UNIX on LAN platforms. This is created either by recompilation or a rewrite of the package. A large proportion of software vendors have selected Novell Netware as the LAN operating platform, strengthening the position of Novell as an emerging standard (Figure 11.10).

Data communications through interconnecting bridge

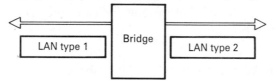

Figure 11.9 Bridge interconnections of different LANs.

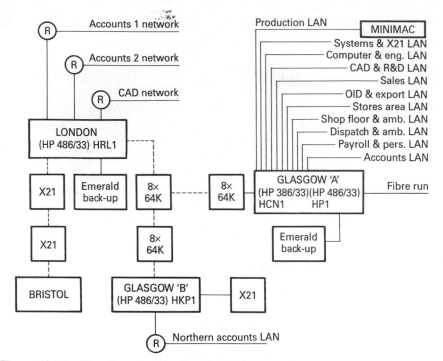

Figure 11.10 Novell networking in a multisite manufacturing services business.

Novell now offers a very wide range of LAN Netware, utility software and hardware products. Replacing part of the DOS operating system, Netware boasts facilities and performance that matches or surpasses conventional minicomputers.

Novell LANs using 486 PC technology

Netware V3.12 and V4.1 and fault tolerant versions are the current top-end products, allowing the full potential of 486 PCs to be exploited. This also supports full disk mirroring for reliability, and allows over 100 Gbytes of hard disk to be supported on the file servers.

The resulting performance of a correctly configured Ethernet/Novell LAN with 386 and 486 nodes is more than adequate to match and indeed outpace many minicomputer platforms. The earlier chart of execution times for a large MRP regeneration on different systems indicates the capability of LAN/486 systems (Figure 11.12).

A PC-based network to support 100+ business users

When you are made aware of a possible alternative solution to computer and communications requirements, thorough checking and investigation must be carried out to discover whether or not the practical reality meets your needs. PC based networking is no exception, and the author was initially doubtful of its capabilities when compared to more established minis and superminis.

However, during several months of investigation in both the UK and the US, it was repeatedly shown that correctly configured PC networks could provide the performance and capacity required by small and medium-sized companies, equalling or bettering minicomputer competition. In the UK there are a growing number of 100+ node business installations running full business and financial applications, including Sky TV, Hussman Manufacturing, and Racal. In the US there are a multitude of large PC LAN sites, including Fox Software (150 node), TRW Technar (120 node), GPS (500 node), Custom MRP (140 node), and so on. An example of a user site is given in Figure 11.12. Users are all more than satisfied with their systems, and are enjoying the advantages of cost and upgrade flexibility.

UNIX and PC networks: upgrade paths

Although a correctly configured PC LAN offers high levels of performance that will meet or exceed the requirements of most medium-sized businesses, there are several paths open for further growth and upgrading.

1. Additional file servers and/or disk storage. Depending on the specific type of upgrading required, this route places additional processing power on the LAN, which can improve response time, particularly if multiple disks are used instead of a single unit. Very high speed disk controllers will allow up to 30 Mbps data transfer rates, many times the rate of pre-1990 devices.
2. Using a Unix operating system, it is possible to connect other (non-PC) devices on a LAN. With Ethernet, Unix-based machines such as the IBM RS6000 can operate side by side with PC machines. This provides a further level of upgraded performance to the LAN route.

Option 2 assumes the applications software can be sourced under Unix. This is now fairly common.

Database configuration architectures

With the full development of LAN-based systems, the way in which they are configured to share and distribute tasks becomes important, offering different strengths for various applications. With the distributed processing provided by LAN systems, it is simple to place computing power where it is required by the user. From a network performance perspective, it may be necessary to consider how best to service the following:

1. database(s)
2. clients (user nodes) interface
3. communications

Client/server and data/server architectures are designed to address these areas,

DATABASE CONFIGURATION OPTIONS

One machine, multiple database
Configured to access two or more databases using the same programs. Useful when running multiple companies that are separate legal entities. A single procedure can access both databases simultaneously for reporting or updates

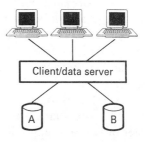

Multiple databases, single machine
Client server architecture can be configured so that one machine on a LAN functions solely as the database server. Multiple machines are then dedicated to client (user interface) functions

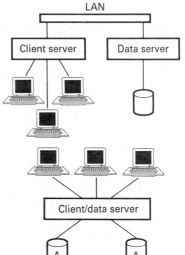

One machine, split databases

A single database can be split into two or more separate files that can then be placed on different physical devices. Useful in increasing system performance

Multiple machines, split databases

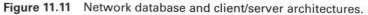
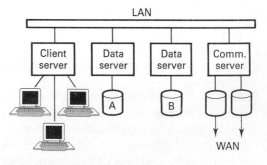

Figure 11.11   Network database and client/server architectures.

describing the provision of dedicated computers (servers) for one or more specific functions that have a high data throughput or require optimum response. These concepts are traditional computer approaches to relieving bottlenecks with centralised systems, but are equally applicable to network systems (Figure 11.11). A database can also be split to provide either mirroring of itself for security purposes, or to give improved access time.

Large corporate networks using PC LANs

GPS network: 12 file servers, 8 Gbyte hard disk, 500 nodes

Great Plains Software (GPS) is a major US software producer, specialising in financial packages. It has a nationwide business, with headquarters in Fargo, North Dakota. To service headquarters and all district branches, they have developed a large PC-based network with 12 servers and 3 Xenix/UNIX 386 host machines. Seven servers run under the Novell operating system, and 5 run under Appleshare. There is a combined total of approximately 8 Gbytes of hard disk capacity on the servers. Five hundred microcomputers connect onto the network of 280 DOS 286 PCs, 150 Macs and 70 remote hosts.

Two mail routers allow email users to connect across the country to the 70 remote hosts located in other cities and towns.

Tape back-up is provided by two Emerald Vast 2.2 Gbyte tape units, using time delay facilities to back up each night from two master workstations. An auxiliary server (AUX) is used as a spare to cover any server breakdown. AUX is also used for testing utilities, new systems work and software load/unload onto the LAN. The network topologies are Ethernet, Arcnet and Localtalk.

Hussmann network: 3 file servers, 10 Gbytes hard disk, 105 PC workstations

Hussmann Manufacturing UK is a good example of correctly configuring and utilising LAN and PC technology to serve the total requirements of a large company (Figures 11.10 and 11.12). Develped over several years, the network has grown from a ten-node system running financial and inventory control applications, into one of the largest PC LAN systems in the UK. Having been restructured and upgraded in the early 1990s, the network was equipped with 486 PC nodes for all major users, such as MRP analysts, BOM and CAD operators. With a file server structure that had disk mirroring and redundancy, and a cabling layout that gave duplicated segment coverage to several functions, the LAN was both rapid and robust. Applications software included full MRP2, office automation, CAD/CAM systems and email. Wide area links were provided to link with other UK sites (Figure 11.10).

Hardware platform

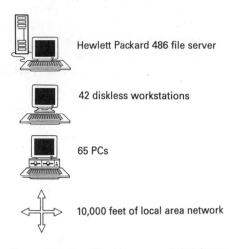

Hewlett Packard 486 file server

42 diskless workstations

65 PCs

10,000 feet of local area network

Figure 11.12   The Hussman LAN (1991).

Shop-floor communications

The need to pass information between computers and other devices within an automated plant has resulted in the provision of a communications facility on all but the most elementary machines and controllers. In the case of small intelligent controllers, the necessary communications hardware and software is incorporated in the computer/controller body, whereas larger controllers and computers have a range of communications modules available to suit different applications. Most manufacturers provide a dedicated network system that may be used for communications between controllers from their own product range:

Manufacturer	Network
Allan-Bradley	Data Highway
Gould	Modbus
General Electric	GE Net Factory LAN
Mitsubishi	Melsec-NET
Texas Instruments	TIWAY

A device may be linked into its manufacturer's communications network by using a proprietary network interface module, if one is available for the particular machine. It will rarely be possible to use equipment from different sources on the same networking system, unless the equipment is listed as compatible with the required network.

Figure 11.13 shows a Texas Instruments TIWAY data network. This is used to transport data from any suitably equipped Texas PLC (programmable

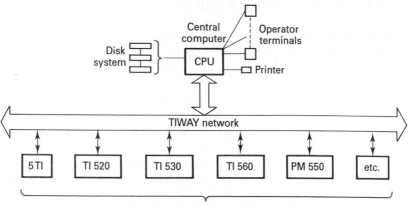

Texas Instruments programmable controllers

Figure 11.13 The Texas Instruments TIWAY network.

logic controller) to any other, or to a central mini or mainframe. The GE NET factory LAN is the General Electric proprietary network, which also conforms to recent world-wide open systems standards under MAP. Before looking at communications networks in more detail, it is worthwhile considering their role (Figure 11.14).

	Business and manufacturing hierarchy	Computer application	Computer system	
1	Enterprise Corporate	Financial, business modelling database application Executive information systems (EIS)	Mainframe Supermini Networked mini/micros	
2	Facility Plant operation	Computer aided design/engineering Computer aided manufacturing Manufacturing resource plan Process planning	Mini or networked micros	
3	Workshop	Production scheduling and control	Mini or powerful micro	Downsizing
4	Work cell	Direct numerical control Cell scheduling/supervision Preventative maintenance Tool management	Mini Micro Custom micros	
5	Workstation	Direct machine control Statistical process control CNC, robotic programming AGV routing	Micros Custom micros	
6	Equipment	Control of sensors and logic devices	Microprocessor PLCs	

Figure 11.14 Communications for distributed control.

Distributed control in manufacturing

Communications facilities also allow the computer or programmable controller to act not just as a dedicated controller on one particular machine, but also as a controller of multiple stations within a large manufacturing area. Thus the PC can become part of a hierarchical control structure, where a co-ordinating computer or programmable controller supervises several dedicated PCs or other intelligent devices, such as robots or CNC machines (Figure 11.15).

This structure is termed distributed control and allows far greater sophistication than a large central computer because the control function is divided between several dedicated processors. A distributed control hierarchy need not consist exclusively of minis and micros; it can include other intelligent devices such as CNC machines, DNC, robots and programmable controllers. Microprocessor-based control panels are small enough to locate at or near the point of final control, simplifying connection requirements.

In large processes it is now common for several microcontrollers to be used instead of a single large mainframe, with resulting benefits in performance, cost and reliability. Each micro can provide optimal local control and can exchange data via other microcontrollers or a host supervisor computer.

Figure 11.14 illustrates the concept of top-down control currently being adopted in the integration of complete companies. It shows the use of mainframes, mini, networks and micros for co-ordination of all levels of plant operation from top management to shop-floor machinery. This may involve communication from one PC to another, or to micros and minis, as well as between different applications software. Acceptable communications standards for hardware connections and data format are therefore highly desirable.

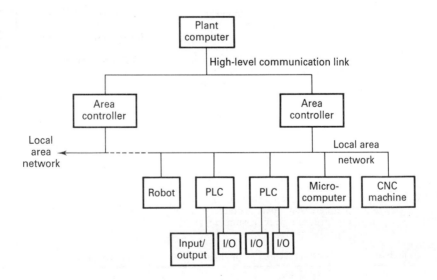

Figure 11.15 Distributed control structure.

Range of requirements

The communicating needs of this control hierarchy differ from level to level. At the bottom of the pyramid, real-time control may be necessary. The LAN carries the control and data signals to and from programmable controllers, robots, CNC machines, etc. and links with an area controller, which may be a larger PC or minicomputer. At this level, communication is mainly small amounts of high speed control data. From the area controller up to the plant computer, most data will be information on plant performance rather than control. The data rate capabilities of the constituent elements may also vary greatly.

This implies a hierarchical network design with stepped levels of capacity and performance, for example, dedicated shop-floor networks to accommodate low and medium data rates and linked to a 'backbone' high speed, high bandwidth communications system that is truly plant-wide.

Variety of LANs

A wide variety of LAN standards exists in both the industrial and business markets, due to the competitive and lock-in nature of the suppliers, resulting in continually changing specifications. Once a firm has purchased a system of this type, it is normally constrained into expanding the system using equipment from the same source for reasons of compatibility.

Unlike business, industry can seldom specify equipment from a single manufacturer. The type and diversity of equipment and plant often precludes single sourcing because the computers, PCs, automatic tools, CNC, etc inevitably come from different manufacturers. Without a standard LAN, each manufacturer would fit its own communications interface, making networks extremely hard to create.

The fourth and fifth levels of the control pyramid shown in Figure 11.14 normally involve real-time control and high functionality. At these levels a company using a single manufacturer of, say, programmable controllers, normally finds it convenient and economic to use a proprietary network for communications.

Ethernet for industry

Ethernet, in its many variations, has become the dominant network standard in both office and manufacturing environments. In the author's experience, Ethernet has proved to be an acceptable LAN platform for both business and manufacture. With the current breadth and depth of Ethernet usage in industry, there is arguably no great need for any other standard LAN. Even when handling several different wideband applications on a single cable, including separate LANs, closed-circuit television and telephones, a fibre optic Ethernet is adequate. Nevertheless, MAP, another LAN architecture, has been heavily supported and adopted by industry.

MAP: manufacturing automation protocol

In 1980, General Motors (GM) in America had a decade to equip its factories with around 20,000 information systems, all of them required to communicate with each other. By 1984 GM had installed tens of thousands of intelligent machines but only 15% could communicate. These quantities were to increase fivefold by 1990. The cost of providing the then standard and bespoke communications systems was estimated at 50% of the total costs (several hundred millions of dollars). GM identified the key element in any factorywide automation, the communications network. It supported a truly open system independent of vendors.

Following this, GM specified a LAN with a manufacturing automation protocol (GM MAP) to integrate all levels of control systems and computer-based products. This was based on the ISO model and the full specification was available to allow vendors to develop compatible interfaces for their equipment. The intention was that GM would, in the future, only buy computer-based equipment that was MAP compatible. All major suppliers of control and information systems committed themselves to MAP compatibility.

MAP used a token bus, broadband coaxial medium at a data rate of 10 Mbps and with its cable bandwidth divided into many separate channels across the frequency spectrum. Two channels are used for transmit and receive paths for MAP network data, with several other channels available for other purposes, such as video and voice communication. Some plants already had broadband cabling suitable for a MAP network.

Plant-wide communications and open systems

The GM MAP was intended to become the standard factory backbone, providing communications between all production units within the factory. Over several years, MAP was developed and revised to version 3. This included subnetworks to provide communications within each production unit; many proprietary LANs already fulfil their requirements. These continue to be used alongside new gateways to the main MAP backbone from the majority of non-standard industrial networks, e.g. Gould Modbus, AB data highway and Digital DNA.

Between different types of network, gateways are available to allow MAP integration of other segmented manufacturing systems without having to replace existing equipment. However, their provision requires network interfaces that are both complex and expensive. For networked communications in office environments, several features or their effects become positive disadvantages. They may be costly or inflexible and sometimes fit awkwardly with the Ethernet infrastructure of many businesses. So, for office-based communications, it can be argued that an ideal LAN would use

- Ethernet as a basis (ISO 802.3) 10 Mbps/100 Mbps baseband.
- thin coaxial cable or twisted pair for ease and cost of installation.
- low cost interfaces.

And, if we assume a MAP-based shop-floor network (or similar), it would need to allow file exchange with MAP.

Wide area communications

Many organisations operate from separate sites up to several hundred miles apart. Similar to single-site companies, multisite organisations require effective national or global communications. This moves beyond the realm of conventional LANs to the domain of British Telecom (BT), Mercury and other suppliers of data lines.

We can think of this situation in terms of wide area networks (WANs) and global area networks (GANs) Wide and global area communication may involve the use of the public switched telephone network (PSTN), using telephone modems to send data down voice channels. The majority of national and international PSTN systems presently offer low quality, low speed data communications performance, inadequate for many applications. Modern PSTN performance can, however, be adequate for infrequent users, or those with requirements for low cost, where slower data rates (up to 28,800 baud) are acceptable.

Where much higher performance is required, for example, to interconnect remote computers or terminals to a central mini or mainframe, the use of high speed data channels is normal. Until recently, users had limited options, mostly leased from BT and based around data circuits between remote premises. Performance varied with cost, and a user could select from kilostream or megastream systems, subject to availability. Figures 11.10 and 11.16(b) show a UK wide area network based on interlinking several Novell/Ethernet sites via Mercury kilostream services.

Packet switching

Packet switching systems (PSS) dedicated to data communications were also introduced by BT in the United Kingdom (similar systems exist throughout the world). Connected to the network of PSS exchanges throughout the UK, users send each message into the network complete with a header containing information on its destination, source and size. The message is broken down into packets by the PSS and each packet is sent over a common network shared by

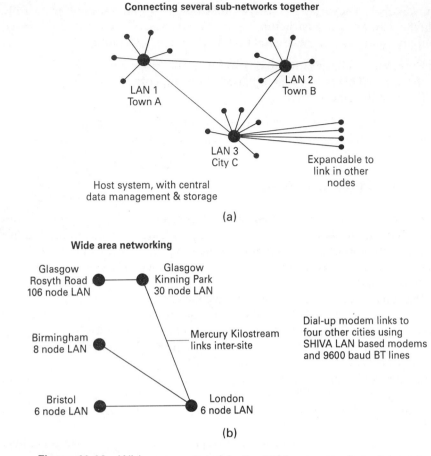

Connecting several sub-networks together

LAN 1
Town A

LAN 2
Town B

LAN 3
City C

Expandable to
link in other
nodes

Host system, with central
data management & storage

(a)

Wide area networking

Glasgow
Rosyth Road
106 node LAN

Glasgow
Kinning Park
30 node LAN

Birmingham
8 node LAN

Mercury Kilostream
links inter-site

Dial-up modem links to
four other cities using
SHIVA LAN based modems
and 9600 baud BT lines

Bristol
6 node LAN

London
6 node LAN

(b)

Figure 11.16   Wide area network in the UK for a manufacturing and
service group of companies.

all other users. Thus each packet is independently routed to its destination by
the header data. If congestion or equipment failure is encountered, the packet
will automatically be re-routed or temporarily stored in one of the PSS
exchanges. Packet switching offers cost and performance advantages over
leasing but only when company premises are close to an exchange. Data
communications facilities from private organisations, such as Mercury, are
now heavily subscribed to, and may offer lower cost services to users.

The integrated services digital network

Much of the national PSTN has been digital for many years. But until recently,
local and rural equipment remained analog, responsible for the slower and

poorer performance of data compared with voice communication. All this is now changing. BT has almost completed the integrated services digital network (ISDN). This offers substantially improved data communications with the rest of Europe and North America, themselves engaged in similar programmes.

ISDN will provide end-to-end digitised speech, data and video transmission at speeds up to 64–128Kbps, compatible with existing digital equipment and within cable limitations. Although video transmissions will be limited to single frame/slow scan pictures, ISDN opens new horizons for wide area data communications at relatively low cost. Once fully established, it will become the default standard for long-haul data communications.

Individual business requirements

When considering the total communications requirements of industry, it is necessary to examine the type of communication at each level of the plant pyramid. Figure 11.2 shows that different levels have different communications needs, impossible to meet using a single medium or network. There is frequently a need for interconnection with other LAN systems at higher and lower levels. At the higher levels, it may be necessary to link with factory minis or mainframes and between computers or networks on different sites. The latter may involve WANs such as the PSTN and other private digital data networks with access to international channels. These levels of communication are far removed from MAP applications but do require interconnection facilities for certain data transfers.

The lower levels of communication are catered for by other systems, including specialist LANs. These form the next layer down in the hierarchy, providing links between computers, cell controllers and the controlling devices in the cells and machines in the plant. Examples are

- Programmable logic controllers (PLCs)
- CNC machines
- Automatic guided vehicle controllers
- Robots
- Transport system controllers
- Bar-code shop-floor data collection

Data communications to and from such devices will often be for control purposes and will be extremely time-critical. A typical factory may require many such interconnections, so the resulting transmission demand may be poorly served by a general purpose LAN, requiring instead a high performance system with emphasis given to very fast response and low mode connection cost.

By selecting a proprietary network that supports a MAP interface or gateway, the user avoids many of the vendor lock-in problems; immediate or future system expansion can be implemented over MAP-based systems, should it be required.

Ⓒ

CASE STUDY 11.1

Strategic options for a business control system, 1991

Hardware background

Previously, all systems at the Hussmann site were serviced by four North 386 Servers, (3 × North Star and 1 × Hewlett Packard Vectra), running under Novell 2.15 operating system. All four servers were interconnected on a LAN providing on-line disk capacity of 2 Gbytes.

Approximately 70 workstations were connected to the LAN, based on Ethernet cable throughout the facility (consisting of several buildings) with several repeaters strategically sited to provide thick-to-thin Ethernet LAN spurs. The LAN topology was improved to increase operational robustness, and allowed simple, low cost extension or rearrangment of network connectors. The workstations were composed of a variety of 286 and 386 PCs and diskless workstations (mainly Compaq or AST). All hardware conformed to standard ISO bus architecture. The options available to the company were as follows:

- Select a suitable MRP package based on a different hardware platform.
- Develop in-house systems to create MRP.
- Select a suitable MRP package based on the existing hardware platform.

After investigation, the project team compiled a comparative list of strengths and weaknesses.

Select a suitable MRP package based on a different hardware platform

During our investigations we considered IBM, Digital and HP minicomputer systems. Each required either a partial or complete re-equipping of our existing network.

Benefits

- Widens scope of range of software to packages such as Bluebird, Impcon, Aquilla, Pansophic and BPCS
- Capitalises on Hussmann corporate experience of minicomputers

Weaknesses

- No existing in-house expertise on other systems
- Recruitment/replacement/training of MIS staff required
- Results in a learning curve delay to implementation
- Benefits will be delayed
- Software and hardware costs are significantly greater
- Consequent dependence on one computer manufacturer

Estimated cost penalty over other options

■ Processor	£80,000
■ Communications	£35,000
■ Maintenance (per annum)	£12,000
■ Delay in achieving MRP and costing systems:	
6 months mininum effective costs	£75,000
Total cost	**£202,000**

Delay to the project is the most significant factor here, since the present lack of these systems was resulting in less than optimal performance of the company. The proposed financial benefits will be delayed. The estimated delay was due to restaffing or retraining MIS, and replacement of existing hardware. Our review found that the cost and inconvenience of changing to another hardware platform makes this an undesirable option, and it should not be considered further.

Develop in-house systems to create MRP

Recent developments of material requirements and related systems by MIS staff had yielded a high quality product and modifications within specified time constraints. This was attributable to recent developments in the use of 4GL tools and formal specification of system requirements with users. However, due to the extent of our business systems needs, timescales and limited MIS resources, it might not be viable to develop these further towards a fully functional and integrated solution wholly in-house.

Select a suitable MRP package based on the existing hardware platform

Hardware

A separate review of the IT systems (carried out by the MRP project team with an external consultant) concluded that the Ethernet/Novell LAN offered a wholly suitable platform on which to build the intended company-wide systems.

One of the main strengths is the ability to upgrade the LAN servers and workstations as required, redeploying lower power units to other parts of the LAN. It was identified that the present server and disk capability was nearing capacity, and would require upgrading either with or without the proposed new systems. During our investigation, we dealt with vendors who offer products on several platforms other than PC/Novell networks. In each case, they confirmed that the basic hardware platform, once upgraded, would offer more than adequate performance against our specification. A number of benchmark tests were conducted on the existing hardware and on sample servers supplied by prospective vendors. These tests clearly ratified the opinions sought from other sources.

Future growth

In judging the suitability of building on the present platform, the group included a perspective on the future potential for growth and development. We concluded that the Novell operating system, being the *de facto* standard in PC LANs (market share approx 70%) offered the greatest stability. IBM's recent decision to adopt Novell as their PC network standard confirmed our judgement.

Alternative operating systems, especially UNIX, were regarded as insufficiently stable at this time (1989/90). We did, however, recognise that UNIX would possibly mature within the next two years to offer a wider growth potential. In selecting hardware and software, therefore, care was taken to ensure that UNIX was not excluded from our growth path options (SCO–UNIX, the most 'standard' version of UNIX available).

MRP2 UNIX-based software

There are over ten high quality MRP systems based on networked PCs at present [1993], and the number is increasing. What was previously the domain of minicomputers has, over the last five years, become both technically and financially attractive using PC networking.

It should be noted that many mini-based systems are now ported onto PC networks, including MTMS, Prodstar and 4thShift. We also found several users who were changing from a minicomputer base to PC networked operation (e.g. J & B Distillers), now an acknowledged route for many corporate and business IT systems. In conclusion, it was decided that this option offers the optimal route forward for the company.

■ Summary

The era of low cost supermicros is here, affecting producers and users alike. Many companies are downsizing from traditional minis and mainframes. Networks are the ideal way to interconnect intelligent machines, with several hundred nodes linking and performing well on standard Ethernet systems. UNIX operating systems also have come of age; in conjunction with supermicros they offer new and better business control at a much lower cost.

■ Further reading

Andersen Consulting (1992) *Trends in Information Technology*, Sunday Times/ McGraw-Hill.

Badgett, T. and Sandler, C. (1993) *Business Software Companion*, Wiley.

Blissmar, R. (1993) *Introducing Computing: Concepts, Systems and Applications*, Wiley.

Jennings, F. (1986) *Practical Data Communications*, Blackwell.

Lecarne, O. and Gart, M. P. (1989) *Software Portability with Microcomputer Issues*, McGraw-Hill.

Shatt, S. (1987) *Understanding Local Area Networks*, H. Sams & Co.

12 | MRP2 Developments and Make or Finish to Order

Things have progressed a long way since MRP1 was introduced – we can now run every function within a business effectively using a packaged Enterprise Resource Planning (ERP) system

Jim Hall, Polaroid UK

This chapter deals with recent developments and trends in MRP2 systems, including hardware independence, and particularly how these benefit the make- and finish-to-order users. Also discussed are bolt-on facilities, such as finite schedulers and forecasting tools. The case study examines MRP2 operation and improvements in a traditionally difficult environment, make- and finish-to-order manufacturing with short lead times.

■ Introduction: business planning and control systems

(Manufacturing Resource Planning – MRP2 plus
Material Requirements Planning – MRP1)

An ever increasing requirement is the effective planning and control of all activities relating to the manufacturing function. Modern manufacturing goals centre around achieving delivery dates, with minimum inventory and other associated direct and indirect costs. This aim is now being realised through the correct application of several separate but related philosophies. For the majority of companies in both the manufacturing and process industries, this means total quality management (TQM) plus just-in-time (JIT) techniques, together with a manufacturing resource planning system (see Figure 12.6).

Perhaps these approaches seem conflicting, but this is not the case. They all aim for manufacturing excellence and competitiveness. Indeed, with ever growing competition both nationally and internationally, it is unlikely that any single approach will give the competitive edge required. But only one is available off the shelf, and that is MRP2.

■ Background to MRP2

In the 1960s, manufacturing companies commonly had unwieldy, paper-based stock and warehouse control systems, which required considerable effort to maintain. Computerised stock control systems were becoming affordable and attractive and basic inventory control was quickly developed. The basic functions of inventory control grew to include facilities for reorder control (two-bin or min/max) and forecasting. It soon became obvious that computers could handle the advance planning and checking of actual material needs, data processing tasks that could not be effectively carried out manually, due to the massive amounts of data processing needed. This brought together the concepts of bills of material (BOMs), material requirements planning (MRP1), and inventory control (Figure 12.1).

Basic MRP calculation

Material requirements planning took the BOM for each required product on the schedule, exploded each BOM into its component parts and calculated the gross material requirements for the schedule. It then compared the gross requirements with the on-hand stock in the inventory system, resulting in the net requirements to be purchased or produced, i.e.

1. Calculate gross requirements from BOMs and end-item schedule.
2. Compare gross requirements with on-hand quantities.
3. Calculate the net requirements to be obtained, and when these have to be ordered to be available in time.

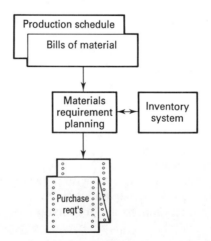

Figure 12.1 Early material requirements planning and inventory control.

The next development was to use the net requirements report to forward purchase orders onto suppliers and works orders onto the in-house production facility. Also, the MRP1 system needed to take into account not only the physical stock held, but also any planned and in-progress materials for both suppliers and production. This *time-phased requirements planning* was then able to generate exception reports to speed up or slow down the flow of incoming materials according to schedule changes. MRP1 could be rerun with amended schedules until a satisfactory outcome was obtained. These batch runs took many hours to run on the computers of the day.

The Materials aspect of MRP1 developed quickly. As more and more businesses bought systems, facilities and features grew into the powerful tools used today. And soon people wondered how other business aspects could benefit.

Capacity planning

Production facilities need adequate capacity for the intended schedule. There was little point in obtaining all the necessary parts if the production facility did not have the capacity or resources to assemble them. So capacity requirements planning (CRP) was developed to run schedules of work against current and planned capacity. Similar to MRP1, CRP reported on all mismatches between schedule load and shop capacity, allowing remedial action to be taken (Figure 12.2). Every manufactured part was assigned a route through the production facility, including time required at each work centre. Similar to a BOM, the routings and operation times were used within CRP to calculate the loading onto each resource. The total relative loading at each work centre was given by multiplying the net quantity of production parts by the routing time for each part. Both MRP1 and CRP can be termed planning systems because they develop valid and achievable schedules for execution within the business.

Closed-loop MRP

So far the MRP1 system described has not extended beyond the manufacturing and service areas: *plan the work and work the plan*.

Closing the loop relates to having feedback on status and progress of each area against the plan or the schedule. Materials was almost a closed loop, via purchase orders, receipts, expedites and goods received notes (GRNs). But production had little feedback, making it very difficult to determine when parts or products were finished. So, to be able to monitor how closely they were keeping to the schedule, shop-floor control and feedback was devised. Execution of the plan and feedback to the planning function formed *closed loop MRP1* (Figure 12.3), adequate for all production-related functions within a business. The master production schedule (MPS) could also be tied and

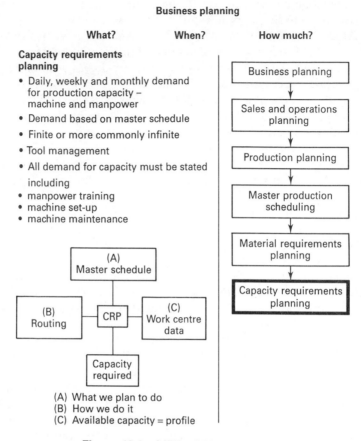

Business planning

| What? | When? | How much? |

Capacity requirements planning

- Daily, weekly and monthly demand for production capacity – machine and manpower
- Demand based on master schedule
- Finite or more commonly infinite
- Tool management
- All demand for capacity must be stated including
- manpower training
- machine set-up
- machine maintenance

Business planning

↓

Sales and operations planning

↓

Production planning

↓

Master production scheduling

↓

Material requirements planning

↓

Capacity requirements planning

(A) Master schedule

(B) Routing — CRP — (C) Work centre data

Capacity required

(A) What we plan to do
(B) How we do it
(C) Available capacity = profile

Figure 12.2   MRP1/CRP operation.

compared to the production volumes and mix necessary to satisfy the financial business plan. *It's not just materials anymore*.

Manufacturing resource planning

Sales and accounts were not generally involved in MRP1. But the system was crying out for them. Data had to be double entered, a major source of inefficiency that led to the development of a complete business-wide system. It catered for all aspects of resourcing and it was called manufacturing resource planning (MRP2).

Sales order processing allowed the passing of order information directly to the stock or schedule for earliest delivery or scheduling. Order information was also passed to financial modules to record the sale. This allowed automatic invoicing and paperwork generation when the goods were shipped

(Figure 12.4). Links between inventory, accounts, costing and purchasing allowed the updating and costing of stock, materials and products, with much reduced manual effort. The developments in MRP1 and MRP2 bring us near to the sophisticated enterprise planning systems available today:

1. Inventory control
2. BOMs, MRP and inventory
3. Materials requirements planning (MRP1)
4. Links to purchase ordering
5. Links to production
6. Master scheduling and capacity planning
7. Forward planning
8. Advanced features in all functions
9. Closed-loop MRP
10. Integrated financials
11. Manufacturing resource planning MRP2
12. Shop-floor data collection
13. Links to CAD
14. EDI Links to external suppliers and customers

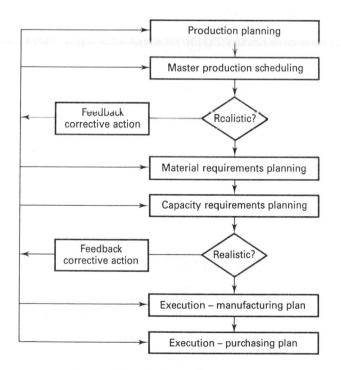

Figure 12.3 (a) Closed-loop MRP1.

337

15. Executive information systems (EIS)
16. Enterprise resource planning (ERP)

Continuing to expand and thrive in all parts of the manufacturing and process industries, MRP2 gained more and more functions and facilities: CAD links, forecasting tools, shop-floor data collection to speed up information feedback and provide job time collection. EDI links to anywhere in the world are now in common use, and high level business planning facilities are provided for strategic planning and rough-cut measurement, using the MRP2 system as a 'what if' tool (Figure 12.5).

Starting out over twenty years ago as a better way to order material, MRP1 has evolved into a company-wide planning system. It is a way to get all the people in the company working to the same information and the same game plan. Using a single system, a company can now plan material and capacity, finance and marketing. It can run simulations to make planning even more effective. And it can determine net time-phased requirements to meet customer due dates. Fully implemented and properly employed, MRP2 is not just *a* company-wide information system; it is *the* company-wide information system.

Typical manufacturing resource planning (MRP2) structure

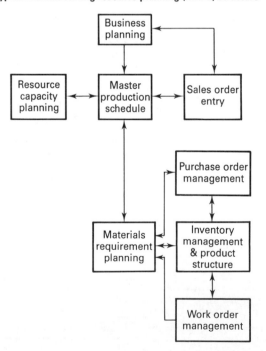

Figure 12.3　(b) Modules for closed-loop MRP2.

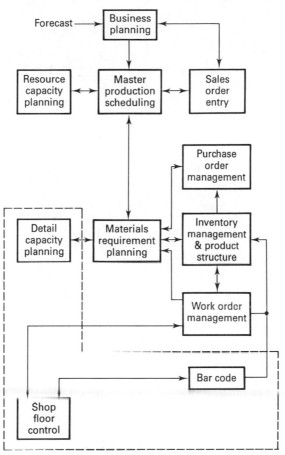

Figure 12.4 Integrating other business modules with MRP1 to form MRP2.

MRP2 includes

- Sales order processing
- Long horizon capacity planning
- Master scheduling for the facility
- Material requirements planning, including made and bought-in items
- Resource planning on a medium or short horizon
- Purchase order processing
- Shop-floor control, including works order generation and WIP tracking
- Inventory control
- Shop-floor data collection to streamline WIP, time and attendance data input
- Job costing
- Financial ledgers and accounting
- Forecasting

339

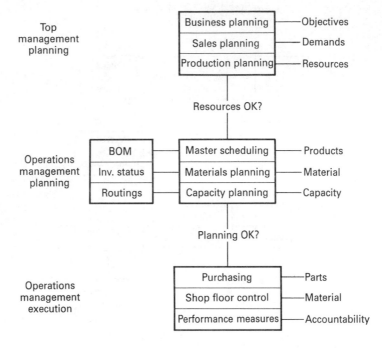

Figure 12.5   MRP2 processes.

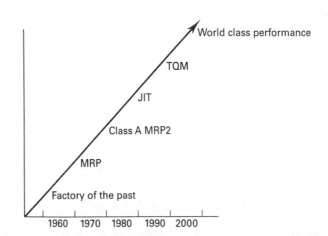

Figure 12.6   The road to manufacturing excellence.

■ Everybody's information system: enterprise wide

Without question, MRP2 meets the everyday needs of *any* modern business, mechanical or electrical, job shop or flow line. From cotton to chemicals, from sugar to steel, process industries may or may not need materials planning, but they are virtually certain to require a corporate-wide information system, in other words MRP2. Many of its principles and features are applicable outside manufacturing. It is truly enterprise-wide and sector-independent. (See Figure 9.13, which shows functional areas of the business information system under modular MRP2/ERP.)

■ Enterprise resource planning

Outside manufacturing, some people were put off by the name MRP, so enterprise resource planning (ERP) was coined. ERP is becoming much more widely understood and it describes the system's full potential. From now on in this text, the terms ERP and MRP2 will be used to describe the *same* company-wide, multifunctional business planning and control systems. And it will be stated when something refers purely to manufacturing.

Advantages

The advantages of ERP are manifold:

- company-wide integration
- shared, accurate and current databases
- greatly reduced duplication of data and paperwork
- reduced NVAs
- very fast access and response
- links to external systems such as CAD and scheduling tools
- ability to perform 'what if' analyses in various areas

The tangible cost advantages of MRP2 are discussed in Chapter 13. They depend on a successful system implementation over many months of continued commitment and hard work at every level of the company. (Chapter 13 also describes success and failure factors of MRP2 implementation, with a recommended approach to all aspects of the project.)

There are vast numbers of modular MRP2 systems available today, and many that provide full ERP coverage, each offering integration between the functions, for companies with needs ranging from one to one thousand terminals. Whilst each system will have family similarities, each will also provide the above functions in a different manner. Several systems have been specifically for chemicals, textiles and food. (Figure 12.7)

341

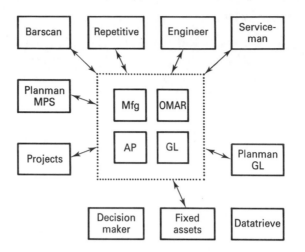

Figure 12.7 MANMAN MRP2 product diagram.

There is thus a requirement for each company to carefully select the preferred system. Because of the importance of this system to the company, considerable care should be exercised during appraisal and selection.

■ Trends in the MRP2 industry

There have been significant changes to virtually all MRP2 products over the last five years, often with the vendor bringing out a completely new or rewritten system to cater for changing needs and sector-specific options. IBM has brought out the MAAPICS DB to replace MAAPICS and other vendors have brought out out major new releases of existing products. But most significant is the emergence of a potentially universal operating system and the downsizing of hardware platforms. These trends are based on

- UNIX growing as a standard operating system
- High power, low cost microcomputers
- Emergence of powerful low cost UNIX and PC network systems

Operating system and hardware platform issues

More fully discussed in Chapter 11, there is now a growing movement towards downsizing corporate computer systems. Networked PCs offer distributed, upgradable performance with the advantages of

- Easy and low cost upgrade path by adding or replacing nodes with more powerful nodes and/or file servers.
- Low node cost versus computing power compared with minicomputer alternatives.

They have major implications for MRP2 business control systems, which have been almost exclusively minicomputer-based for the last two decades. Digital, IBM and Hewlett Packard minicomputers have been the preferred hardware platform for the most popular and successful MRP2 packages, including MAAPICS and MAAPICS DB (IBM), Manman (Digital), Impcon (Digital), BPCS (IBM), MM3000 (HP), MTMS (various). Many of these system developers have now ported or redeveloped their MRP2 products onto UNIX.

Workstation LANs: distributed performance

The effective performance of a correctly configured Ethernet/Novell LAN with 486 nodes has been shown to outpace many minicomputer platforms. Figure 12.8 compares execution times for a large MRP2 regeneration and indicates the capability of LAN/486 systems.

UNIX networks can obtain substantial speed advantages by using RISC processors as host or workstation. For PC LANs, 486- or Pentium-based file servers and workstations/PCs can also offer highly impressive performance.

Processor	Thousands of transactions/hour	MRP regeneration time
PC/Network		
486/20 MHz processor	5.4	4h 13m
486/33 MHz with cache	10	2h 5m
486/50 MHz	15	1h 25m
Pentium P120		
Mini		
DEC VAX/VMS 3800 (24Mb RAM)	N/A	6h 29m
NCR 9500	N/A	10h 32m
UNIX/RISC		
RISC/20 MHz	6.0	3h
RISC/28 MHz	8	2h 6m
RISC/30 MHz	12	1h 40m
RISC/41 MHz	20	1h 24m (est)

These test results indicate the relative performance and suitability of micro and RISC processors during (the same) actual MRP runs, using live data.

Figure 12.8   Some computer benchmarks.

343

Downsizing: hardware test benchmark

One benchmark is a full regeneration explosion for a typical materials requirement. This commonly takes several hours to complete and would be run overnight for large applications. Figure 12.9 shows results for a database with

> 15,000 part masters
> 200 items per BOM
> Bills of 2 or 3 levels
> 300 end-items per week
> working on a 10 week planning horizon

The run covers one week's requirements for a worst case involving 84,000 allocations in the database.

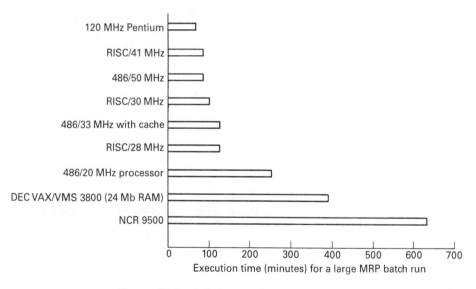

Figure 12.9   Relative performance of computers.

■ MRP2 software products

Until recently, very few MRP2 systems were available on UNIX or PC/network platforms. With few exceptions, those that were available tended to be rather limited in functionality and/or unsuitable for large installations. Popular PC MRP2 was provided mainly by

- Fourth Shift (UK and US),
- Micross (UK),

- MicroMRP (US)
- TSIpro (US)
- MicroSafes

All the popular PC MRP2 systems could be run on Novell or PC LAN networks. Several UNIX MRP2 systems were available, but were again perceived as better suited to the smaller, less demanding business or installation.

New MRP2 products

In the late 1980s, as awareness grew of the potential power and flexibility of distributed PC or workstation platforms, several MRP2 products were created or recompiled for UNIX or PC LANs.

UNIX options

In the late 1980s UNIX came of age as the sought-after open system needed to convert many established MRP2 packages, such as MTMS (BEC), Impcon, and Manuflo.

Most major OEMs now market comparable UNIX systems offering wide choice of hardware, MRP2 software under UNIX is now plentiful. Among the new UNIX MRP2 systems are

- MFGPRO (QAD Inc.)
- Bluebird Software
- MANMAN/X
- System 6000

And new PC LAN MRP2 systems include

- TSIPRO V.2 (Thornapple Software)
- Micross Professional (Kewill)
- MFGPRO (QAD Inc.)
- Swan MRP2

A large proportion of PC-based systems selected Novell Netware as the LAN operating platform, strengthening the combination of Novell and Ethernet as an emerging standard. Microsoft's LANManager offers similar functions, but has much smaller market penetration.

■ Additional features in modern MRP2 systems

Over the last five years, MRP2 and ERP have developed to include:

- User interfaces
- Configurator/features and options
- Functionality changes, including the MRP net change
- Expert systems
- Finite scheduling

User interface

Many of the recently developed MRP2 systems have been created using modern, fourth-generation languages and have provided new and useful features not found in traditional systems. MRP2 has exploited the power and graphics facilities of the modern computer workstation to produce

- pop-up windows,
- mouse-driven operation
- user-defined report generators, menus and screens
- relational databases
- integrated graphics displays
- on-line help screens and manuals
- multiple languages, currencies and methods

Particularly useful, on-line help screens usually refer to their immediate context and are easily updated by system managers (Figure 12.10). This is a major step forward from the many manuals formerly required for larger MRP2 systems. *Onscreen procedures and work instructions for BS 5750 or ISO 9000 can also be easily accessed and updated from disk.*

Pop-up windows provide greatly enhanced guidance to users, displaying the range of options available at any specified screen field. For maximum tailoring, some systems allow the user to define where and how windows are used.

The windows and mouse combination can be extremely powerful in use, allowing the user to open one or more window(s), then to select one or more data elements for use in the next operation. Features such as this are now becoming standard on even the lowest cost system, rendering the older and larger MRP2 products lumbering and unfriendly (Figure 12.11).

Report and screen generators

Extending the use of the data dictionary provided in earlier MRP2 packages, many new systems provide full creation and edit facilities for report calculations and for screen field calculations. This allows the user to adjust how to calculate a field value, e.g. safety stock, building up a formula from both numeric and data name variables (Figure 12.12).

```
,              ÖÄÄÄÄÄÄÄÄÄÄÄÄÄÄÄÄÄÄÄÄÄÄÄÄÄÄÄÄÄÄÄÄ·
,                 ▪  Batch Process parts (L/R)    ▪
, ÛÄÄÄÄÄÄÄÄÄÄÄÄÄÄ¿ ÓÄÄÄÄÄÄÄÄÄÄÄÄÄÄÄÄÄÄÄÄÄÄÄÄÄÄÄÄÄÄÄÄ½
, , User Help   ,
, ,    Screen   , System: Requirements Planning.
, ÄÄÄÄÄÄÄÄÄÄÄÄÄÄÄÙ
,                 This option scans all parts with a policy code of "L" or
, < Topics   >   "R" and if, through the analysis of the part, an order
,                 is required then the part is written out to the MRPORD
,                 database with the corresponding quantity required plus
, < Previous >   the date it is required for.
,
,                 The user can select a range of parts to process so the
, <   Next   >   full range of parts can be done in sections. But please
,                 note that all other orders for that week are cleared as
,                 the option is run (even for a small subset of parts).
, < Tech Info >
,
,
, <   Exit   >
,
,
```

Figure 12.10   Onscreen help.

Integrated graphics arc not a new idea or feature; many long-established systems provide excellent links to PC-based graphics and spreadsheet packages, allowing the effective display of data in various ways. State-of-the-art MRP2 takes this a stage further; its built-in graphics handlers offer single-key selection in some or all functions.

Sector-specific software modules

Led by the Oliver Wight organisation, most systems provide the essentials, and many have been created for a particular market sector or provide sector-specific modules. There are significant differences in the systems needs of process, batch and make-to-order sectors. If only a single fixed set of planning and control software was available, it would either be so restrictive as to be of limited value, or have so many configuration options as to be unusable. Sector-specific modules are both understandable and desirable (Figure 12.13). **All MRP2 systems are not the same**.

347

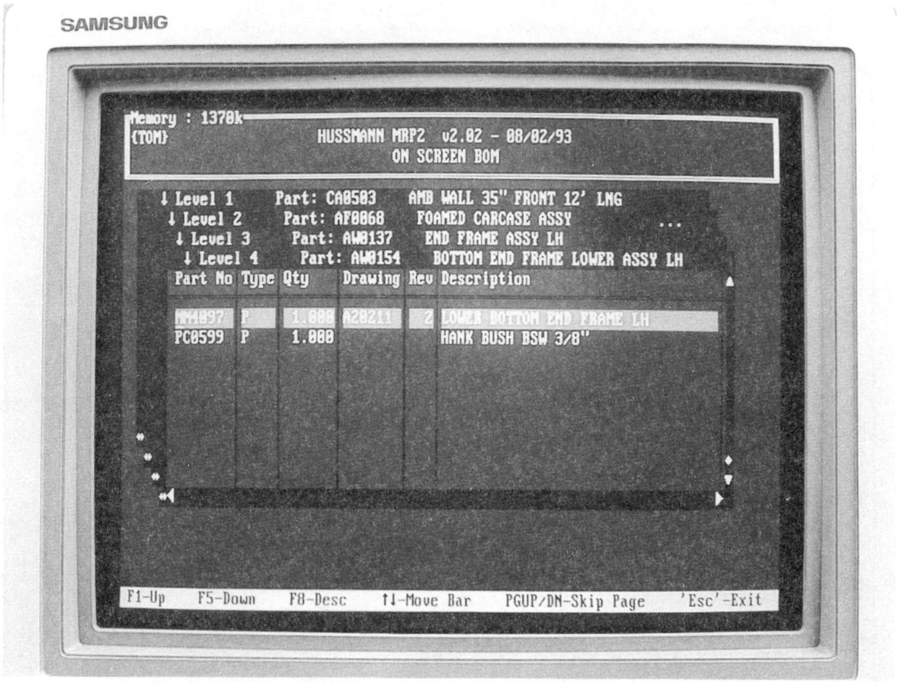

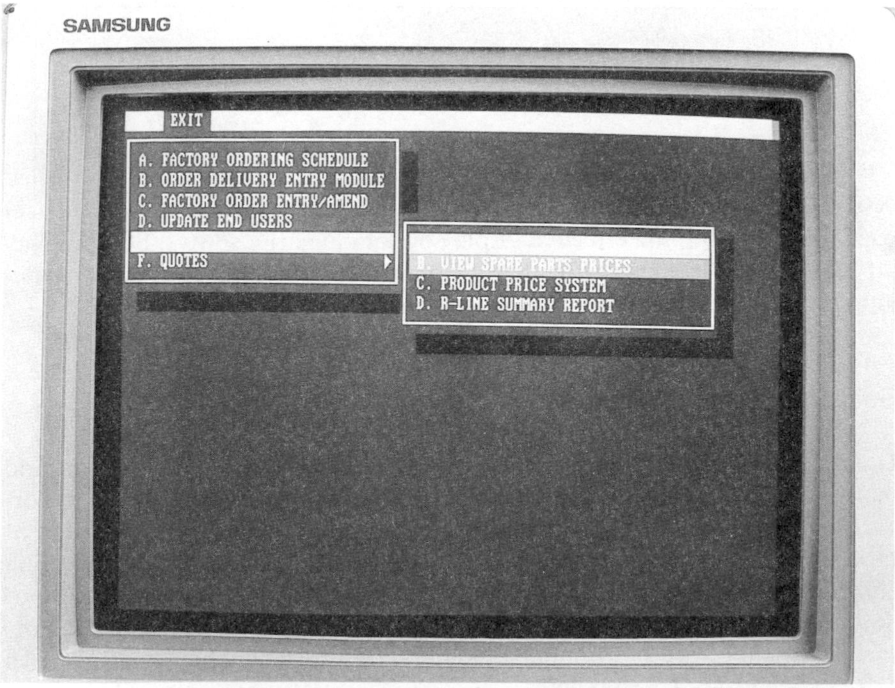

Figure 12.11 Some Windows features used in MRP2 screens.

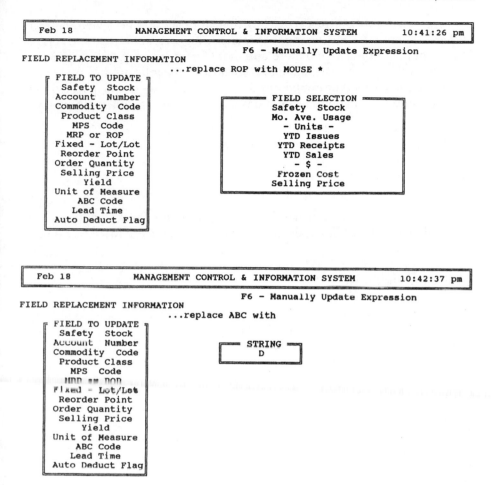

Figure 12.12 MRP2 field definition using data names and user-defined equations.

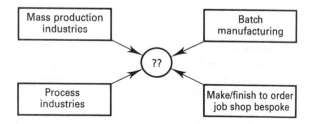

Figure 12.13 MRP2 sector-specific products and/or modules.

The same criticism of excess functionality can also be levied at many system vendors, who, in attempting to increase the sector coverage and vertical market of their product, add more and more features and options, where the end-user will only use a maximum of 30 or 40% of the total product functions.

In addition to module options for different sectors, more and more systems are being offered with optional modules for other more general requirements:

- JIT-based production
- Full traceability for product quality assurance
- Finite scheduling
- BS 5750 /ISO 9000
- Features and options for custom product manufacture

This is the trend in system development, providing a pick-and-mix set of optional modules that sit beside and integrate with the core MRP2 modules.

MRP2 example: MFG/PRO

A modern MRP2 system that provides optional modules for several industry sectors is MFG/PRO from qad.inc North America (Figure 12.14). Addressing virtually the whole manufacturing spectrum from repetitive to configure-to-order, it provides facilities for

- process
- batch
- repetitive
- make-to-stock
- configure-to-order
- multiple site
- multinational, multicurrency and multilanguage

This is achieved through optional modules, including

- Configuration of custom products
- Service/repair orders for after-market work
- Repetitive manufacturing for high volume/large batch production
- Works order for mid-range WO-based producers
- Formula/process module for recipe and process-based industries
- Forecasting module, including seasonal build/sales facilities

MFG/PRO was designed using the manufacturing MRP standards set by *The Standard System*, published by Oliver Wight and Gray Research, and by the APICS societies in North America. It was developed using the Progress flexible relational database manager, with data dictionary, 4GL application generator and editor. A fast development environment, it allows easy modification and customisation of the system. Multi-user database locking to the individual record level is supported, with automatic database recovery from any failure

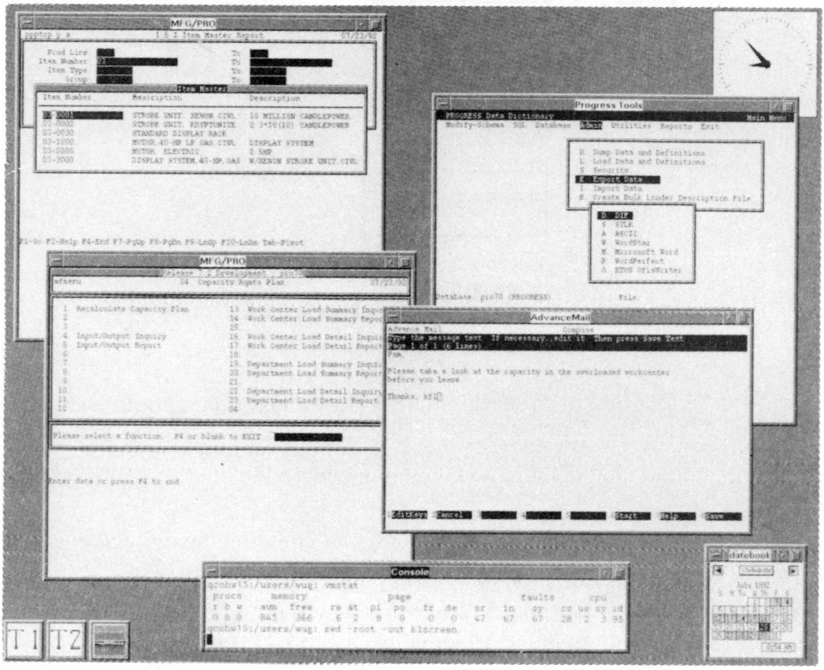

The PROGRESS 4GL allows MFG/PRO to run unmodified under X-Window-based graphical user interfaces (GUI's) such as Motif®, Open Look®, Open Desktop™ and DEC Windows® as well as Microsoft Windows®.

This capability allows an end user to become more effective in managing a business. On one screen, a user can run MFG/PRO, access mail and word processing, and use remote communications capabilities simultaneously.

Figure 12.14 MFG/PRO, a modern Windows based MRP2 system.

in the hardware or power supply. Progress also allows the software to be transported between operating systems without the need for code alteration.

MFG/PRO is thus available for a very wide variety of operating systems and hardware platforms supporting Progress, including UNIX, Ultrix, HP-UX, Ultrix, Xenix, VMS, and MSDOS. Networking is supported on Novell and 3COM systems. This allows users to start with a system of only one PC, growing to virtually any size. MFG/PRO is used worldwide.

Systems for specific sectors

Some MRP2 systems remain heavily biased towards a particular industry due to the long involvement of the software supplier. Several systems are available purely for clothing manufacture, which largely precludes their use elsewhere. And a few systems are so industry-specific they cannot transfer between sectors as closely related as clothing manufacture, and textile production.

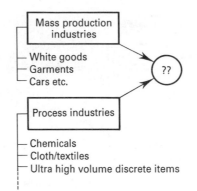

Figure 12.15 Sector breakdown into often dissimilar requirements.

No need to reinvent the wheel

Many users, particularly larger companies with a data processing/systems department, continue to try and develop bespoke, customised software to perform these very functions. We must question why. They are readily available off the shelf, at relatively low cost. Hundreds or thousands of man years were devoted to producing the package MRP2 solution; how can a company systems department hope to produce something even half as flexible in even twice the time? Most systems need to be fine-tuned to reach the levels of commercial packages, and most companies cannot spare the time.

Even corporate giants are turning to commercial packages, fine-tuned by their own information system resources. The early 1990s saw several new MRP2 products that use advanced 4GL languages on smaller, more cost-effective hardware. Often more customisable than their predecessors, they offered new levels of flexibility and performance (Figure 12.16).

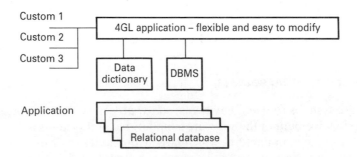

Figure 12.16 Customised business control system concepts.

Customised MRP2

An extreme example of customised MRP2 was a product resold by Hewlett Packard third-party software distributors. Not a standard suite of modules, it was tailored by the distributor to meet a detailed user requirements specification. And according to vendors, state-of-the-art development systems make the total time and cost involved comparable with conventional packages. The bespoke approach also achieves a better fit.

So, this customising approach is becoming more practical as the power and development speed of software development tools continues to grow, coupled with the data dictionary and database management systems now in widespread use.

■ MRP for the make-to-order sector

MRP1 and MRP2 offer a high degree and ease of fit to businesses, making mainly standard, well defined products, from canned food to cars. This is not surprising, since their origins lie in materials planning and control (MRP1) in assembly and make-to-stock industries.

It is arguable that the majority of traditional MRP2 systems were not particularly well suited for the make-to-order sector, nor did they provide as standard the flexibility needed in finish- or assemble-to-order industries. Inevitably this led to systems modifications, which in turn resulted in cost penalties and maintenance difficulties.

The basic MRP1 equations and calculations are based on defining end-item (saleable product) requirements, exploding the BOMs for these end-items, then comparing the gross requirements against the planned stock in a time-phased manner. This results in recommended purchase and works order actions. Where the BOMs are firm and current for standard products and the available options, MRP1 will operate well, provided the data accuracy and currency is maintained.

BOM limitations: a major reason for MRP failure

Chapter 13 discusses the failure of a high proportion of MRP installations. Poor and/or inappropriate implementation accounts for the majority, but failure also stems from BOM limitations with respect to rapid configuration from product orders to produceable items.

In the make-to-order sector, the range of product options and the frequency of change can make conventional BOMs almost impossible. Even in certain assemble-to-order situations, market pressure and customer requirements can lead to end-items being requested as configured from non-standard component parts.

In many companies this is carried out by creating unique BOMs for each order. Depending on the underlying BOM structure and the MRP BOM system, this can vary from being a time-consuming task to being an almost impossible one. BOMs are one of the most important data sources that feed MRP and the factory. They must therefore be maintained and controlled to the highest standards. But in the situation where unique BOMs are created for each job, control and maintenance become virtually impossible, and accuracy suffers.

This need for unique or modified BOMs is largely due to the limited features for BOM configuration in many MRP systems. Where no features exist for providing product options at various levels of the BOM, the only options available to the user are

■ Create new or modified BOMs for each order
■ Attempt to create a very wide range of BOMs and/or lower level kits that define all product options

The latter option may be feasible for some assemble-to-order situations, but cannot meet the needs of the majority of make-to-order situations. However, in both cases the net effect is the need for continuous staffing to be devoted to the task. This is both costly and ineffective, and can be a weak element in a company's operations.

Depending on the level of technical expertise in the estimating or customer order services function, it may be able to use the BOM system as a pick-and-copy tool to create each detailed job estimate. However, this is uncommon, and standard BOM systems often contain only basic copy and edit facilities.

Many companies have had to set up separate functions to perform two-stage order definition:

■ Initial estimate and outline definition (estimating)
■ Second-stage detailed job definition (BOM)

Significant duplication exists between such functions. This inefficiency shows up in extended overall lead time and in frequent conflict over what can or cannot be built. An example of this front-end overlap is given in a case study later in this chapter.

■ Links to CAD from MRP2

Most modern MRP2 systems provide software options for linking to a range of CAD systems to transfer part and product information. Either provided as a product-to-product link, or as a more general EDI module, the CAD data is used for

■ BOM generation in standard product applications

- BOM construction in make- or assemble-to-order applications
- Product/part viewing from within the enterprise-wide system

In the make- or assemble-to-order applications, automatic BOM generation is rarely possible or desirable, due to the large amount of variation and expert knowledge that has to be used for detailed specification. However, the specification experts may be able to usefully select and build up components and/or assemblies from the CAD and BOM databases, saving configuration lead time.

CAD drawings are commonly viewed using shop-floor and office monitor screens assisted by high resolution PC displays on many MRP2 systems. This facility can be of immense value to employees for part identification and detail information.

■ MRP2 order definition and configuration facilities

The customised product problem asks

- What is to be made? – Can it be made?
- How is it to be made?
- What is the estimated cost?
- What should be the sale price?
- What profit margin will be made on the order?
- What delivery date can be given?
- Will it be designed and completed on time?
- How do we describe and invoice the job?
- Who is responsible for each of these functions?
- How are they to be managed?

In addition to these problems and questions, there are other compounding factors, including limited resources and time, and the potential for any mistakes to be very costly to the business.

Where the product definition process requires selection of options at various stages of a product, we require a BOM/definition system that offers comprehensive configuration facilities. This must cover not only the engineering aspects of product definition, but also the estimating and costing requirements of the business. Having accurate costs rolled up as a job or quote is prepared is a major advantage compared to having to rely on best guesses. Several established MRP2 products offer configurator modules that integrate with the SOP and BOM modules, for example, BEC's MTMS configurator, and the MFGPRO options module.

MTMS configurator

The MTMS configurator provides attributes that are defined and used to mark any number of optional parts and/or kits that can be built into a product. Users can then configure the system to display the available options at each appropriate stage of order estimating and definition. A reasonably effective configurator, it can handle many of the requirements for option handling. It does not provide facilities for building in additional selection rules, except by customising the software.

MFG/PRO configurator

A second-generation MRP2 system, MFG/PRO contains many of the newer functions and facilities required for custom product manufacture. The features and options module of MFG/PRO allows the user to maintain data on product configuration options and to process customer orders for assemble-to-order products.

Customer sales orders are entered by selecting a particular product configuration from a predetermined list of configuration options. The system then generates works orders for assembly of the specific product configuration. The module includes the following functions:

For products/items

- a built-up purchase/manufacturing code identifies configured products
- an automatic interface to SOP prompts for configuration
- a purchase/manufacture code identifies families of configured items

Features and options

- product structure identifies all product features
- mandatory feature specifications cannot be omitted
- default options can be selected
- options may call other options (nesting)
- forecasts can be specified for options or accessories; planning bill preparation
- master scheduling can be done at many levels, including option levels
- production of option quantities is derived from forecast percentages
- costs, prices and margins are calculated automatically from configurations

Sales order processing

- automatic prompt for product configurations
- rapid entry of SOs by selecting default options
- separate product structure for each line item in the sales order

Works orders

- the final assembly WO is created from SO configuration
- WOs are pegged to the SO

The following screen examples from MFGPRO indicate how the features and options configuration is carried out (Figure 12.17).

MANMAN/x, the UNIX version of the well-known MANMAN MRP2 system from ASK, also has a powerful product configurator (Figure 12.18). This is a rule-based tool that allows the user to specify customised items at any stage in the order process. These rules can be user-defined to allow the creation options, features, dependencies and constrains. This operates at all levels of the BOMs and allows the configuration of product from any mixture of standard and customised components. The system also generates costs, prices and documentation based on the specified rules.

MANJOB: Make to Order and Job-shop Control System

Traditional MRP2 systems largely ignored the job-shop sector of manufacturing, which had very few standard parts or operations, requiring great flexibility and ease of creating one-off products and services.

A new package – *MANJOB* – was launched in 1995 to address this need. Written by the BRIM Group in Microsoft ACCESS and Windows, *MANJOB* offers a flexible, easy to use jobbing and contracting system with selected MRP2 functions integrated to key job-shop requirements. For example

- product and service configurator
- one-off parts
- simple MRP
- offcuts
- job costing
- auto faxing of purchase enquiries

See Figure 12.19.

MANJOB is now widely used in the United Kingdom and Europe, in manufacturing and commercial sectors. Because of its configurable nature, it has proved to be an ideal operation and control system for a variety of businesses, including project design, sub-contract, medical, building, electronics and plastics, as well as general job-shop applications. It will cater for between one and 50 users, networked using Novell.

(The BRIM Group is a leading UK supplier of manufacturing and business control systems, including *MANJOB* and *MRP Expert*. *MRP Expert* is a bolt-on analyser to assist in improving the use of materials planning in a range of popular MRP2 systems. The Group is based in Glasgow.)

357

CONFIGURED PRODUCTS

The Configured Products module is used to enter product configuration information, review product configurations on sales orders, and release work orders to assemble configurations ordered. Configured Products is most often used by manufacturers who configure their products uniquely for each customer order.

■ ■ ■ ■ ■

Option Definition: Configured product bills are defined in this module and used to prompt for option selections during order entry and sales quotations entry.

Work Orders: May be created directly from these sales orders with a work order bill for the exact product configuration specified during order entry.

User Specified Price: When entering a sales order or invoice for a configured product, it is possible to enter a different price and discount on each of the configuration options selected.

Backflush Option: When the sales order is shipped each of the configuration options selected may be backflushed from inventory. The sales order does not need to be released to create a final assembly work order and issue components.

KEY FEATURES

- Mandatory features
- Default options
- Production forecast percentage specified for each option or accessory
- Multi-level master scheduling
- Production forecasts calculated from forecast percentage
- Creates final assembly work order to create a unique configuration
- Lot/Serial numbers may be assigned to component issues and sales order shipments
- Final assembly work orders are pegged to the sales order

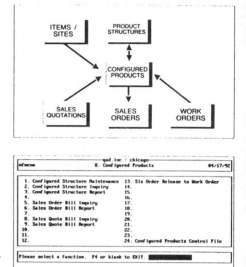

MFG/PRO GLOBAL SUPPLY CHAIN MANAGEMENT

Figure 12.17 MFG/PRO option handling.

Fluent in all business dialects. MANMAN/X's multi-lingual abilities are only the beginning. The system is designed to accommodate the full range of international variables — enabling companies with several subsidiaries to localize MANMAN/X to regional business practices, and to each country's currency and taxation requirements. The system converts among currencies, handles exchange rate variances, and consolidates financial results worldwide through a multi-tiered, multi-company fiscal structure.

These powerful capabilities are integrated across all MANMAN/X modules. They can be adjusted and modified to fit changing needs and market conditions so that, as you expand across the globe, MANMAN/X can grow with you.

A SYSTEM WITH UNPARALLELED FLEXIBILITY AND VERSATILITY

Full Support For All Manufacturing Methodologies
Whether you are using make-to-stock, assemble-to-order, made-to-order, or engineer-to-order manufacturing methodologies — alone or in any combination — MANMAN/X fits your business practices.

... And For Customer-driven Manufacturing
Since the time of Henry Ford, manufacturing has operated according to a mass production model in which everyone bought an identical product. Nowadays, intense global competition is obliging manufacturers to adopt a more customer-driven stance. What customers are asking for are individually tailored products at mass production prices.

MANMAN/X enables you to respond to your customers' needs without compromising your economies of scale. Whether you build products one at a time in a project-oriented mode, run a mass production operation, or employ some combination of both, MANMAN/X's exclusive Product Configurator provides you with a solution — and a competitive edge.

The Product Configurator is a rules-based tool that lets you specify customized items at any stage in the process, and feed them into the production pipeline in the manner best suited to your methodology.

User-defined rules enable you to determine your own options, features, dependencies, and constraints. By operating at all levels of the bills of material, the Product Configurator lets you tailor individualized products that can be fabricated from a mixture of standard and customized components. Product costing, pricing, and documentation are automatically generated in accordance with the rules you have specified.

Bills of material for a product, such as a personal configurable computer, present all the parts necessary to build all versions of the product. Product feature lists present all possible variants in producing the personal configurable computer.

Figure 12.18 MANMAN/X product configurator.

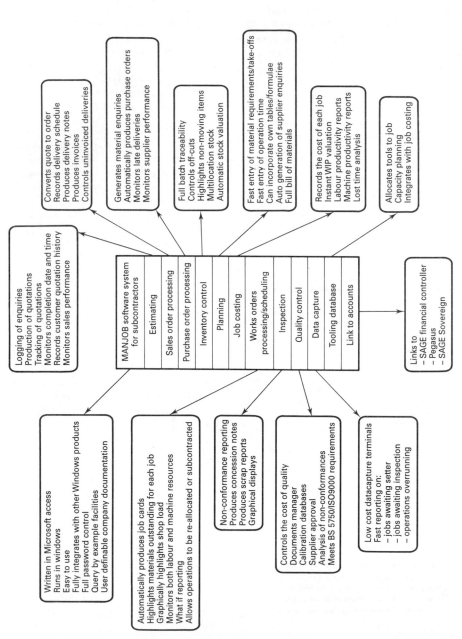

Figure 12.19 MANJOB job-shop control.

■ Expert systems for order options configuration

Whilst these type of configurators are useful and even effective in some applications, there are many cases where a more powerful and intelligent tool is required, for example, where selection of one option affects what further options should be displayed to the user. Standard configurators cannot readily deal with such a problem. For this reason, much resource has been devoted to using expert systems in order definition by both MRP2 producers and third-party software vendors. Expert systems are discussed in Chapter 9.

Expert systems in configuration

Of the many applications involving expert systems within manufacturing, configuration is proving to be one of the most challenging and worthwhile. In 1989, the product configuration system at Henderson Doors won the DTI's award for best manufacturing intelligence application in Britain. Henderson's 'Easy BOM' was primarily a database application incorporating appropriate decision and selection rules. Further details can be found in *Manufacturing Intelligence*, Spring 1990, published by the DTI.

Exports: a package export configurator

A handful of quality MRP2 systems are now available with an expert configurator. Written in a form of BASIC, EZXPERT is a front-end expert configurator fully integrated to Interactive's Infoflo MRP2 system. EZXPERT is described as easy to use and able to solve custom product configuration problems. Its approach is removed from the more traditional ways of tackling custom product definition:

■ intelligent part numbering
■ modular BOMs and kits
■ manual definition by engineering department

Instead, EZXPERT uses an expert system shell for defining and maintaining a custom product configurator (Figure 12.20). As with most available configurators, the system objectives are concerned with both product definition and accurate cost build-up:

■ the configuration of a complete build specification
■ calculation of complete manufacturing costs
■ calculation of recommended selling price from margin and pricing policy rules
■ rapid collection and presentation of information for management reporting

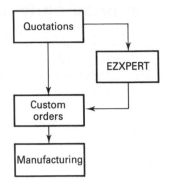

Figure 12.20   EZXPERT custom products configurator.

Output from EZXPERT is a costed and priced BOM structure that completely defines a specially configured product.

Operation

The system is first configured through a question-and-answer session to define the actual requirements to be controlled. The engineer/configuration expert defines a set of rules that govern the configuration of a product. Then step-by-step question and answer sessions are developed to guide the end-user through to full custom product configuration. This in turn generates a build specification to customer requirements, including manufacturing costs, suggested selling price and profit margin. By allowing only the selection of buildable configurations and highlighting exceptions to the rule base, EZXPERT can lead to improved customer service, more accurate and standard build programmes and better delivery performance. It also encapsulates a wide and deep set of knowledge and experience from any selected industry sector, reducing a company's dependence and vulnerability to the loss of key members of staff (Figure 12.21).

EZXPERT can be used either as a standalone module for a variety of applications, or as a complement to the Infoflo business information system. Its major benefit will be as an integrated part of a total system, where quotation and customer requirements information can be fed directly into EZXPERT, which is then used to derive the expert-based definition of each custom order. Sitting beside a conventional orders system, EZXPERT may be selected or invoked only when necessary, passing custom orders through to the planning manufacturing modules of MRP2. (Further details of the EZXPERT and Infoflo MRP2 systems are contained in Appendix H)

System overview

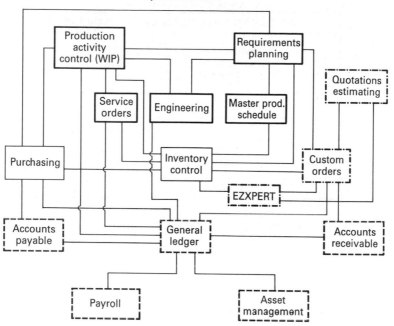

Figure 12.21 Infoflo MRP2 and EZXPERT configurator. (Courtesy of Interactive.)

CASE STUDY 12.1

Configuration of the single office in Chapter 6

A commercial white-goods company manufactures a wide range of food display cases for supermarket chains. Each end-user has its own specification, which contains variants for colours, shelving, lighting and electrics. The massive variety of products and options, and the level of finishing demanded by the market meant there was extreme difficulty in any one person defining a detailed requirements list for estimating and ordering the job. Estimators did not have a technical background, and this would frequently result in incorrect job quoting and description. This could in turn lead to wrongly priced jobs and lower profit margins than expected.

This difficulty in job definition contributed significantly to the total lead time within the company, as all subsequent activities were either delayed until full definition was achieved, or carried out with partial and/or inaccurate information (Figure 12.22). The results of this were frequently late and changing schedules, material shortages and inefficient working.

The company had already undertaken a wide and lengthy review of MRP2 systems in their search for a suitable solution for their finish- and engineer-to-order business (Figure 6.18). A vital area in this search was product configuration, but no suitable package system that met their requirements was found.

363

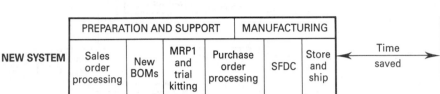

Total lead time reduction – attacking the early stages

	PREPARATION AND SUPPORT LEAD TIME					MANUFACTURING LEAD TIME	
OLD SYSTEM	Quote Prepare sales order (job line)	Full order definition (job sheets)	Plan and schedule shop and materials	Procure and make parts	Manufacture		Store and ship

	PREPARATION AND SUPPORT			MANUFACTURING			
NEW SYSTEM	Sales order processing	New BOMs	MRP1 and trial kitting	Purchase order processing	SFDC	Store and ship	Time saved

Figure 12.22 Total lead time diagram.

Subsequently, MRP2 was implemented in the company, but with the knowledge that an additional configuration system was to be developed and integrated to the MRP2 suite.

Knowledge-based system solution

From a feasibility study, it was proposed to design a configuration system (either knowledge-based or heavily option-driven) to use directly and to integrate with the existing BOM system, which was to be restructured for this purpose. The configuration system would be used by estimators to define each quote and order, using screen menus showing available options at each stage of the definition process. This would give a guided but rigid path through each necessary stage in the process, ensuring no important areas or questions were left unanswered (Figure 12.23).

Where the selection of an option affected other parts of the case, the configurator would offer an amended list of options or flag the choice as unavailable. Technical options would be configured automatically once the question-and-answer session had completed all relevant data (Figure 12.24).

Advantages

Using this system, turnaround times for quotes and job lines were reduced by over 50%, with the level of errors and incorrect configurations dropping to almost nil. This directly contributed to improved response time to customer enquiries and reduced lead time. Staff efficiency and morale improved significantly, as did control over job quotation pricing and profit margin.

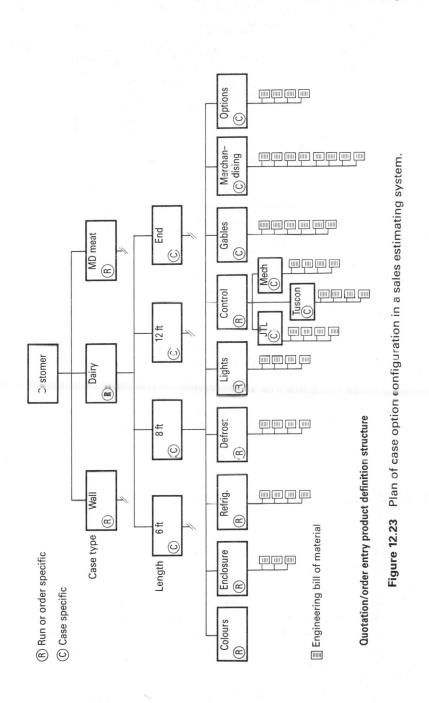

® Run or order specific

© Case specific

Quotation/order entry product definition structure

▦ Engineering bill of material

Figure 12.23 Plan of case option configuration in a sales estimating system.

365

```
                        Check the fields to view
                                                       BoM
        [ ]  Reset all            [X]  OHSS             [ ]
        [ ]  Defaults             [X]  Gables           [ ]
                                  [X]  Bumper protection [ ]
        [X]  Colours     BoM      [X]  Merchandising    [ ]
        [X]  Cases       [ ]      [X]  Options          [ ]
        [X]  Defrost     [ ]      [X]  Bumper colour
        [X]  Lights      [ ]      [X]  Parts prices
        [X]  Enclosures  [ ]      [X]  Total price
        [X]  Controls    [ ]      [X]  Case features
                         <Continue>
```

```
                    5 message(s) received

        Errors in quote:  SA00120

        Part: PA1325 has no agreed selling price
        Part: KA0705 has no agreed selling price
        Part: CZ0523 has no agreed selling price

        Received from user Colin   At 15:26:56 on 22/01/93
```

```
        Do you want to view the message(s)?

                <Yes>     <No>
```

Figure 12.24   Example screens of selection options.

■ MRP materials and production planning problems

MRP1 and MRP2 can run into trouble of many shapes and forms, particularly where the basic product design and/or sales strategy are less than adequate.

In some make- and assemble-to-order sectors, the product lead time (LT) will be long enough to plan and source materials to meet the delivery date required by the customer. However, in a large number of sectors, the market-led lead time for products will be less than the longest lead time for the component parts. Where there is a forecast build plan, this does not present significant problems. However, where forecasts and production plans are largely unavailable, material and production planning become very difficult, as there are few end-items scheduled early enough for MRP1 to be effective.

Make-to-order manufacturing

Product LT (order to delivery) = 4 weeks

Average component LT (0 to 13 weeks) = 6 weeks

The problems here could be largely overcome if it were possible to build or part-build a product/carcass and assemble to order, however, if the product has not or cannot be designed and built in this way, then the problems are significant. Without major changes in the way the product is designed and/or made, and in how the firm takes orders, we are faced with major problems at the material planning level. The first approach is to investigate the master scheduling of batches or units of production against the sales forecast for product and option sales, using a percentage MPS approach.

Figure 12.25 shows how Sales is able to break down the total sales quantity by percentage of sales with particular options. This is not an easy task and sales figures are usually inaccurate, though by how much depends on the market sector.

For material procurement, however, this approach is reasonable, as it allows the MRP1 system to plan quantity information on parts against the option spread given. MRP1 doesn't really care about the end-item at this point. It knows enough to get early orders to suppliers so that stock is available for meeting customer orders with short lead times. Feeding this forecast data as planning bills into the MPS module allows MRP1 to produce requirements to cover the option mix.

However, if these sales forecasts are unreliable, material planning will suffer and substantial shortages will result. To combat this problem, many firms plan material against other forecast information, assembly and part level forecasting. Although this is wholly against the MRP philosophy, measures such as these are regularly used where shortages of vital stock will result in

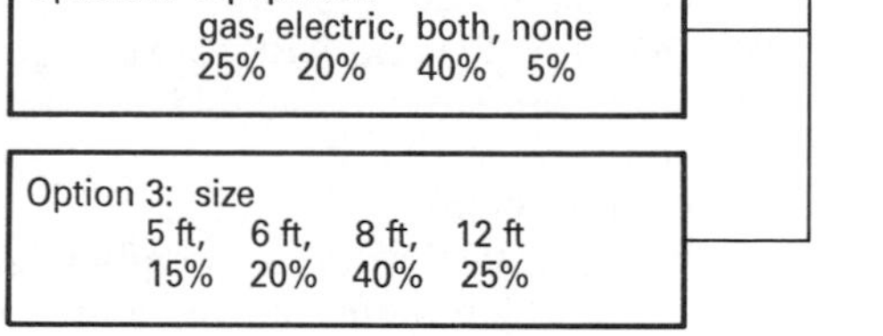

Product A forecast sales for next 4 months: 500

```
Option 1:  colour
              red, blue, green, white, black
              10%  5%   20%    25%   40%

Option 2:  equipment
              gas, electric, both, none
              25%  20%      40%   5%

Option 3:  size
              5 ft,  6 ft,  8 ft,  12 ft
              15%   20%    40%    25%
```

Figure 12.25 Sales breakdown of a product into option groups and percentage splits.

missed due dates, loss of customers and destruction of a market niche. Climbing stock levels can be tolerated as long as the company is actively reviewing and improving its sales, design and modular build for longer-term solutions. Where lower level forecasting is needed, this may be done internally in the package or externally via a PC spreadsheet or forecasting tool (Figure 12.26(a)).

Several MRP2 systems allow real-time generation of changing material requirements, showing onscreen the effects of high and low level demand for products and components, together with the consumption of forecast volumes (Figure 12.26(b)). This can be very effective when used for analysing MRP2 recommendations for ordering and expediting, and to observe the effects of moving jobs and orders.

Forecasting and forecast adjustment are largely automated in modern MRP2 systems, with smoothing factors and exception reputing available to the user (Figure 12.27). With these tools, it is possible to fill in for the lack of better product and sales data, but not to appreciably improve business performance and response. For this, the root causes of the problem must be solved.

Assuming the business strategy is to maintain or improve on the four-week lead time, longer-term steps have to be taken to achieve this in an effective and controlled manner. The following case study illustrates this problem and its solution.

CASE STUDY 12.2

Modify to order versus customer delivery date problems

In a commercial white-goods manufacturing enterprise, market demand is for a product that can be delivered complete to site in less than four weeks from order. It is extremely difficult to define an order; its content, detail, etc. are almost impossible to tie down, and frequently change several times during the period from order to delivery. To compound the problem, the product designs are not in any way modular or flexible, closing off any real opportunity to build in advance and then assemble and finish to order. Building a stock of popular end-items has a very low success rate, with most stock build requiring major rework to fit a customer order.

From a materials planning and control point of view, the four-week lead time is a major problem, since the average component lead time (LT) is six weeks, ranging from JIT to thirteen weeks. MRP1 and MRP2 are up against the odds from the start. Even if the products were more flexible for assembly build, the problems of order definition, the amount of order change and the LT requirements make this a tough problem.

This problem was encountered during a company-wide review (see Appendix B), and it was quickly identified that several front-end problems (including lack of modular design and imprecise order definition) were the root cause. These problem areas were addressed by an improvement team, with the following actions:

TEST.XLS

	jan	feb	mar	apr	may	jun	jul	aug	sep	oct	nov	dec	
89	20	20	17	20	22	22	20	9	20	22	22	17	
90	20	20	22	16	22	21	20	10	20	23	22	16	
91	19	20	21	17	23	20	23	12	20	23	21	16	
92	20	20	22	18	20	22	23	6	21	22	21	15	
Average	20	20	21	18	22	21	22	10	20	23	22	16	235
Total	75343	76267	80958	77337	77943	90924	74379	31535	89560	77369	87835	35606	875056
work days /mont	20	20	21	18	22	21	22	10	20	22	22	16	
Ave/work day	3767	3813	3855	4297	3543	4330	3381	3154	4478	3517	3993	2225	
Smoothed	3790	3812	3988	3898	4056	3751	3621	3671	3716	3996	3245	3109	44654
trend	4060	3994	3928	3861	3795	3729	3663	3597	3531	3464	3398	3332	
month nos	1	2	3	4	5	6	7	8	9	10	11	12	
smth as % of tre	93.36	95.446	101.5	101	106.9	100.6	98.87	102.1	105.3	115.3	95.49	93.3	
month factor	0.079	0.0812	0.091	0.077	0.1	0.09	0.093	0.043	0.09	0.108	0.089	0.064	1.005162

Part No	PEO418			Description		Defrost Heater 12ft							
Annual Total	1982			Year	1992								
Forecast	157.5	161	179.9	153.3	198.3	178.2	183.4	86.08	177.5	214	177.2	125.9	1992.231
Actual	183	231	199	210	191	320	286	120	138	258	76	0	2212

Forecast for Part No — PEO418

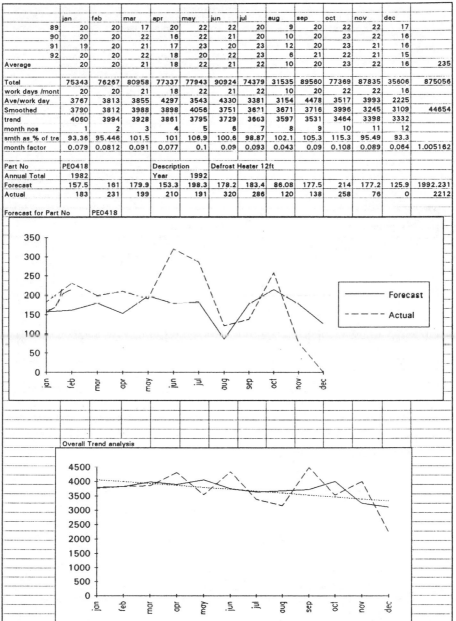

Figure 12.26 (a) Forecasting parameters in typical MRP2.

```
                      HUSSMANN MRP1 SYSTEM - 03/02/93
PART : PE0097               FORECASTING DISPLAY
```

Description :FAN MOTOR 220V 7W 50HZ CCW   UNIVE

	MRP ALLOC	FORECAST	GROSS REQUIRED	PURCHASE DELIVERY DUE	FREE STOCK	PLAN ORDER FOR	PLANNED FREE	PLAN ORDER RELEASE
FEB 93	212	600A	362	1050	1039	0	1039	0
MAR 93	0	600A	600	0	439	0	1489	461
APR 93	0	600A	600	1050	889	0	889	570
MAY 93	0	600A	600	0	289	461	750	120
JUN 93	0	600A	600	0	-311	570	720	0
JUL 93	0	600A	240	0	-551	120	600	0
AUG 93	0	200A	0	0	-551	0	600	0
SEP 93	0	600A	0	0	-551	0	600	0
OCT 93	0	600A	0	0	-551	0	600	0
NOV 93	0	600A	0	0	-551	0	600	0
DEC 93	0	600A	0	0	-551	0	0	0
JAN 94	0	600A	0	0	-551	0	0	0

[RET] - Zoom into month   [F2] - Detailed Info.   [F3] - Purchase Orders

```
                      HUSSMANN MRP1 SYSTEM - 03/02/93
PART : PE0097               FORECASTING DISPLAY
```

Description :FAN MOTOR 220V 7W 50HZ CCW   UNIVE
——————————————— ZOOMED IN TO WEEK LEVEL ———————————————

WEEKS	MRP ALLOC.	FORECAST	GROSS REQUIRED	PURCHASE DELIVERY DUE	FREE STOCK	PLAN ORDER FOR	PLANNED FREE	PLAN ORDER RELEASE
5	58	150	58	1050	1343	0	1343	0
6	110	150	110	0	1233	0	1233	0
7	44	150	44	0	1189	0	1189	0
8	0	150	150	0	1039	0	1039	0
9	0	150	150	0	889	0	889	0
10	0	150	150	0	739	0	739	0
11	0	150	150	0	589	0	1639	461
12	0	150	150	0	439	0	1489	0
13	0	120	120	0	319	0	1369	0

[RET] - Zoom into month   [F2] - Detailed Info.   [F3] - Purchase Orders

Figure 12.26 (b) MRP2 on-line materials planning/forecasting display.

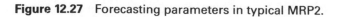

		MONTH
Seasonality Y/N ___	Seasonality pattern 1 – 10	1 Annual, peak in JFMAMJJASOND trough in JFMAMJJASOND
		2 Annual, two peaks, etc.
		3 Quarterly ....
		etc.

Forecast smoothing Y/N ☐
Smoothing factor 1 – 10 ☐ (Preset effects selected by system/inventory analyst)
Display graph Y/N   ☐
Graph style 1 – 5   ☐
Period to display (M)   ☐

Figure 12.27   Forecasting parameters in typical MRP2.

- Design for manufacture was established
- Change control established to limit/plan change to customer orders, with a charge system for change within four weeks of delivery
- Estimating and order interpretation functions merged to give improved identification of order content

However, these improvements were going to give medium- or long-term relief; more immediate answers were needed to provide effective materials planning in the short term. These improvements were to be achieved using specialist MRP1 tools and good inventory management practices

- Inventory accuracy was improved via cycle counting/perpetual inventory using ABC classifications.
- Material planning was moved onto MRP1.
- End-item forecasting was used to derive MRP1 requirements for components where possible.
- Dependent part and subassembly forecasting was used where no end-item data was available or reliable. Auto-adjustment of forecasts against usage was also used.
- Part-level forecasts were used to drive MRP1 and reorder control systems.
- A sophisticated trial kitting module of MRP1 was implemented to give rapid on-line testing of customer orders against LT and material requirements, producing a go/no go and problem parts report.

Trial kitting (Figure 12.28) was a virtual necessity if the master scheduler was to be able to give firm promise dates to the sales function for delivery of product. When the estimator asked for a customer delivery date, the BOM for the job would immediately be trial kitted against the requested date. The process took only minutes to run, with actual times depending on gross number of component items. The materials analyst could then advise whether the job

- could be taken with the requested LT
- would be feasible with specific part exceptions, and their delivery status
- should not be accepted on the specified date, but could be delivered for a (tested) different date

Depending on the priorities of alternative jobs, there may be a degree of reshuffling that could take place to insert the requested job. Similar to a net change MRP1 run for a single or small group of products, the trial kitting module does the following:

Calculates gross requirements for the job or product
Compares gross requirements with availability of components prior to the due date
Where material availability is not okay, checks open orders due and reports on each part
Allows a test job to be moved forward or backward in the schedule to find a feasible slot
Allows auto-allocation of the job if kitted okay
Determines the knock-on effects of loading the test job on other later jobs
Provides-allocation (backflush) facility for stripping out a job
Makes net changes instantaneously to produce shop and supplier documents

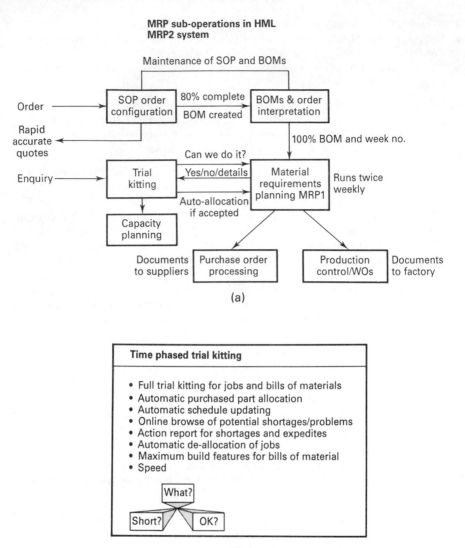

**MRP sub-operations in HML
MRP2 system**

Maintenance of SOP and BOMs

Order → SOP order configuration — 80% complete BOM created → BOMs & order interpretation

Rapid accurate quotes ←

100% BOM and week no.

Can we do it?

Enquiry → Trial kitting — Yes/no/details → Material requirements planning MRP1 — Runs twice weekly

Auto-allocation if accepted

Capacity planning

Documents to suppliers | Purchase order processing | Production control/WOs | Documents to factory

(a)

Time phased trial kitting

- Full trial kitting for jobs and bills of materials
- Automatic purchased part allocation
- Automatic schedule updating
- Online browse of potential shortages/problems
- Action report for shortages and expedites
- Automatic de-allocation of jobs
- Maximum build features for bills of material
- Speed

What?

Short? | OK?

(b)

Figure 12.28 MRP trial kitting operation and facilities.

Material ordering and/or expedite information would be generated at each normal MRP1 run, either net change or regenerative. In effect, this trial kitting is an ultimate net change MRP1, allowing instant 'what if' testing and reporting on any size of batch, job, product or subassembly (Figure 12.29).

In support of these MRP1 tools, new inventory policies for bought-in material were developed. Standard procedures where implemented, such as ABC coding and statistical analysis of monthly usage of bought-in parts was carried out.

```
┌Memory : 642k══════════════════════════════════════════════════════════╗
│(IANW)                    HUSSMANN MRP2   v2.02 - 03/02/93              │
│                          TIME PHASED TRIAL KITTING                     │
╚═══════════════════════════════════════════════════════════════════════╝

        ┌───────────────────────────────────────────────────────────┐
        │                    TRIAL KITTING MODULE                   │
        │   Ignore Parts with lead Time less than Build Lead time - 1 wks │
        └───────────────────────────────────────────────────────────┘


     Kit Part No: KA0001 Quantity : 1                    Build Week: 5
     Part No│Description          │Req Qty│Plan Free│Diff │Short│Exps│Xords│Xrep│▲
     ═══════════════════════════════════════════════════════════════════════════
     PA1640 │PIN FITTING SUPPORT  │   1   │    0    │ -1  │  √  │    │     │    │
     PA1641 │PIN FITTING 10 3/16" │   8   │    0    │ -8  │  √  │    │     │    │
     PA1643 │PIN FITTING SUPPORT  │   1   │    0    │ -1  │  √  │    │     │    │
     PA1644 │PIN FITTING SUPPORT  │   1   │    0    │ -1  │  √  │    │     │    │
                                                                              ◆
  ◄◆                                                                           ▼
                                                                              ►
         ENTER -View Purchase Orders          ESC -Exit
         F3 -Shortages Report                 F4 -Potential Orders Report

     Select ALL / Shortage Parts or QUIT



         ENTER -View Purchase Orders    F2 -Max Build      ESC -Exit
         F3 -Shortages Report           F4 -Potential Orders Report
```

Figure 12.29   (a) Trial kitting report.

Cycle counting was fully employed in the stores. Priority was given to critical parts, high cost items and items which have a critical effect on production capacity. JIT sourcing was developed for a large proportion of items. These and other techniques gave significantly better performance under the modified MRP1 system.

Parallel improvements
To further accelerate the 'sales order to schedule' cycle, which had been impeded by the difficulty in fully defining each order on receipt, a form of knowledge-based front end was developed for MRP1, to function as a new SOP and definition function.

Date: 21/08/92

HUSSMAN
TRIAL KITTING POTENTIAL ORDERS REPORT

Page: 1

Job No: X1078 No Cases: 2

Build Week: 39

Part No	Description	L/T Weeks	Safety Stock Req	Alloc Quantity	Qty Short	Expedite Qty	Date Required	Xord Quantity	Date Required	Xrep Quantity	Date Required	Problem Jobs	Job Qty Alloc	Build Week Date
PA1805	FAN GUARD 6" PAPST LZ 27	5	100	6.00		0.00	/ /	328.00	17/08/92	0.00	/ /			
PC0628	SCREW SET HEX M20 X 100	5	1000	8.00		0.00	/ /	1285.00	17/08/92	0.00	/ /	F1322	9.00	07/09/92
PE0172	CAPACITOR 6MU F GC2429	5	10	2.00		40.00	17/08/92	105.00	17/08/92	0.00	/ /			
PE0685	ELECTRONIC TEMP PROBE 12FT TAILS	5	1000	8.00		0.00	/ /	0.00	/ /	21.00	24/08/92	A0335	40.00	24/08/92
												F1388	4.00	24/08/92
PE1056	SWITCH ROCKER 16AMP DBLE POLE HIGH INRUSH C1350 VQ	9	350	2.00		0.00	/ /	0.00	/ /	60.00	17/08/92	A0335	10.00	24/08/92
												F1690	13.00	05/10/92
PE1321	FAN MOTOR 240V 36W 50HZ 150MM X 55MM PAPST 7855N	5	100	8.00		350.00	17/08/92	0.00	/ /	18.00	17/08/92			
PE2350	FAN MOTOR 240V 24W 50HZ PAPST TYPE 6350S	5	100	6.00		410.00	17/08/92	0.00	/ /	56.00	17/08/92			
PE2360	ELECT TRAY E,D TUS,C M&S LOW-5.5	5	0	2.00	2.0	85.00	17/08/92	13.00	17/08/92	0.00	/ /			
PG0459	GLAZED UNIT DUAL 44 11/16" X24 1/2" SHAPED	5	5	4.00		150.00	17/08/92	67.00	17/08/92	0.00	/ /			
PR0571	CONDENSER BEEHIVE BC3022/110	5	0	4.00	4.00	100.00	17/08/92	86.00	17/08/92	0.00	/ /			
PR0572	COMPRESSOR TECHCUMSEH AK5512E CSR ELECTS 240V 50HZ	5	0	4.00	4.00	46.00	17/08/92	140.00	17/08/92	0.00	/ /			

Figure 12.29 (b) Trial kitting report.

CASE STUDY 12.3

SOP product configuration at Hussmann UK

The existing SOP system was basically a simple order entry screen that allowed the user to enter item type, quantity, delivery date and sales price. All detailed breakdown was carried out external to this system, using reference catalogues, notes and a WP system. A second order interpretation function (OIF) then defined the order technically, preparing a BOM for loading into the MRP1 system; virtually every BOM was unique (Figure 12.30).

The new SOP system was developed to contain logic and rules that would assist an estimator in rapidly selecting all the features and options required for full end-item definition, without the need for later technical interpretation, (essentially NVAs that extended the internal lead time). Whilst not a true expert system, this new SOP module had a large amount of rules built in to guide and route the user towards allowable options at each step of the definition process. It also rolled up costs and sales prices as the definition was built.

This rule-based SOP system started by prompting the user for the end customer. There were approximately ten major customers, each taking different types and finishes of product. Choosing the customer narrows down the options by eliminating all other customers and their options.

This is a similar concept to modular BOMs, except the modules/features and options are presented or hidden from the user under preprogrammed rules. Each successive selection from a range of options builds up the main BOM for the chosen product, and, where a selection would conflict with a previous or additional selection, these are flagged as unavailable, and cannot be selected.

Facilities were also provided to enter non-standard item requests, with estimate costs, to allow for customer needs peculiar to this market sector. These custom items are appended to the defined BOM for later definition by the OID. In summary, the new SOP system (Figure 12.31) provided the following facilities:

- Rule-based user prompts for each selection
- On-line help

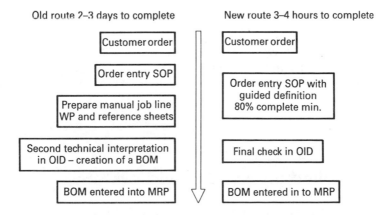

Figure 12.30 Old and new routes to BOM development.

Order configurator

- Automatic BOM part selection
- Job/run/case level defaults by model
- 'Knowledge based' parts inclusion/exclusion
- Pre-defined end user colour schemes
- On screen quote build up
- On screen cost build up (margin display)
- On screen graphical help on part selection
- High level 'sales' BOMs selecting engineering BOMs
- Automatic creation of bill of manufacture for MRP

Figure 12.31 SOP configurator functions.

- Cost and sales price roll-up as the definition progresses
- Selection logic for AND, OR and EX-OR
- Custom item inclusion
- Powerful copy, modify and replace features
- Discount tables and profit margin display

See Figure 12.24 for examples of SOP screen options.

Maintenance of the customer-specific products and logic options was a known overhead, but required substantially less effort than demanded by the previous system. The successful implementation of this system cut out several time-consuming and error-prone sections of work, with a number of tangible benefits:

- Throughput time reduced from 2–3 days to 3–4 hours to fully process an order
- Job lines and BOMs on time for MRP1 runs rose from an average of 75% to over 92%
- Improved BOM and jobsheet accuracy
- Savings in labour through removal of duplicated effort and ineffective practices

An alternative approach would have been to develop a more 'expert' system, such as described in Chapter 9. But this may have taken longer and a short timescale was a prime objective. Expert systems were considered for future developments.

■ Finite scheduling and expert systems

Production planning and scheduling can often have requirements that are too complex for a standard MRP system.

Scheduling technologies and tools

At the sharp end, schedulers allocate work to labour resource, and machines according to customer orders or a demand pattern. They juggle changing due dates and job priorities, shortages of staff, machines and materials, not to mention scrap, rework and breakdowns. Schedulers have few aids; their planning boards and management software are merely extensions to memory and notepad.

In many businesses, there are sizeable savings to be made from improved scheduling. Capacity can depend on the type and mix of job, and the varying work content at any given time. This should be remembered when asking a planner to schedule many different jobs, with changing dates and priorities to meet the customer's requirements. Computerised scheduling may be able to offer the planner more powerful tools and techniques to deal with these high levels of variation and complexity.

Using MRP2 for scheduling

Perhaps MRP2 seems an obvious scheduling tool, but is it? This is a grey area, for several reasons. MRP2 means different things to different people; it also does different things, in different ways, depending on the MRP software chosen and the way it is used within the company.

Typically, MRP2 systems are not good for scheduling shift operations. Some MRP2 systems have very basic capacity and resource planning facilities, which may be adequate for long- and medium-term planning but are completely inadequate for any form of dynamic scheduling in the short-term shop-floor area. At the other end of the scale, a few MRP2 systems, if fully and correctly implemented, provide shop-floor scheduling facilities that include suggested rescheduling and exception reports.

Finite capacity

MRP2 carries out capacity and resource planning on the basis of infinite capacity. It is possible to plan against finite capacity of staff and machines, but this requires entry and regular updating of data concerning actual capacities per time period, time fences, staff availability, absenteeism, etc. Finite capacity planning is much more than just machine capacity. Data needs to be accurate with respect to both detail and time, if the plans are to be at all realistic.

Once accurate data is regularly input to the MRP2 system, we can begin to ask the 'what if' questions we want to answer. This implies frequent running of MRP2, not unusual but often time-consuming. In large installations it can take several hours, so some systems are still run overnight. If the run time is short, e.g. several minutes, the planner will make appropriate changes selected from the relevant reports or screens then rerun the system. It may require several iterations to achieve a good schedule, but this will not be interactive, nor measure the cost of alternative schedules in terms of money, time or effective utilisation.

To provide fast run times for company models with different schedules, it is common to create a separate model of minimal size. This can then be used exclusively for scheduling trials. It is often preferable to link the MRP2 system to a scheduling optimisation package, specifically designed to allow rapid schedule amendments and comparisons.

Scheduling products

Scheduling products range from software packages on personal computers through to multi-user systems across a complete corporation. In most cases the user is required to build a model of the production facility in terms of machines, staff, materials and quantitative data. Once the model (or models) are completed and validated, the scheduler can enter production requirements to see their effect on the available resources. Most scheduling tools offer 'what if' facilities, allowing the comparison of different schedules in terms of capacity, due date performance, cost, etc. Typically, a graphics display of the resources is provided, with Gantt chart information on utilisation and sequencing of jobs. Some scheduling packages are

- Cimitar/Factor (BAe)
- Moses (NEL)
- OPT (Scheduling Technologies)
- Provisa (AT&T Istel)
- Schedulex (Scheduling Technologies)
- Optimiser (Sanderson Computing)

Some newer MRP2 systems, such as MANMAN/X, are now equipped with an integrated scheduling module, removing the need for an additional third-party package to be purchased and linked in (see Figure 12.32 (b)). The illustrations in Figure 12.32 (b, c) are from PROVISA by AT&T Istel. (a) is a view of a schedule simulation as it is played forward in time. The white band across the bar by each machine shows its cumulative utilisation level at that moment in time, and where the area is below the acceptable standards set (c) is the PROVISA planning board display, with a detail screen that allows problems to be traced through the model. This is a typical PROVISA Results window, showing a summary of the number of items that will be early or late in production.

Description and examples of the PROVISA Finite Scheduling Systems are given in a case study at the end of this chapter.

A SYSTEM THAT CAN CHANGE AS YOU DO

When you are running a system as sophisticated and all-inclusive as MANMAN/X, you need the power to control and manage your environment. You also will need the freedom to adapt it to evolving conditions and customer demands.

With three layers of functionality, MANMAN/X Tools allow you to customize your system at whatever level is appropriate to meet your business needs. System changes are language-neutral, and will not impede MANMAN/X's multi-lingual capabilities. Modifications can be made in whichever spoken language or languages your system is configured for.

The MANMAN/X Foundation module contains the facilities required for managing system configuration and security — concerns such as assigning user access to specific functions and determining which printers are available to the system.

Customizer lets you modify screens and reports to fit your needs, and it lets you create completely new formats to satisfy various criteria. You can adjust screen, report, and menu formats for individual users, and make system-wide changes dictated by corporate guidelines. MANMAN/X's Release Control feature ensures that all modifications will be preserved when you move to future system releases.

Developer provides a full 4GL development environment, through which your information system specialists can modify system logic, develop new functional extensions, and create stand-alone applications. Once you define the data dictionary for new applications, MANMAN/X will automatically generate maintenance and reporting programs, which you can then modify to your specific requirements.

YOUR BEST INVESTMENT IN A BUSINESS MANAGEMENT SYSTEM

Only ASK offers you a system with MANMAN/X's raw power, flexible functionality, worldwide versatility, and freedom of choice — all fully integrated into a complete and seamless package. No one else provides you with our all-encompassing support services, or the confidence that can come only from the size, experience, and reliability of an acknowledged industry leader.

Your ASK sales representative will be pleased to sit down with you, discuss your needs and objectives, and show you how an individually tailored MANMAN/X system can position you to take full competitive advantage of all the opportunities that lie ahead.

◄◄◄

With the Planning Board, you can determine the most efficient schedule to build the personal configurable computers. All shop floor operations are graphically presented, enabling you to make changes and perform simulations through simple manipulations of the elements. MANMAN/X's extensive planning and forecasting capabilities give you the power to optimize your materials and resources.

Figure 12.32 (a) MANMAN/X screen display.

Provisa-integrated finite scheduler

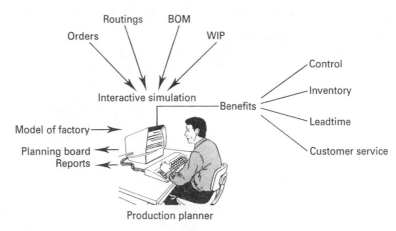

(b)

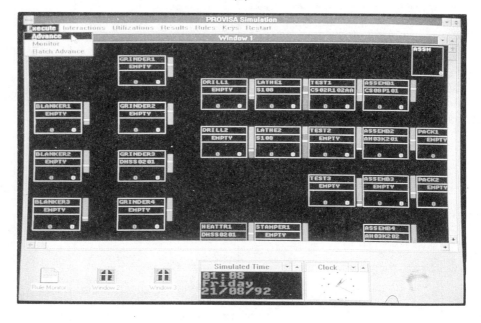

(c)

Figure 12.32 (b) PROVISA scheduling system; (c) PROVISA screen display.

Principles of computer scheduling systems

At the lower end, the scheduling tool may only provide facilities to manually move jobs around the Gantt chart on screen, requiring the scheduler to try out alterations manually. However, there are packages that can automatically perform many of the comparison activities required, often based on predefined rules and goals.

Finite, computer assisted scheduling can be thought of as placing jobs in the optimal sequence on each of the available resources in the company. This requires that a realistic model of the enterprise is constructed and used within the finite scheduler, similar to that developed within an MRP2 system. This contains

Resource information: capacity, overtime limits, shift patterns, sub-contract options, operations, times, set-up times, queue sizes

Job information: priority, due date, operations required, batch/lot sizes

Next, define the rules and priorities within the scheduling system, for example:

1. Schedule not to exceed 10% overtime
2. Schedule to fit in all priority 1, 2, and 3 jobs
3. Reschedule only to accommodate priority 1 jobs

Using rules such as these, a scheduling tool will attempt to satisfy the programmed rules, in order of priority, by moving and rescheduling jobs on all appropriate resources. This usually involves shuffling jobs to minimise and flatten the loading on particular work centres to levels within the available capacity.

Scheduling systems will normally link directly to an existing MRP2 system, allowing data to be passed back and forward for scheduling activities, and for monitoring progress using the MRP2 shop-floor systems. The creation, entry and maintenance of the database for a scheduling system is a major task, and will often require substantial resources for a lengthy period of time. Once loaded, the data and rules require testing to ensure validity and usability within the scheduling environment. Parallel testing and line-at-a-time implementation is strongly advised (Figure 12.33).

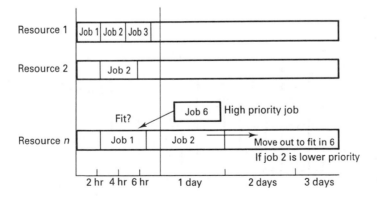

Figure 12.33 Basic outline of scheduling jobs across resource capacity.

A basic difference between this type of computer assisted scheduling and MRP2 capacity planning is that where MRP2 highlights and advises the planner of overloads and the need to replan job priorities, tools like PROVISA and Optimiser automatically carry out a large proportion of the rescheduling driven by the programmed planning and priority rules (Figure 12.32). The planner is required to make final adjustments and to accept a proposed schedule.

Figure 12.34 indicates the type of job movement that would be automatically undertaken by a scheduling tool, provided this did not conflict with a second rule to complete job F100 by the end of period 1. If so, an exception may be generated, or a second set of rules brought into play, and other lower priority jobs moved out to later periods.

The major take-up of scheduling systems has been in the process industries, such as chemical and refinery operations, where rapid scheduling and rescheduling of complex product mixes is common and cycle speeds are high. There has been some penetration into discrete product sectors, although there is often a need to simplify the process and product before using this level of computing power to solve a complex problem.

Again, the application of one of the available scheduling systems is not going to be of value unless the management and culture of the enterprise is attuned to its operation and use. The complex task of scheduling is often poorly understood by senior management, who view the production and scheduling functions as the root cause of late deliveries and poor customer service. A scheduling tool requires discipline and understanding from all those involved.

Optimised production technology

Optimised production technology (OPT) is more than just software; it comes with an alternative approach to production and inventory control, an approach that can be taken up with or without the software products. OPT is the creation

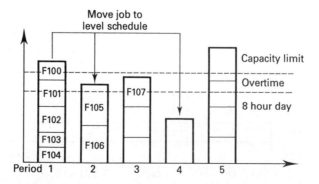

Figure 12.34 Capacity and loading profile for a resource.

of Eli Goldratt, who developed rules for identification and control of bottleneck resources within a facility, tying production to the speed of these constraints. The OPT drum, buffer, rope and capacity constraint management approach is outlined in Figure 7.5. The rules of OPT fully explain the principles of bottleneck management and can be extensively applied to many production facilities with very worthwhile results.

The OPT software goes deeper still in identifying and controlling bottlenecks, and is intended for retrofit to existing manufacturing information systems such as MRP2. OPT software would provide the shop-level control and scheduling mechanism. At around £100k, this could be viewed as a fairly expensive route, particularly if the company has already invested a similar or greater sum in MRP2. A new PC-based version of OPT is available, potentially offering OPT analysis and scheduling facilities at a lower cost and incorporating improved control logic.

Expert production scheduling tools

When a scheduling package has to select from several different rules and priorities, expert systems can offer an additional dimension. Expert systems are rule-based computer systems with the knowledge and reasoning ability of a human expert (or experts) in the chosen field. When applied to production scheduling, expert systems can contain knowledge and rules relating to particular products and/or parts and their effect on the shop floor. For example, the system could contain a set of different batch sizes for different shop-floor loads and part mixes. The expert system would analyse current and planned production loading, then apply selected rules and strategies to the problem.

The expert scheduling system uses two main sets of information: the production database and the knowledge or rule base. The production database contains static data on

- each part
- each product
- routing
- resource needs (labour and plant)

Dynamic data is held and loaded for analysis against

- job priority (against other jobs and/or due date)
- current schedule and work in process (WIP)
- machine and labour status (variations)
- allowable tolerances and variances in the above

The knowledge or rule base contains all defined scheduling rules, elicited from scheduling experts, management and any other appropriate source. Typically it will contain

- optimal line and product loading mixes in various situations of capacity and material status
- ranked desirable and undesirable mix combinations
- rules for maintaining or amending an existing schedule

From the results of processing all relevant production data with the knowledge rules, the expert system can run and compare several iterations using different scheduling strategies (Figure 12.35). The scenario that ranks highest against a predefined set of measurements is indicated to the operator, who can then execute it. The system also gives reasons for its selection. The plan is executed by transferring the control data to the company MRP2 system. The expert system can also learn from each operation, storing results as solutions to particular scenarios, for later use. Examples of scheduling systems that contain expert systems are DAS (Strathclyde University) and KEE (IBM).

MRP2 Expert Analyser

A very different but equally valuable application of expert systems to manufacturing is the *MRP Expert Analyser*; developed by the BRIM Group. This is a 'universal' performance analysis software package for attaching to a range of MRP Systems, with the aim of improving the management and materials control of the target system.

With DTI support, they have developed a fully functional expert system for assisting in the material planning and analysis functions of MRP in manufacturing operations. This was inspired by the findings that many MRP

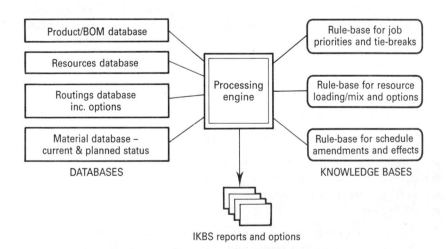

Figure 12.35 IKBS and scheduling tool operation.

systems have relatively poor, unchecked configuration of key parameters for the MRP materials control, often resulting in poor performance of the system in total, due to shortages and/or high inventory costs.

It was found that few Materials managers actually had time to patrol the system operating parameters, often leaving these tasks to subordinates. Most MRP system applications have a fairly large parts database to be controlled (thousands of part nos.) by few staff, working near to capacity. The material planner's job involves a variety of tasks which often results in inadequate time being available for MRP analysis and fine tuning, i.e. for

- management reports
- fine tuning the system – its parameters and exception reports
- defining and redefining MRP control parameters for each part

These are areas where more 'effort' and intelligence/assistance was seen as necessary. These task areas are the core part of material planning, but are also very time consuming to do, which can in practice mean they are not carried out fully or well enough to gain the potential maximum benefits from material planning and control.

It is here that the *MRP Expert* system has significant benefit, as it can

- integrate to the existing databases in several MRP systems
- interrogate the way each part is classified, coded and controlled in MRP terms
- analyse past and forecast part control performance based on certain criteria
- compare/evaluate alternative MRP scenarios for each part, giving ratings
- provide recommendations for amending part control parameters, with reasons
- run on demand, or automatically at set intervals.
- have selectable options for the user to set and adjust to alter how the system considers each part and control method.

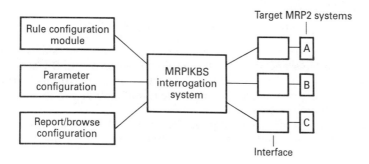

Figure 12.36 MRP Expert.

Modules

Data fields and parameters include:

- Part number
- Class A B C etc.
- MRP control category LFL, ROP, consuming forecast, etc., JIT
- Lead time
- Order interval
- time fence
- average usage/various
- standard deviation
- forecast and smoothing factors
- safety stock
- stock performance over last n months etc.

Facilities

Within the *MRP Expert Analyser*, there are facilities that include: parts within/outwith n% of forecast, forecast inaccuracy and/or adjustment recommended, parts with stock/s. stock in excess of/below 'm' (which will indicate over-stocking or under stocking/stockout potential.) incorrect part classification – ABC, recommendations for part review based on various criteria, etc.

The analyser is configurable to suit particular sector and company needs, is Windows based, and easily linked to a range of MRP systems. It provides management with a powerful tool to assist in improving materials control, and has a growing user base from various sectors.

(The BRIM Group is a leading UK supplier of manufacturing and business control systems, including *MRP Expert* and *MANJOB*, a make to order/sub-contractor business management system. The Group is based in Glasgow.)

CASE STUDY 12.4

PROVISA scheduling in an aluminium manufacturing plant*

Aluminium Corporation (ACL) is a specialist producer of non-standard aluminium rolled products and is currently within the Enterprises Division of Alcan Aluminium (UK) Ltd. It has production facilities at Dolgarrog in North Wales including a casting operation, hot breakdown mill, a wide range of cold rolling mills and finishing equipment.

The company produces a large variety of products to order, for example, circles for saucepans and light fittings, sheet for satellite dishes, patterned

---

*Courtesy of AT&T Istel.

sheet, superplastic alloys and high purity products. The operation is necessarily one of batch production and a number of process routes through the factory can be identified (Figure 12.37(a)).

In the late 1980s ACL's management team began to make a critical appraisal of the way in which orders were scheduled through the plant. In addition to high work in process (WIP) the company faced the problem of high manufacturing elapsed times as a result of the inherent queuing in the production system. The decision was taken to purchase the optimised production technology (OPT) scheduling software and to incorporate the bottleneck management philosophy into production management and control within the factory.

OPT implementation

The first step was to develop a simulation model of the factory and its manufacturing processes. On this basis a number of critical processes

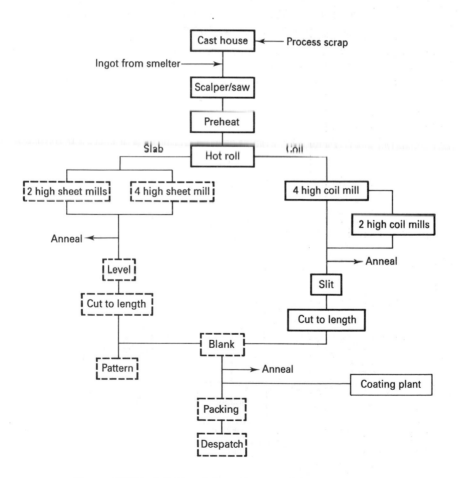

Figure 12.37 (a) Aluminium Corporation process route.

(bottlenecks) were identified which determined the output for the factory operations as a whole. In terms of scheduling, the key was to load orders to the critical resources which would maximise their activity, whilst pulling work 'as required' through preceding activities with excess capacity. This ensured a continual flow of work but avoided WIP build-up in front of non-critical resources. The OPT software, in scheduling orders, could also limit the total amount of work in circulation in the factory at any one time, and so exert a positive influence on manufacturing elapsed time.

The second initiative, linked to the company's programme of manufacturing excellence (Figure 12.37(b)) entitled Gwella (Welsh for improvement) was to seek to increase output, improve machine reliability and product quality at the critical operations. After all, any improvement in productivity at the bottleneck would result in improved productivity for the system over all. In some cases the resultant improvements had the effect of turning critical into non-critical activities and in a sense moving the bottleneck.

In addition to the direct benefits of a 50% reduction in WIP and manufacturing elapsed times, ACL also realised the following as a consequence of the new approach to manufacturing management:

■ Improved product quality through process improvement groups, reduced handling and storage damage.

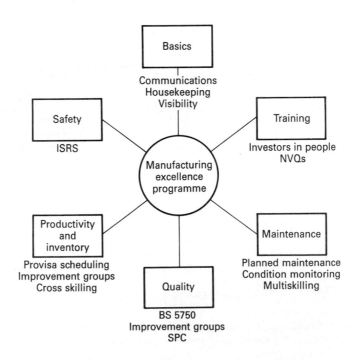

Figure 12.37 (b) Manufacturing excellence programme.

■ Improved customer service in terms of delivery lead times and on-time delivery reliability.
■ More effective maintenance with resources focused on maximising uptime on critical resources.

By the end of 1990 it was becoming apparent that the ability to further reduce inventory and manufacturing elapsed time was being limited by the modelling capability of the OPT software, in particular in the area of continuous preheat furnace modelling. The company set itself a new objective of further reducing manufacturing elapsed time by 50% to four days, and carried out a further review of scheduling software packages to assess their capability in achieving this measure.

Whilst maintaining the philosophy of bottleneck management (i.e. finite capacity scheduling) the ideal package would also be

■ more user friendly in terms of displays and reports
■ require less processing time for scheduling runs
■ allow 'what if' simulations within real time to allow the effects of changes in production routings to be evaluated before any process batch was committed to the shop-floor.

PROVISA implementation

In April 1991, after an evaluation period lasting six months, ACL had shortlisted three packages as being capable of dealing with its particular scheduling difficulties. There was a need to cope with a constantly changing product mix, which could demand alternative routings for the same product, dependent on machine loading. Machine set-up times and sequencing rules, particularly relevant to rolling mills (in terms of gauge, width and temper variation), had to be modelled satisfactorily. There was also an unspecified but recognised need to manage labour availability, in terms of variable shift patterns, skill categories and the facility for overtime, since under certain product mix conditions production labour itself was a critical resource.

The package which was ultimately chosen was the PROVISA scheduling package from AT&T Istel based in Redditch. PROVISA was selected following a workshop period during which the capability to accurately model all constraints within the production process was proven. In addition to meeting the user requirements listed previously, PROVISA offered multifunctional scheduling capability based not only on customer due date (the usual sort rule) but also taking into account set-up losses, customer priorities, user rules, etc., in a fast and easy-to-use database. Importantly, it allowed interactive simulation for the scheduler to view the schedule, stop it and edit or amend rules as necessary. This promoted user confidence in the decision-making processes at each stage of scheduling, greatly enhancing the user friendliness over black-box report generators.

The PROVISA finite forward scheduling package requires certain inputs from other standalone or integrated computer systems in terms of orders, WIP, bills of materials (BOM) and routings giving manufacturing control over such derivatives as inventory, elapsed time and ultimately customer service in terms of lead time and due date.

At ACL these inputs are supplied from a PC-based Novell Network system using an Advanced Revelation database (Figure 12.37(c)). The ACL manufacturing system uses Provisa (box 5) to output machine schedules to the shop-floor, determining the optimum processing sequence given variable plant workloads and capacities to achieve due date. Allocation of crude plant and process capacities in terms of due week to customer is done separately at a load balancing stage called rough cut capacity planning (RCCP).

In order to understand the success of Provisa in reducing WIP inventory and manufacturing elapsed time, the complex production routings outlined in Figure 12.37(a) can be simplified. Initially there are two production routes for cold rolling and two finishing routes, which normally determine production capacity (bottlenecks) and require inventory buffers. Before a process batch is launched into the plant, Provisa checks overall constraints such as due date, availability of raw material and a suitable preheat furnace schedule. An order will then be released subject to a dependency matrix whose values are set by the levels of inventory (both current and simulated) at five different stages in the routing

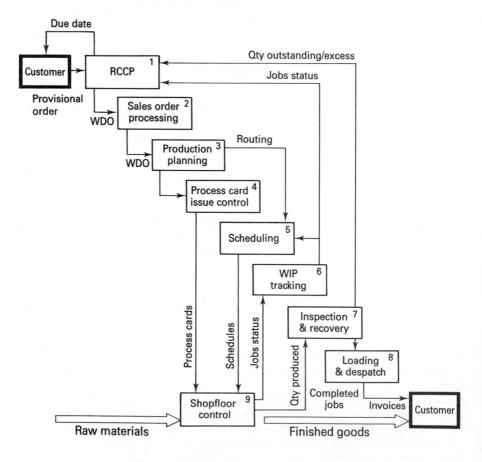

Figure 12.37 (c) Scheduling function.

(Figure 12.37(d)). Unnecessary WIP queues can be avoided simply by releasing an order into the plant as long as certain critical inventory levels will not be exceeded.

The main output screen from the simulation package is a Gantt planning board configured to show customer orders by batch number and their lateness by bar colour. This readily provides the scheduler with a visual image of machine utilisation and potential customer service problems. Other outputs are histograms of labour utilisation, and product throughput and manufacturing cost predications.

At ACL the availability of this data on a daily basis to the manufacturing director, production managers and production controller allows them to utilise their combined flexibility to expand resource capability (e.g. overtime, additional labour, alternative routings) before a predicted capacity problem manifests itself as a queue in WIP.

Aluminium Corporation began live running with the PROVISA system in June of 1992 and has seen a considerable improvement in its key success

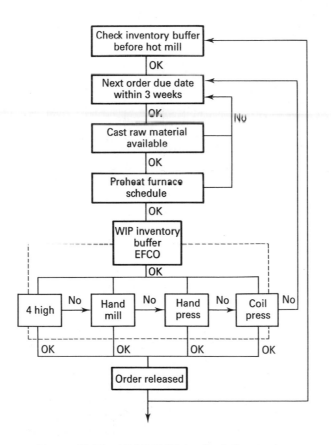

Figure 12.37 (d) PROVISA scheduling logic.

parameters of total inventory and manufacturing elapsed time. Inventory levels are now down to 685 tonnes with similar weekly outputs to those achieved using 1,985 tonnes prior to OPT (Figure 12.37(e)). Manufacturing elapsed time has similarly reduced from fourteen days to less than four (Figure 12.37(f)) and in the same time direct labour productivity has increased by 15% due to better matching of resource to product requirement. These improvements are just part of an overall manufacturing excellence programme which has allowed a rolling mill of 10,000 tpa capacity to survive in the Conwy Valley (when European competition has production:scale economies an order of magnitude greater) from where they expect to be rolling out the schedules for several more years.

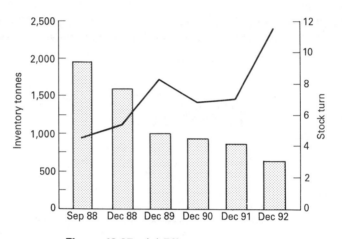

Figure 12.37 (e) Effects on inventory.

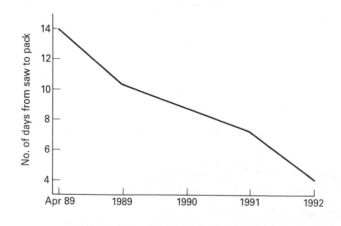

Figure 12.37 (f) Effects on lead time.

CASE STUDY 12.5

Furnace scheduling using expert systems*

Figure 12.38 shows a knowledge-based scheduling system developed by Standard Oil and Pfalder Balfour to link into their on-site MRP2 system. It proved very effective at applying operational rules to the scheduling processes.

Summary

A rule-based, forward chaining expert system was developed to schedule a multipass glassing and furnacing operation for glass-lined vessels. Due to the nature of the manufacturing process and unexpected disturbances, such as rework, the fabrication time of each part cannot be tightly controlled. This results in the need to schedule the glassing and furnacing operation manually. But the large number of parts, allowed part mix, and product constraints make manual scheduling very tedious and time-consuming.

The furnace scheduling advisory system (FSAS) captures the expertise of the shop-floor supervisor and uses heuristics to account for resource constraints, such as availability and capacity of the furnaces and availability of firing tools, and operational constraints, including allowed part mix, firing temperature, and allowed thickness difference. The dynamic schedule generated by the expert system is in a heuristic sense the best solution to meet delivery date requirements, optimise utilisation of multiple furnaces and minimise energy consumption.

Comparison of the actual manual furnacing schedule with the furnacing schedule generated by the expert system using a month of data reveals that significant improvement in furnace utilisation can be achieved with the furnace scheduling advisory system. In addition, we expect significant improvement in on-time delivery, energy savings and staff savings. The heuristic optimisation and scheduling strategies used in the furnace scheduling expert system can be extended to scheduling of any batch manufacturing production process.

Figure 12.38(a) illustrates the overall manufacturing process of the glass-lined vessels discussed in this report. The process consists of several major operations: fabrication, sandblasting, glassing, furnacing, and final assembly. Each vessel may include vessel body, cover, agitator, access hole, supporting legs and other miscellaneous items. All parts initially go through the fabrication process then each part requires different machining operations, which take different amounts of time. Parts are accumulated in a temporary storage area.

The expert system schedules sandblasting, glassing, and firing for all the parts. Parts are scheduled for sandblasting only when there is sufficient time for the firing schedule to accommodate the sandblasted parts. All parts undergo several glassing coats, each of which has to be fired in a furnace. Some coats can be fired in more than one furnace. A number of parts can be fired in one furnace at the same time provided they have the same temperature, firing cycle and meet certain thickness requirements. Each part is classified according to its type and weight. Furnace loading is governed by the preferred combinations of the items of different categories.

---

*Extracts from *A Scheduling and Planning Expert System for Multiple Furnaces* by J. Hall, F. Litt *et al.*

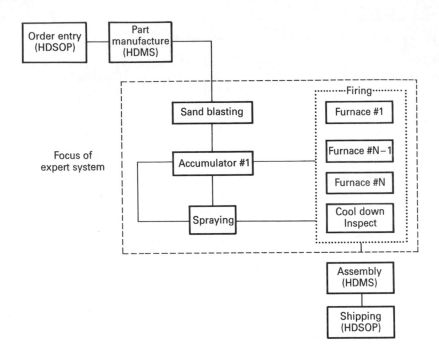

Figure 12.38 (a) Furnace scheduling expert system.

Knowledge acquisition

The expert system was implemented in DEC OPS5 on a VAX 785 and then transported to a Micro VAX II. The system was also tested on an IBM PC running the Common LISP version of OPS5. The rules were generalised to cover any furnace. This meant the parts data set was very large and many working memory elements could match one rule. The conflict set also became very large, to provide system flexibility, increasing the execution time. Extra rules were also required because OPS5 permits only one vector attribute per element class. Thus, in cases where several vector attributes were needed, several data structures had to be implemented. Other minor disadvantages of OPS5, which made things more difficult, were unavailability of COMPUTE (except for matching existing working memory elements on the left-hand side of rules and logical OR (except for constants). Again, extra rules were required, slowing down the system. Use of external routines callable from OPS5 could solve this problem.

A pilot test was done using actual production data collected for one month. Results of the scheduling obtained using the expert system compared very favourably with the manual scheduling. There was a significant productivity improvement. The throughput and number of completed parts were higher and more parts were closer to the delivery date under FSAS than under manual

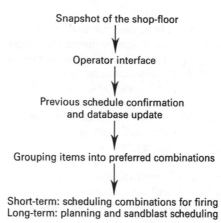

Snapshot of the shop-floor

↓

Operator interface

↓

Previous schedule confirmation
and database update

↓

Grouping items into preferred combinations

↓

Short-term: scheduling combinations for firing
Long-term: planning and sandblast scheduling

Figure 12.38 (b) Flow diagram of overall expert system strategy.

Furnace:	8
Batch number:	2
Combination:	D
Firing code:	N following N
Firing code:	900
Priority:	1
Combination ranking:	4

Thickness:	18 mm
Total weight:	7,400 kg
Starting time:	1987/4/1 2:38
Preheating time:	0 min
Firing time:	1 hr 45 min
Handling time:	20 min
Completion time:	1987/4/1 4:43
Next Ready time:	1987/4/2 16:43

There is no combination with higher priority or higher
combination ranking. Also, this firing sequence has better
energy utilisation by minimising the heating and
cooling of the furnace

Object ID	Object type	Coat type	Category
SB00011V	Vessel	G1	D
SB00627_42	Vessel	G1	D
SB00627_41	Vessel	G1	D

Figure 12.38 (c) Example of a scheduled batch.

scheduling. The main reason is that FSAS looks for an optimal set of
combinations to schedule and thus limits the very large set of feasible solutions.

Ⓒ

Conclusions

The rules were generalised to cover any furnace. This meant the parts data set was very large and many working memory elements could match one rule. The conflict set also became very large, increasing the execution time.

The expert system for furnace scheduling is modular and flexible. The rule base is so general that it does not have to be modified when the user needs to add categories or combinations, change calculations of firing times, add glassing sequences, change times allocated for sandblasting, change numbers of firing tools, change temperatures associated with glassing coats or change rankings of allowed combinations. And the system is easy to maintain because the rules are clustered according to context. FASA can evaluate a very large set of feasible solutions in a short time then recommend in a heuristic sense, the best solution to the scheduling problem. Significant improvement in furnace utilisation, on-time delivery, energy savings and staff reductions are expected.

■ Summary

What was essentially a tool for manufacturing and material control has now grown up to become the most comprehensive business planning and control software. Where conventional packages for accounts, sales and purchases stop, ERP systems carry right on. Offering configurable and selectable modules for virtually every business need, ERP is a desirable and essential tool for world class business performance. Where the enterprise is in the manufacturing sector, improved features and options provide even greater planning and control facilities, combined with additional specific software tools for detailed scheduling and product configuration.

■ Further reading

Goldratt, E. (1988) *The Race (OPT Principles)*, Productivity Inc.
Oliver Wight Co. (1981) MRP2: Unlocking America's Productivity Potential.
Peterson, R. and Silver, E. A. (1991) *Decision Systems for Inventory Management and Production Planning*, Wiley.
Wallace, T. (1990) *Manufacturing Resource Planning 2: Making it Happen*, Oliver Wight.
Wantuk, K. (1989) *JIT for America*, The Forum.

Also look at the *Knowledge at Work* newsletters produced by the IKBS group of the DTI.

13 | A Guide to Successful MRP2

> *MRP(2) is a good system, but doesn't always meet our expectations . . . because it accepts our manufacturing processes as if they were right, but are they?*
>
> Kenneth Wantuk, 1989

■ Introduction

Making MRP2 systems work effectively remains a very difficult task, usually unrelated to the actual hardware and software. Adequate preparation and careful implementation within a total quality philosophy will enable the potential benefits to be realised.

This final chapter addresses two main areas: the considerations and steps involved in implementing an MRP2 system, and the key elements or prerequisites for successful system operation. While discussed from the point of view of implementing MRP for the first time, this material is equally useful for rejuvenating an existing system that is not performing up to expectations.

■ Management considerations

Many MRP2 installations fail – why?

There have been many MRP2 installations over the past twenty-five years, which have met with widely varying degrees of success. Many have been studied carefully and the reasons for success or failure analysed. While each company is unique in its methods of operation, it is nevertheless clear that there are a number of basic principles common to all users of MRP that must be adhered to if the system is to work.

Experience has shown there are certain methods for implementation and operation that lead to rapid success. Conversely, if these methods are ignored or violated, the system results will be degraded, and its operation will become an exercise in futility and wasted effort, leading to possible failure. Overall, the two major and most consistent elements contributing to a successful program are the *commitment of the top management* and the level of *dedication to company-wide education* and involvement in the system.

A majority of MRP2 implementation attempts may be seen to have failed, or fallen well short of the expected benefits, because these people, policy and procedure aspects were largely ignored. An MRP2 system is just a tool that can be used extremely well or extremely badly.

However, success or failure cannot be determined unless we have measured the progress against objectives. Performance goals have to be determined in a quantifiable way so the project has something tangible to achieve.

A B C D classification

MRP2 systems have been formally classified into categories of measurable implementation success. Classes A, B, C and D, can be used as a guide to setting goals and as a way to measure the degree of success.

Details are given later in this chapter of how these measures are defined and calculated.

Note that MRP2 success is also closely related to the effective operation of all functions, requiring them to be correctly organised and operated through procedures and policies that use the capacity and power of an enterprise-wide system in an effective manner. **It is clear that achieving world class performance will normally require Class A MRP performance as a key component.**

Studies have shown that while 50% of the *cost* of MRP2 is due to computer hardware, software and the consulting costs associated with implementing the software; only 25% of the eventual benefits can be attributed to the computer path. Of the benefits, 75% will arrive via the management path which has educated, trained and reorganised the people in new methods and procedures established by management to make the manufacturing process more effective

---

Class performance characteristics

A	Complete closed-loop business planning system. Top management use the formal system to run the business. All functional areas average between 90 and 100% in measured effectiveness.
B	Formal systems are in place but not all functional areas are working effectively. Top management approves of the mission, but may not actively participate. Functions average 80 to 90% effectiveness.
C	The MRP1 manufacturing schedule is not well planned; functional areas are not working together well. Several subsystems are not in place. Functional areas average 70 to 80% effectiveness.
D	Formal system is not in place or not working. Poor data accuracy or integrity, little management involvement and low employee confidence in the overall operation of the business. Functional areas are 50% effective or less.

(Figure 13.1) This matches the findings of Chapter 9. **You don't install MRP2 or ERP, you implement them as a different way of running the business.**

The first thing we must do when developing an MRP2 project plan is to consider it as having two distinct pieces or sections, the computer and systems section and the management and operation section. The computer section provides tools and processing power to speed up and automate the handling and presentation of data. This is particularly important here since it is frequently obvious in at least the initial stages of project planning that most people think of MRP2 simply as the implementation of a computer software system. This attitude is still encountered in many installations.

Why projects fail

MRP2/ERP systems are (computer-based) planning and control systems which track the total resources of a business or manufacturing organisation to improve results in meeting that organisation's business and financial object-ives. The key to success is planning and control. Projects fail when top management and middle management are not enthusiastically involved throughout. There are several fundamental reasons for this.

Lack of clearly defined objectives

Our aim is not to install an MRP2 system, instead it is to lower our inventory costs, increase our inventory turns, improve productivity or improve customer service. But it needs to be clear; vagueness leads to failure.

Lack of total involvement and/or commitment

If only the project team is serious about the success of the project, it's doomed to fail from the start. MRP2 is a very complex procedure which impacts on almost every functional area of our manufacturing operation. It requires serious involvement and willing co-operation from all levels of the organisa-tion. We must all be prepared to climb out of our respective ruts and seriously

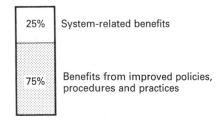

Figure 13.1 Proportion of benefits from MRP2.

consider whether the way we've been doing things over the past twenty years or so is appropriate to carry us into the next century.

Lack of leadership

A post-mortem on a failed project often reveals that no single person had overall responsibility for the project. Someone was given the nominal title of project manager but not the authority to effectively manage it. Project management by committee can also be a common problem, resulting in constantly changing goals, switched priorities and a project that slowly dies.

Lack of structure in the project plan

Many projects get into trouble because critical tasks are overlooked in the planning phase or are not scheduled in the proper sequence. The project plan must be well structured so it is more than just a long list of tasks with no discernible interrelationships. We should not overlook the role of microcomputers in the development and subsequent implementation of an MRP2 plan. Project planning software on a micro can introduce structure and establish the interrelationships between tasks.

Lack of detail in the plan

There will always be several unknowns at the beginning of this type of project, and several tasks won't be apparent until an unforeseen problem arises, e.g. lack of availability of a key piece of data or when to schedule a task with no idea of its timescale. Breaking down each major task into its subtasks and planning them back from each defined milestone will normally result in the level of detail necessary for success. However, it must also be flexible enough to be easily modified as unforeseen tasks and/or schedule changes develop. Microcomputer-based project planning can handle the knock-on effects of these changes.

Lack of financing and cost overrum

We will be required to establish an estimated total cost of the project in our capital plans to head office. Significant overspends can occur without careful budgeting. There are several unknowns at this point, so a contingency should be built in. Whether at the time or in advance, management must decide how to correct overruns early enough to prevent major damage to credibility and eventual success. Back to the project management again. A detailed budget and spend record must be kept throughout the life of the project.

Lack of resource commitment

Resource allocation of time, staff, equipment and space is vital. Critical to success is dedication by top management, the people with authority to allocate the resources. This probably means accepting short-term pain for long-term gain in some instances and the application of strong persuasion at other times, to secure the co-operation of staff who may be reluctant to give enough of their time to the project.

Lack of tracking the project

After the project starts, planning stops. No, for any project to work, feedback from the people involved is needed to compare real progress with planned progress. It may be necessary to change the plan to reflect the real situation. Without feedback, the plan will become out of step, counter-productive and eventually fail. Of primary importance is the establishment of a feedback mechanism for the project team to use in reporting actual data on a project: start dates, completion dates and other actuals, those amounts of work actually executed on tasks.

Lack of communication within the project team

Chaos results when members of a project team neglect to inform each other of the status of their work. The project manager must ensure that team members have the information needed for their part of the project, e.g. changes in plans, shifting of responsibilities, progress reports and unantici pated problems. Regular meetings – daily to weekly – between manager and team are a basic necessity.

Lack of continuing direction

The project can stray from it's original goals. Despite starting a project with clearly defined objectives, it will often stray because of changing circumstances, altered allocation of resources or the impact of a strong personality in a functional area insisting things be done a certain way. Any major redirection of the project must be analysed for impact on the original objectives and how the new objectives are related to the original plan. It is the steering committee's responsibility to ensure the project remains on track, with the correct objectives. The project team regularly reports to the steering committee and receives direction from it.

Lack of accurate data

The lack of an accurate database will always result in errors and ineffective use of the systems, leading to poor running of the business. Data inaccuracy will greatly nullify the benefits of any integrated, formal system.

Success stories: Hussmann Manufacturing

In the late 1980s, harsh economic realities forced change on Hussmann, a Glasgow manufacturer of commercial refrigeration equipment. 'We were both the source and the solution to our problems. To turn it around, we had to take control of the way we managed our people, information and other assets. This led us to class A MRP2,' recalls Tom Whyte, director of quality. In July 1991, Hussmann management drew up and signed a statement of commitment to achieve class A MRP2 by August 1993. To reach class A within two years requires concentrated effort, but Hussman made it (Figure 13.2).

Performance measurement focuses on twelve functional areas: business plan, sales plan, production plan, shipment plan, master schedule, materials plan, bill of materials, inventory control, purchasing, shop-floor control and delivery performance. 'For example, when we started this system, our delivery performance was only 70 or 80% on time,' says Whyte, team manager for the project. 'Our approach at that time was to deal first with our biggest customers, or whoever shouted the loudest.'

When Hussmann began measuring delivery performance, management

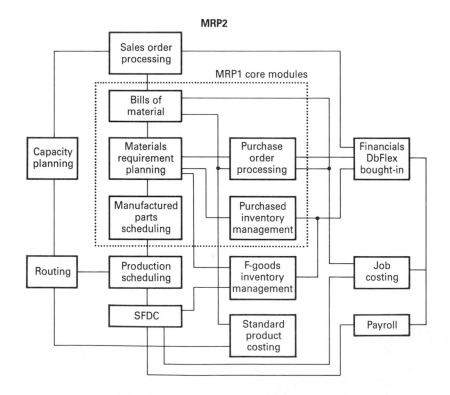

Figure 13.2 Hussmann MRP2 system.

witnessed rapid improvement just by focusing on the problem; first to over 80%, then to 90% delivery to the hour, based on the delivery date requested on the customer order. Within a year, Hussman consistently shipped over 98% on time. Materials planning shortages, first measured at 58% accuracy in September 1990, quickly rose to 85% by December 1991, following implementation of MRP1, and to 97% in late 1992.

Throughout the project all of the employees involved were exposed to appropriate education programmes, ranging from a minimum three hours per person to a maximum of forty hours. Other communications media were used, including project newsletters and prestige commemorative items, including MRP2 pens and mugs.

Project team members with responsibility for key areas of the system received even more extensive outside education. Whyte recalls

> If there's one thing we learned, it's that you can never underestimate the importance of education and involvement for all the people on an ongoing basis, both during the project and afterwards. Instituting Class A MRP2 was our highest priority and has proven to be the saviour of Hussmann. It is a lot of hard work to get there. Furthermore, if you are fair and honest, you will uncover a lot of problems you won't like, and you must just deal with them. They won't go away without work. But all the things you are working on will make you a much better company. *You'll see the results in performance, productivity and profits, together with improved morale across the company.*

Hussmann also created a tougher set of class ABCD measures than is normal, with the breakpoints set higher (class A was 85% overall rather than 80%) This was done both to make the final class A target a little harder, but also to adjust the C/B threshold to be a meaningful target in the early years. Hussman has successfully maintained these high levels of performance, and profits have grown significantly over the life of the project.

Success stories: Polaroid

In 1988, Polaroid formed an MRP2 implementation team led by the managing director. After an extensive education programme, the team launched the project for class A, working its way through achieving accuracy in inventory, bills of material and forecasts

'We realised the importance of management commitment and education,' remembers Jim Hall, consultant member of the original team and then made director of manufacturing. The project team attributes its success not to the system but to the people within the company.

Once the class A project was under way, progress was rapid. When the first performance measurements were taken in January 1988, inventory accuracy was 60%; by 1989 it was 99%. Since then it has not dropped below

97%. At the end of 1990 Polaroid was performing at an average of 97% accuracy in all key areas, arriving at class A in two years.

Management commitment

Education will help top management to understand and subsequently define how MRP2 will support the overall business strategy. It will allow them to set and align the project objectives with the business goals, avoiding the frustration of not knowing where you are going with MRP2. Top management must communicate this throughout the company by leading project launch sessions and contributing to the project update sessions and news-sheets. This will secure commitment and transfer enthusiasm and support throughout the company. Top management commitment also means support and acknowledgment of change if it is right for the company.

MRP2 is the ideal vehicle for change if top management creates the correct environment. During the implementation process, the need to change the way the business is run, both inside and outside the application of MRP2, will undoubtedly be highlighted. Top management must grasp this opportunity if the implementation is to avoid the classic mistake of simply automating the way the business is currently run.

Do not attempt to develop a fully detailed plan covering the complete implementation timescale, as the required tasks for a successful implementation will emerge as you progress.

Successful MRP2

Many companies fail to understand the opportunities presented to them from implementing MRP2. Often seeing MRP2 as 'just another piece of software', managers simply automate the way the business is run. Companies with vision realise that MRP2 is an excellent vehicle for change and a foundation on which to build utilising techniques, such as JIT and TQM, on the journey to becoming world class.

Without doubt, implementing MRP2 is one of the largest projects a company will undertake. Many companies are enjoying significant benefits, but there is a great risk of failure or underachievement. Getting it right from the start requires a company to have Management commitment, a strong project organisation, an achievable implementation plan and educated people.

MRP2 education and training

A properly executed education and training programme will build enthusiasm and confidence, help people to identify with the benefits of implementing MRP2 and allay their fears about their new roles within the company. It will develop an environment for change and create a workforce hungry to learn more. Ultimately it will correctly assign responsibility for making it happen, and ensure the optimum benefits from the MRP2 process.

Educated people are essential to successful MRP2 implementation. Always remember that MRP2 makes it possible but it is your people who will make it happen.

Moving from informal to formal operation

Achieving Class A performance takes far more than just measuring current performance and saying, 'Now lets all work harder to get better.' It is a process of determining objectives, assigning responsibilities, measuring performance, identifying and solving problems to improve results. Through this process a company can develop a much improved system of standards, expectations and accountability, with responsibilities clearly defined and communicated to all employees.

For many companies, this process requires changing from an informal environment, where there is little accountability, to a more formal environment. In a formal environment performance is measured and the results from each functional area are posted for review by *all* employees. This makes each functional area accountable not just to management but to the entire organisation.

Informal system symptoms

Schedules
- Many orders in arrears
- Orders in arrears that are not needed
- Overload in the current period
- Invalid shop and vendor due dates
- High level of shortages

People
- Don't believe the schedules
- Perceive no real accountability
- Lack co-ordination and teamwork
- Operate in crisis mode

Results
- Less than best customer service
- Excess inventories
- Reduced productivity
- Excessive costs
- Uncompetitive product lead times

Formal system characteristics

- Maintains valid schedules
- Plans orders for release and tracks status
- Work is done to schedule
- Corrective action takes place
- Systems simulate reality
- Formal policies, procedures and disciplines are, fully documented and adhered to
- Continuous improvement occurs

Appendix D gives examples of several basic formal procedures that are required in most production environments. To make them lasting, everyone in the company must recognise that they are operating in a highly competitive environment where change is now the norm, not the exception. Every aspect of business is changing, and in response we must change the way we run our business. Adopting a **management of change philosophy** is essential, not only to move from an informal operating environment to a formal performance measurement system, but to establish the foundation to strive for continuous improvement. Change management provides the attitude and direction to introduce new systems non-threateningly, so they are acceptable to all employees.

MRP2 can be a catalyst, forming a company-wide focus on the need for change and its benefits. This will also be the most difficult and testing part of the total project – people, not systems, are the key resources. If you fail in this stage, the MRP2 implementation is very likely to fail.

Cost/benefit justification

Installation of an MRP2 system is a major project and should be preceded by a thorough cost/benefit analysis to understand the real magnitude of the various elements of the cost and to involve top management, securing its full commitment to the financial investment and major effort required to reap the benefits. The fact that the actual costs and benefits may be hard to pin down is not an excuse for avoiding this important first step. Implementation costs include

1. **Costs of the MRP2** software and hardware; costs of expert system tools and any other supporting systems.
2. **Maintenance costs** of the hardware and software.
3. **Costs of educating and training** management and all users, including software supplier support, consultancy as needed, outside agencies, in-house training and payment for employees' time and expenses. Training often costs more than software.

4. **Costs of a full-time project leader**, and possibly several part-time assistants (for work on procedures, training, etc.), and for additional staff to carry the normal workload while part-time team members are working on the project.

Some of the expected benefits:

1. **Reduction of inventory** as a one-time gain of released capital plus the continuing savings in annual carrying costs, applies to purchased stock, work in process (WIP) and finished goods inventories. Reductions of 20 to 40% are to be expected. A goal should be set so that this reduction is fully realised.
2. **Reduced manufacturing costs** from improved efficiency through better planning visibility (at least 5 to 7% of manufacturing cost is reasonable). Again a goal should be established.
3. **Reduced purchasing costs** through increased visibility of long-term needs. A goal of between 5 and 10% should be set.
4. **Improved customer services**. This may be hard to estimate financially and it depends on the shortfalls in the existing service. There should definitely be a competitive advantage.
5. **Reduced costs** of expediting shortage list review meetings, overtime, line stoppages, premium freight costs and employee time consumed in wasted effort.

Cost/benefit studies may show large costs but the benefits may be even larger. Senior management will understand good investment opportunities and payback, and will be more likely to approve the project and be committed to its success after a careful cost/benefit study.

Setting goals in the company

'to measure is to know'

What can we expect to achieve when we implement MRP2 effectively? First, we have to identify and quantify our expectations. The best way is to do a performance audit of our current performance levels in all areas of concern. Company performance audits were described in earlier chapters, and measures of particular relevance to MRP2 are discussed later on. With pre-MRP2 performance percentages established, we then must decide target years to achieve each classification, A, B, C and D.

Return on investment (%)

	Class A	Class B	Class C	Class D
First year	250	100	50	25
Thereafter	1,000	500	250	100

Functional benefits (%)

	Class A	Class B	Class C	Class D
Inventory reduction	35–50	25–35	10–25	5–10
Productivity increase	20–30	10–20	10	5
Quality rating	95 +	90–95	85–90	75–80
Customer service rating	95 +	90	85	80
Profits increase	25	15	10	5

The following charts developed by the Strathclyde Institute, a leading firm of UK management consultants, shows the experience of four of the companies they have worked with in the achievement of class A MRP2.

Company return on investment

	Company			
	1	2	3	4
Year of class A	1989	1989	1990	1990
ROI in first year (%)	280	300	250	350
Annual growth (%)	330	900	1,000	625

Company functional benefits

	Company			
	1	2	3	4
First year inventory reduction (%)	40	25	20	50
First year productivity increase (%)	10	5	15	9.5–10
First year quality improvement (%)	+50	+10	+15	+15–20
First year customer service level (%)	94	90–95	91	95
Turns at start/turn now	3/8	3.5/10	2/4.3	2.4/5

CASE STUDY 13.1

Cost/benefit analysis of a small pharmaceutical company

Through improved planning and scheduling of production and purchases, we expect a reduction in the average lead time/process time ratio, from 5.3:1 at present. This reduction in lead time (LT) will allow finished goods stock to be reduced, since the production facility will be able to respond to demand much more quickly. Improved scheduling of raw material will reduce this inventory, with work in progress (WIP) being more precisely controlled. Less WIP will be pushed into the system, hence reducing the WIP level. Conversely, the reduction in WIP will also reduce queue sizes and queue time, hence reducing product lead time.

For completeness we have compared three different levels of saving: 25%, 50% and 75% reduction in lead time. This is expected to be achieved over two years.

£ millions

Scenario	Ratio	Raw materials	Work in progress	Finished goods stock	Total saving (£m)
Now	5.3:1	1.2	0.6	1.1	0
25% reduction	4.12:1	0.9	0.45	0.82	0.725
50% reduction	2.75:1	0.6	0.3	0.55	1.45
75% reduction	1.4:1	0.3	0.15	0.27	2.175

This saving would be one-off in nature. Cost of carrying the current level of inventory at 15% is £430k per annum.

Intangible benefits

In many cases the intangible benefits can be more rewarding than those that can be readily quantified.

- Ability to grow without a proportional cost increase
- Reduced management fire-fighting therefore more time to plan
- Improved accuracy and currency of management information
- Improved quality of service between departments
- Improved employee morale and job satisfaction
- Improved customer service
- Ability to react quickly and change effectively

Cost of the project team

Assuming an average salary of £18k, the five full-time equivalents cost £90k per annum. Over two years this equates to £180,000. Other costs:

Software	£140,000
Hardware	£200,000
Training	£110,000
Consultancy	£90,000
Subtotal	£540,000

Total cost including internal teams £720,000. The 50% reduction scenario gives a saving of £1,450,000 over two years. Depending on their phasing, this would give a payback in approximately 1 to 1.5 years.

What are we going to do differently and what benefits will this give us? According to Tom Whyte of Hussmann, we should first look at the much improved bottom-line results, then talk to people in our company who were here before we instituted MRP and are still with us. They will tell us whether life it is much better and whether their jobs are more satisfying, whether they are in control and whether they feel more involved with the business.

The Sloan School of Management at MIT warns that significant technological investment without organisational change is a prescription for failure. Businesses must use sophisticated information gathering systems as part of their corporate decision-making or else they will lose ground to those that have learned to do it. It is no longer sufficient to think of information technology as a way of reducing costs or improving productivity, and successful companies will be those who understand early the strategic potential of new technologies and act to ensure their own competitiveness – and survival.

Potential pitfalls

Most problems can be attributed to inadequate and ineffective implementation, people and procedures. But unless they are carefully selected, the impression given by performance measurements may be inaccurate. 'Leaders and Laggers in Logistics', a survey carried out in 1990 by A. T. Kearney Management Consultants, found that businesses using MRP2 were achieving some of the expected MRP benefits, but possibly at the expense of other important measures. Across all industry sectors they found that leaders carried seven weeks' stock whereas laggers carried twenty. Leaders offered improved customer service for make to order and ex-stock delivery (95% verses 90% service).

Of the businesses surveyed, 70% used MRP1 or MRP2. The survey compared their performance with the non-MRP companies and found that their stock levels were approximately 15% lower but their customer service was poorer by approximately 5 to 10%. A.T. Kearnay concluded that introducing MRP1 or MRP2 will neither necessarily decrease customer service nor necessarily increase customer service. MRP guarantees nothing.

Nevertheless, the results suggest that the companies involved were not of class A status; they were not measuring or driving for the improvement of acknowledged key business performance criteria. Customer service must be very high, if not the highest on our list of desired improvements. It is likely that many of the surveyed firms were examples of poor implementations of MRP1 and MRP2.

Used incorrectly, the formal nature of MRP2 can sometimes lead to problems in giving customer delivery dates, particularly in any industry sector that is competing on lead time. Used correctly, rapid delivery date processing can be achieved, surpassing that of non-MRP operations (Figure 13.3).

■ Managing and running the project

The formation of the team begins with the project leader. This individual should be someone from within the company who has a good, solid working

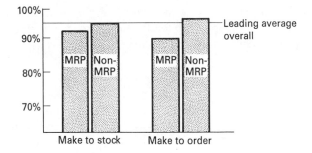

Figure 13.3 Effect of MRP2 on customer service in make-to-order and make-to-stock sectors.

knowledge of the manufacturing function of the company and of its products. Computer knowledge is not required for an MRP project leader, since it is largely about people, not technology. Characteristics of suitable leaders are further described in Chapter 6.

Once chosen, the project leader must report directly to top management, either to the management team or managing director of the company. Throughout this project, the project leader will be proposing and making changes to the way the company does business, which will influence every functional area of the manufacturing operation. For this reason, the project leader should normally be drawn from the ranks of the manufacturing or operations area, so the project is not perceived as just another task being imposed by distant bureaucratic personnel.

Full-time project leader

A full-time project leader

- does not need to be a computer expert
- does require to have in-depth knowledge and experience of the business and the products

Managing the MRP2 project should be defined as a full-time job. Otherwise, the project leader's regular responsibilities will receive first priority and the project will fail. The search for a project leader should begin as soon as management commits to class A MRP2 so that this individual can become involved in the initial education and training process. The project leader should pursue detailed education through outside classes to quickly become the in-house expert on the functions and requirements of the new MRP2 system.

Once the steering committee has selected from the candidates within the company, it works with the new project leader to select the project team, to develop the initial communications programme for the MRP2 project and to develop the education and training programme plans. As the project is

411

implemented, the function of the steering committee is to receive regular progress reports from the project leader (e.g. once a fortnight or once a month), resolve problems and to keep the project on target.

Project team members

The project team should have representatives from each of the functional areas involved, or from each business unit or cell in a restructured enterprise (Figures 13.4 and 13.5).

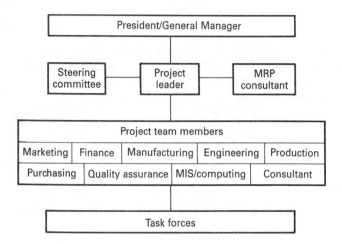

Figure 13.4 Conventional project structure.

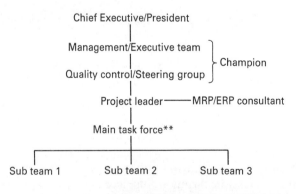

Figure 13.5 Resources and structure for ERP projects.

Detailed project team – an example

For successful implementation of **MRP2** across the company, we recommend that a project team be established, reporting to a high level steering committee. The proposed structure is shown below.

STEERING COMMITTEE

Chairman – chief operations executive
Torchbearer – technical and operations director
Project manager Financial director
Commercial director
External manager or consultant

PROJECT MANAGER ———+——— CONSULTANT

PROJECT TEAM

FULL TIME	PART TIME	AS REQUIRED
Production Planning	Sales and marketing	Key users
Production	Data processing	Vendor
Purchasing	Information systems	Temps
Finance and accounts	Personnel	
	Quality control	
	Marketing	

The project team will ideally have a core of full-time members plus part-time and as-required members. For the company this would mean the following people full-time:

Full time	Project manager
	Production
	Purchasing
	Finance and accounts
	(Consultant)
Part-time (20 to 50%)	Data processing
	Information systems
	Estimating/SOP
	Quality control
	Personnel
	Marketing: internal and export sales

Full-time or part-time team?

The steering committee must decide whether it is advantageous to create a full team from specific areas at certain stages of the project. MRP2/ERP is an integrated system just as the company is an integrated organisation, and it is more effective for the team to act together as a single unit. However, a part-time or revolving team places a reduced strain on the company's human resources, and may be the only approach possible.

As the project is progressing, the project team should meet at least once a week to develop the project plan, review progress, identify problems and decide on the appropriate actions. The team's first action is to create the project plan, which should include an education and training plan.

Train the trainers

Once the MRP2 project team is established, the project leader should take the team through the education programme and additional detailed training. The members of the team can then become trainers, educators and promoters of the project for their respective task forces.

Each task force is composed of the person or group of people who are assigned the responsibility for carrying out each task in the detailed plan. For example, it could be the engineer assigned the job of bringing up the accuracy of the bills of material of a particular product line. The task force reports to the project team member responsible for that area of the plan. The skills required can be one or more of the following:

- Project management: identifying tasks, quantifying and ordering those tasks and assigning them to others: the ability to use tools to assist in this process, such as microcomputer-based project management software packages.
- Application skills: in a computerised MRP2 environment this means knowledge about computer applications in general, and specifically
  - how computerised MRP2 systems function
  - how the modules are interconnected
  - how best to use the features and facilities of each module
  - where the problems and likely pitfalls are
- Software package knowledge: most people do not fully use the facilities of their software package. They reach the minimum level of competence to meet the original objectives. They suppress their desire for experimentation in order to get on with the job. Detailed specific package knowledge is critical.
- Technical skills: needed by new users of computer equipment and software, e.g. workstations, printers and screen/report generators.

Consult the consultants

Class A MRP2 must begin with a commitment from top management, to lead the project then to follow its implementation plan. Unfamiliar problems will be encountered – problems which cannot effectively be handled in-house. These are the times to seek advice from a consultant. Find someone who has deep experience of MRP2 and a track record in Class A. (Chapter 4 gives further guidance on selecting and using consultants. Check references of MRP projects. Avoid playing guinea-pig to a 'rookie'.

Call upon the consultant when there is a need for an experienced outside perspective. A consultant can help management to define potential problem areas and to develop the corporate objectives, policies and plans required for MRP2. And a consultant can help project team members to prepare for the various software modules related to the functional areas.

After the implementation is accomplished, continuing effort is most important to identify and implement the remaining disciplines and to establish the levels of accuracy needed for class A. We are much more likely to accept direction from an unbiased consultant in this touchy area than from our own co-workers. Perhaps a consultant is a person who will tell us the things we need to do, whether we want to hear them or not, and make us do them too. Consultants are catalysts for change, not just back-up management. But it is important for us to take responsibility ourselves. But a consultant can also be extremely useful when opposition is encountered, for two different reasons.

1. The consultant usually has greater experience of business improvement, and may be able to inform any opposers of cases where MRP2 has overcome similar opposition and improved similar operations.
2. Since your company is paying for the services, consultants tend to be listened to, otherwise the management team would feel they were not getting value for money. We've paid for this guidance, let's follow it.

Fortunately, most businesses are swayed by the first reason, not the second.

The apprenticeship concept for MRP2

A large proportion of MRP2 projects may be deemed to have failed or dropped below targets. Several reasons for this have been highlighted. The use of proper education and training programmes has been recognised for some time as being an essential ingredient for success. Yet MRP2 projects continue to fail, providing none of the stated benefits within the agreed schedule of deadlines.

One explanation may be the complex nature of the knowledge to be acquired then transferred throughout an organisation. Another may be the high cost of making errors during an MRP2 project. These could be confronted by a more traditional method of skills transfer, *apprenticeship*. College and textbook training provides valuable knowledge, but real working knowledge and experience is best learned person-to-person. Since MRP2 projects are lengthy and complex, an apprenticeship should be considered for the assistant

415

project leader. The project leader has more than a full-time job and may need a full-time assistant.

Education and training

Effective leadership and education are vital factors in establishing a successful MRP2 programme. They are just as important when initiating any change. This is especially true of a company planning to achieve class A MRP2, sustained performance of 95% or better performance in all functional areas. Hardware and software can't deliver this performance, people do. But they can't if they aren't knowledgeable about the system, don't fully understand its objectives and don't see management support for it.

An MRP2 system is not just an end-user's tool, not something only hands-on operators need to learn. It is not sufficient to educate people using the cheapest instruction manuals and a rented video course. MRP2 education requires long-term investment in people, investment that allows people to master the system and improve their performance.

For a company simply wishing to implement MRP2 to computerise existing functions, these two 'quick and dirty' approaches may be adequate. But the company wishing to achieve and sustain top performance must commit to a different way of running the business, a new management philosophy and a corporate culture which strives for continuous improvement and manufacturing excellence.

Class A, B, C, D and relevant education

Class	Performance (%)	Characteristics
A	90	Totally balanced plan. Top and middle management, and all functions participate in the education process using the full range of methods available
B	80	Essential components in place but no motivation. Middle management and the project team participate in education programme that top management approves but does not participate in.
C	70	Education is picked from several sources but is confined to people who will be operating the system.
D	50	Lowest budget programme. Doing minimal education and concentrating on training provided with the system by hardware and software vendors. No real involvement or linking to operate the business in a different way.

In some businesses, you may be limited in the available options for education and training. However, you can use what is available to the fullest extent. Sources include the established MRP2 training companies (Oliver Wight, Mike Salmon, Aurora, etc.) and several specialist manufacturing consultancies, including the Strathclyde Institute, Coopers and Lybrand Deloitte, CSC and Price Waterhouse.

Many colleges and universities offer standard or custom courses on all aspects of manufacturing and business improvement. This may be a cost-effective way to cover a large proportion of your training needs. Education is defined as general manufacturing management and specific MRP2 theory and practices. Training is defined as the specifics of using the software package. Top management will be restricted to education but middle management and operational levels will be involved in both categories (Figure 13.6).

During educational workshops, we can use videotapes marketed through various agents. They are not to be viewed as movies and they should be appropriately oriented for top management, middle management and workers. Depending on the complexity of the subject, top management spends one or two days per group, middle management five, and workers two or more.

Software-specific training

For training on the chosen software, the supplier should have a programme in place which will show each person involved how to interface successfully with the package and train middle management in what to expect out of the system and how to handle it. The training involves hands-on experience with the actual hardware, the software menus, displays and reports.

Your chosen consultant could be used to do the initial performance audit, track progress toward performance goals, define the causes for success and failure, and hold specialty training sessions for the particular groups which need to improve performance. Without them, implementation will take longer and you may never achieve your performance goals.

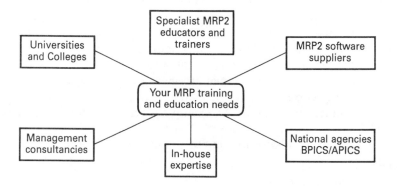

Figure 13.6   MRP2 education and training sources.

Project team training

The project team is in the best position to lead the project astray, so its education and training must provide an even clearer picture of what the company is trying to accomplish and how. Continuing education is important. If the project is generally received with enthusiasm, a number of people will be interested in learning more about their own areas. People from any level of the company should be encouraged to take local college courses in manufacturing management, such as the BPICS course.

■ Organisational approach

Ownership involvement and accountability

As described in Chapter 8, one of the most effective ways to manage change is to take a team approach, involving the entire company in the definition of objectives. Involvement leads to ownership, and ownership leads to accountability.

Once the company, working as a team, has identified its objectives and established appropriate performance measurements, management must develop the performance measurement report described above. After reviewing the report, each functional area nominates someone to regularly carry out and report its measurements to management.

Throughout this process the goal is to push the responsibility for measurement down to the level of the work. Consider bills of material (BOMs). Accountability for BOMs should not automatically be given to the design department manager or director of engineering. Instead it should be divided into groups by product or product line, and each line engineer should be responsible for the BOM accuracy of their line.

Again, formal procedures for the auditing of BOMs need to be established and carried out on a regular basis, for example, weekly auditing of fifty BOMs by a cross-functional team producing the sample accuracy figures for the company performance report.

Organisational accountability is an important aspect of manufacturing excellence, but accountability alone is not sufficient. A formal system must be put in place to ensure that people are measuring the performance of the right things in the correct manner with an effective feedback system in place for top management to monitor. An example of such a system is described in Figure 13.7.

One path through Figure 13.7 might go like this:

■ Define objectives
■ Set goals and parameters
■ Develop an action plan

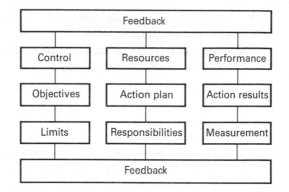

Figure 13.7 Feedback system for measuring progress.

- Measure performance
- Provide feedback
- Achieve objectives

Class A performance measurement must begin with a personal commitment to manufacturing excellence, which, through concentrated effort, can begin to flow through the entire organisation. In the beginning, an MRP2 project team, representing various functions of our operation, works to develop the performance measurements. But the goal eventually, is for employees to take responsibility for their own performance and the success of the company. *The real reason for performance measurement is to identify and solve problems by continuous improvement and feedback*

Under a formal MRP2 management system, business performance is measured against planned performance. Continuing education for employees, as well as a constant feedback, supports a drive to achieve ever higher performance levels. Throughout the company, performance improvement replaces promises and failure as a way of life and culture.

Ideally, this system begins with leadership from top management. Inspired operating managers will in turn communicate this leadership to other employees.

- **Start immediately**: avoid the temptation to overstudy and overjustify commitment to management excellence. Begin with one operation, then build on its success.
- **Educate and empower your people**: understanding is an excellent first step towards improved performance. Give employees the tools and training they need to succeed.
- **Lead by example**: Memos, videotapes and flip charts have their place, but the most powerful motivators are leading by example, and getting people involved.
- **Focus on performance objectives**: Don't set goals on anything less than

Class A. Ensure you communicate an unfaltering commitment to improving performance up to and beyond the planned objectives. Understand and communicate how the shared accountability for productivity and profit performance affects all employees.

Is it working?

You will be starting to succeed when you notice the shift toward personal involvement and responsibility. This happens naturally when employees understand how their work affects everyone else and the company as a whole. People begin to respond with 'I understand,' 'I'll do my part,' and 'I'll deliver,' instead of 'That's not my job,' or 'Our department did our part right.' *Total Quality – motivation and involvement.* By working together, we can all make a difference in productivity and profit performance. In the long run this will ensure the business survives and prospers. Much of this approach is similar to the management of change, described in Chapter 8.

Planning the project

The project plan is an integral part of successfully implementing MRP2 and ERP. In the early stages of planning, intensive, full-day sessions are needed to identify the major tasks for each of the two areas involved (management and computer/MIS). The steps in each of the tasks and their interactions have to be established and a schedule set with deadlines.

The early stages provide an opportunity to directly involve others. Successful MRP2 implementation is highly dependent upon user participation and co-operation. If resistance can be overcome in the very early stages, the probability for success is greatly improved. When users are not consulted until a project is fully defined and ready for implementation, it is not unusual for subtle forms of sabotage to occur (bad-mouthing, refusal to use the system, lip service instead of co-operation). Users need to feel they have made a valuable contribution to the planning and implementation of the MRP2 system and have a vested interest in its success. Co-operation can be encouraged by

- **MRP newsletters**: the project leader can periodically publish project status, changes and interesting information related to the MRP2 project. Any employee interested in the project should have access. The informal discussions that wide readership generates can help in keeping the user interest at a high level.
- **Review meetings for users**: users should have the opportunity to talk to and questions other users, the project team members and the project leader.
- **Decision meetings**: when a compromise has to be made, the project leader

should hold a meeting with the affected users, where each must have an opportunity to state their case. It is important that no user feels they have lost out because decisions were made in a vacuum. Not only must compromises be good for the organisation, they must be seen to be worthwhile.

■ **Public relations**: the project leader and champion should take every opportunity to boost the concept of MRP2 and its benefits across the company. People tend to lose interest in projects which take a long time to produce benefits, so make regular efforts to jog people's memories and keep interest high. Project pens, coasters and mugs actually work!

Implementation planning

The project plan is a lengthy document prepared by the project team. It lists all the tasks that must be accomplished, from start to finish. In reality, it is a work schedule that briefly describes each task, estimates duration and lists the person responsible for its completion and when. Interrelationships between activities should be identified.

Preparing the plan is a time-consuming process because every aspect of the program is examined and its work defined. A useful ground rule is to allow no task to exceed forty hours. Larger tasks should be divided. This assists progress reporting at review meetings. Often it is advisable to have an experienced consultant or MRP expert review the plan to ensure its validity and inclusion of all critical tasks.

The project plan is a firm guide, a work-to list, but to remain realistic it may sometimes need to be rescheduled. Review meetings of the project team monitor progress each week, handle problems and replan future activities. The following describes the kind of planning which can be used for management activities and computer system activities. The project leader must develop an implementation work plan which defines each task in the following way:

■ Name of task
■ Description of task
■ Name of individual responsible for on-time completion of the task
■ Start date of task (based on prerequisite task completion dates)
■ Resources required
■ Completion date of the task
■ Milestone definitions of the groups of tasks and dates
■ Review

The plan and its monitoring should follow the rules of sound project management (Figure 13.8) Some considerations are

■ Tasks/activities should be scheduled in parallel where possible.
■ Task completion deadlines should be based on realistic resource commitments; company operations must continue throughout the project.

Thirteen golden rules for implementing systems

1. **Appoint project champion**
   Business oriented not technical boffin
2. **Top-level commitment**
   Continuous and apparent
3. **Ownership**
   Implement with end-users, not for end-users
4. **Define the objectives**
   Measure, audit and improve (RAT, relevant, attainable and testable)
5. **Define the requirements**
   In terms you understand: keep technical data to appendices
6. **Try for package software**
   Accept best fit; don't be first; insist on reference site visit
7. **Time scales**
   Project plan must be realistic
8. **Don't computerise chaos**
   Simplify and improve first
9. **Working practices**
   At least as important as computers
10. **Quality of information**
    Garbage in, garbage out
11. **Education and training**
    Define levels; include awareness; senior management must attend
12. **Openness on your terms**
    Portability, choice and integration; look at tools used

○━━ AVOID CHEQUE-BOOK MENTALITY

Figure 13.8 Essential rules for successful project implementation.

■ Problems will occur frequently during the implementation; some time should be established in the plan for coping with them and determining changes to operational methods and procedures.
■ Early and late conditions should be identified rapidly by the project leader so as to make the appropriate corrections in the plan.
■ Dry runs of parts of the system should always be made before full-scale testing is undertaken (parallel runs with the old systems remaining in operation). This will test the procedures, disciplines and controls of the systems. Failure to do this will make it very difficult to separate people problems from computer problems and will delay implementation.

Bad planning

The success of MRP 2 is just as dependent upon the detail and quality of the implementation plan as it is on the system itself. It will be obvious in the early stages of implementation whether the plan is good or bad. If the plan is good, schedules will be met, users will have enthusiasm, communications will be open and positive.

If the plan is bad, slippage in the schedule will occur at the first milestones. Tasks will be constantly redesigned and rescheduled. Users will be confused. The organisation will lose confidence in the whole MRP approach. More MRP2 failures occur because of inadequate policies, procedures, disciplines and controls than for any other reason. Formal education and training contribute to the recognition of this fact by top management.

Project plan development

The development of a project plan must be approached in stages. The first stage is to develop an **overall project plan** (Figure 13.9) which covers the full scope indicated below:

Public relations promotion
Education and training plan
Business plan
Sales plan
Production plan
Master production schedule
Bill of materials
Inventory accuracy
Material requirements plan
Purchasing
Operation card routings
Capacity requirements plan
Shop-floor control

In this first stage, the project team must develop benchmarks which will be used to measure progress throughout the project and then to set out the first quarter's targets. These points should be reviewed by the steering committee. Some sample benchmarks are listed below.

Develop overall project plan
Develop major milestones
Define detailed tasks by area
Assign responsibilities by tasks
Establish timetable
Steering committee review
Steering committee approval

Once this **master plan** and the overall benchmarks are created, the second stage begins. In the second stage, the project team must review each task area shown as the thirteen points above and develop the **detailed project plan,** the sub-task-forces for each area and the specifications of each task. This involves identifying the key issues for each area. Some examples are listed below.

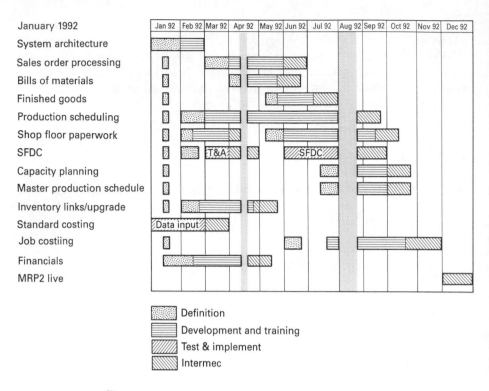

	Jan 92	Feb 92	Mar 92	Apr 92	May 92	Jun 92	Jul 92	Aug 92	Sep 92	Oct 92	Nov 92	Dec 92

January 1992

System architecture

Sales order processing

Bills of materials

Finished goods

Production scheduling

Shop floor paperwork

SFDC

Capacity planning

Master production schedule

Inventory links/upgrade

Standard costing

Job costiing

Financials

MRP2 live

Definition

Development and training

Test & implement

Intermec

Figure 13.9 Hussmann overall MRP2 project plan.

An example is cycle counting for inventory accuracy. Beyond the first-stage guidelines, the spin-off task force must specify the areas to have controlled access areas, the frequency of counts and the tolerances. Then it must define who will do each task and who will be accountable. In this stage, the project plan's development begins to converge along three aspects; education, management and technical (Figure 13.10). Education requirements have been described in the previous sections, and we now move onto the management and technical aspects of implementation.

Management aspects

Management aspects of implementation must be addressed to get performance up to desired levels while software is being implemented. On the technical side, the inventory module of the software may be installed and implemented but that does not make the inventory accurate. The project team and inventory management task force must ask what changes and cycle counts are required to get 99%. They must decide on the control groups, frequencies and

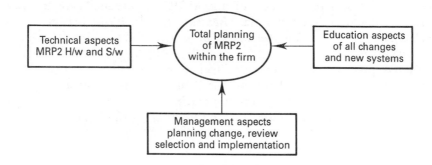

Figure 13.10 The three aspects of implementation.

tolerances. And they must assign staff to collect the data and take responsibility for its integrity. The full range of management and people issues might involve stockroom layout and materials movement, as well as feedback audit trails to identify and correct errors.

The same management questions must be asked of each area to be integrated into the system. How can we use this ERP tool to achieve the desired performance from forecasting, sales, master scheduling, bills of material, accounts, purchasing, shop-floor control?

Under the total quality approach to MRP2, the company operations and systems are scrutinised thoroughly to identify problems, opportunities and potential ways of improving the operation in all functions (Chapter 5). In conjunction, we must consider how best to use the necessary MRP2 facilities. This relates to total quality management, where improved policies and procedures are developed in parallel with the definition and selection of the required MRP facilities and functionality. It is no use developing a simpler and more effective procedure for sales estimating, only to try and fit that in to an overcomplex and largely irrelevant SOP front end. Strive for mutual harmony is the objective. Several questions require to be asked about the corporate plan:

- What significant changes need to be made in our methods and procedures to get the full benefit from the project? Do they knock on to other areas?
- What are the company's current performance measurements in top management's planning and forecasting, and operational performance of each area involved?
- Which class A performance levels are to be attained, and when?
- Who will be responsible for each functional area?
- What are the specific objectives and results expected from each of these functional areas? And what is required to achieve them?

The answers to these questions will come from the findings of the detailed business review that should precede this project. A total quality approach will define the objectives, changes and business structure needed to become competitive. This in turn reflects on the MRP2 requirements for the new enterprise, functions and definitions. *Remember, 75% of the benefits derive from the management portion of an MRP2 project, only 25% from the computer portion.*

This is equally applicable to the rejuvenation of an ailing or unused MRP system, which, after the TQ review has defined the way forward, can be reimplemented using a total quality approach. This will frequently involve using the systems, facilities and procedures in different ways than before.

■ Prerequisites for success with MRP2

Several basic foundations must come before MRP, as system performance is in direct proportion to their implementation (Figure 13.11).

■ Start with the business plan and proceed to the next level only when good plans have been made.
■ Operate the business using formal policies and procedures that address weaknesses identified in the review.
■ Preserve the integrity of data, especially

Bills of material accuracy   98%
Stock/inventory accuracy   95%

■ Train people to use the system capably. Retrain regularly.
■ Ensure management is committed to the expense and effort of the implementation and is willing to use the system.
■ Encourage all users to work with and through the system.

The detail and importance of achieving these prerequisites is addressed in the following section.

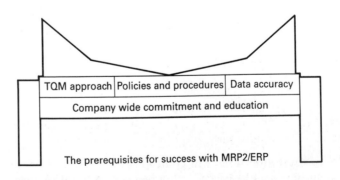

TQM approach	Policies and procedures	Data accuracy
Company wide commitment and education		

The prerequisites for success with MRP2/ERP

Figure 13.11   Prerequisites for success – building a bridge.

Review and existing system study

The ideal preparation for the selection and implementation of an MRP2/ERP system is to completely review the way the business operates, then define and create the changes necessary to give overall improvement. As part of this, the need for business-wide information and control systems is examined, either as a review of existing systems, or as a new requirement altogether. Again, the relationship and interaction between manual and computer-based systems must be understood, if we are to develop an effective operational systems platform for business development, i.e. both system areas are closely linked and must be developed and operated in sympathy.

Documenting the current methods of operating is a necessary first step in any review, to have a sound basis for planning required changes. Existing written procedures should be audited to see if they are appropriate or being followed. Where procedures are missing, drafts describing current methods and practices should be prepared. Graphical tools, such as IDEF diagrams and/ or flow charts should be made to understand the origin, processes and transactions involved in

- Customer orders
- Purchase requisitions and orders, including receiving and inspection
- Work order release, monitoring and receipt
- Inventory transaction, issue and receipt, kitting, random issues, rejections and material review, cycle counts and stock adjustments
- Engineering reviews and implementation responsibilities.

Prerequisites for success

- Formal, total quality approach
- Policies and procedures
- Data accuracy

Policies and procedures

The best computer system in the world will not yield the desired results without proper policies and disciplines. Many companies have cut corners in this area. Proper policies, procedures and disciplines do not simply evolve as required. They are difficult to formulate without interfering with operations, which has led to people avoiding the issue. Correct formulation and implementation of proper policies and procedures goes alongside disciplines to make them work. Procedures must be an integral part of MRP2 management, not trial and error patches applied after problems occur during implementation. In formulating policies, procedures and disciplines for an MRP2 project, the following steps should be taken:

■   Use the software package's concept design and operation, list all the things which could prevent the system from delivering the desired results. This is an excellent application for brainstorming the production area.
■   Give each problem a relative importance.
■   Formulate a control framework for policies and procedures then use it to address each potential problem.
■   Explain how policies and procedures will be monitored and enforced.
■   Formalise policies, procedures, disciplines and controls for all the potential problems in each area, and obtain the required approvals from local management.
■   Distribute policies and procedures to everyone involved.
■   Train affected personnel to ensure understanding and compliance.

The project team should see that complete written procedures are operating the new system, both for planning functions and for daily transactions. These are best if they provide step-by-step instructions for preparing the plans, for making transactions or for preparing data inputs. Actual examples should be included in the procedure. The procedure should explain:

Why the procedure is required.
Who performs it.
How and when it is to be performed.
What to do about exceptions and errors.

The procedures should be maintained and audited for correct use. Procedures should also be the basis for training, to show how the step-by-step instructions are to be followed. It is very important for users to understand why the procedure is needed and the effect of any departure. Having good procedures and adequate training is the best method for securing the data integrity and accuracy that is essential to an MRP2 system.

Record accuracy

Accurate data in all sectors

The major focus of our implementation plan should be to ensure that our data is, and remains as accurate as possible, as the system will only be as good as the data it has to work with. If top management is unable to provide the necessary planning and forecasting accuracy, then this will have a ripple effect all the way through the performance of the organisation. This is also true under a manual system, but with MRP2 the impact is devastating.

Immediate steps should be taken to determine the level of accuracy in all records. In a manufacturing environment, audits of bills of material, routings, and open order balances are necessary. If cycle counts of inventory are not being taken, they should be started. More important than just correcting errors is the investigation of their sources and development of procedures to prevent them in future.

Part/item master accuracy

Begin by assigning responsibility of each of the major data groups. These groups are composed of data elements such as

- engineering/design
- purchasing
- manufacturing/production
- sales
- purchasing
- costing
- finance

The task force assigned to each major data group begins by making sure that all the data is at least present. Missing data elements, such as material cost or purchasing EOQs, must be created then entered into the computer. Once these data elements are filled in, each associated task force must also make the effort to determine their accuracy. The objective is to have accurate information contained in each data element.

Production and inventory control is normally responsible for accurate item master data – lead times, codes, lot-size rules, etc. – and for the continuing accuracy of inventory balances (with cycle counting). Purchasing is responsible for developing minimum but realistic lead times with suppliers, for monitoring open order quantities, and maintaining accurate delivery dates.

Inventory accuracy

Set up or expand the cycle counting procedures. The control group of test parts can be expanded to a new set of parts once the previous group maintains the necessary levels of accuracy. As these control groups reach 95% accuracy then an all-out cycle count effort can be made on the rest of the inventory. The real aim is to identify and rectify inventory control problems, not merely to reach high levels of accuracy. This occurs automatically as more of the control problems are found and dealt with.

As part of overall inventory control, manufacturing is responsible for accurate reporting of production activity, work in process, movement of goods and the reporting of scrap and loss.

Bill of material

A bill of material defines the product. Design engineering provides accurate bills of material that are formatted to work with product option/feature planning, and reflect the actual methods of production. If just one mistake is present in the bill, it cannot be considered as accurate since production is actually building it differently than the bill. The engineer who is responsible for the product should perform a bill of material audit by checking all items that go into the product, verifying for parts and materials, making sure of the correct quantities and making sure that each level is accurate to produce the product. The bills should be compared with a production audit to confirm they are in agreement with real production practices.

Operation and routing cards

Operation and routing cards should be audited using a similar method to the bills of material. When problems or deviations are identified by the task force responsible for the production process, it should investigate and correct the error. Routings can also be checked with a production audit to confirm that the operations agree with the present production process in the plant. Again, the objective is to identify and fix problems. If the project team encourages people to identify problems then fails to fix them, it sends out the message that data accuracy doesn't really matter much to the company. Manufacturing engineering provides accurate production routings and time standards that will be used for production scheduling and costing.

Order definition

Particularly in the make-to-order sector, accurate, punctual and complete order definition is essential. Sales and marketing must provide this accurate order definition and order entry. Effective, not perfect, sales forecasts must also be made, and marketing should work with the master scheduler to manage and satisfy demands without making unreasonable requests on the stability of the master schedule.

Data accuracy pervades all areas in the MRP2 project. Data integrity within the system is the responsibility of every functional department, every unit and every system user. Inaccurate data must not be considered as a failure of the MRP computer system, but it is a critical milestone in the project.

The current master production scheduling (MPS) should also be reviewed. A critical examination should determine whether the MPS tends to be realistic and stable, and a correct representation of the management strategy. Steps should be taken to correct any deficiencies in this critical input to MRP.

Data input, and to a large extent data accuracy, is user controlled. It depends on training, discipline, willingness to follow proper procedures and accountability for errors. There needs to be a willingness on the part of users to check on themselves and audit their inputs after entry. The system itself can also contribute to data integrity by editing transactions and data inputs as they are entered, and by providing clear audit trails when investigation of the source of errors is required.

User profiles

A study should also be made of the people who will work with the new system to see if they understand the systems and their capabilities. Appropriate education and training should remove inadequacies.

■ Technical aspects of implementation

MRP hardware and software

Activities related to hardware and software can be identified by questions such as

- What are the user requirements? How are they determined?
- What software and hardware should be chosen?
- How do we implement the MRP/ERP software?
    What are the tasks involved, their size and importance?
    Should we implement module by module, or as a big bang?
    What are the timescales involved?
    Who is responsible for each module?
    Should there be a pilot?
- How and when will the company convert existing systems and procedures to the new MRP2 system?
- What extra DP/MIS resources are needed? When, and for how long?

As the study of the current system continues, the project team begins to define general and specific needs expressed by the user community. This leads to detailed specification of the features required in the new system. Once complete, the team can proceed to evaluate and select the software and hardware to be purchased. Chapter 12 described current developments in MRP2 and many of the downsizing options now available. However, several basic guidelines apply to every selection and implementation:

    Use a standard package as much as possible
    Keep it simple
    Stay in control
    Sophistication is difficult
    Obtain vendor back-up
    Put people first

Use a standard package

Use a standard package as much as possible. Virtually all current MRP2 systems are based on good business practice and it is very rare to find a business that cannot operate within the procedures and facilities of MRP2. So, do not attempt to reinvent the wheel, and minimise the amount of modification done to a chosen system. Extensive modifications are often a way of transferring old and poor ways of operation into the new system, usually by avoiding the discipline of good manufacturing practice.

Keep it simple

Keep the system as simple as possible. Very high sophistication in techniques reduces the probability of success.

Stay in control

Remember the purpose of the system is to enhance human decision-making by providing accurate and up-to-date information to the people who must run the business. The computer is an information management tool; it will never run the company. People will always control the system.

Sophistication is difficult

The operation of the system should be transparent to the users; but they need to understand its logic. If the system is not understood, users will be tempted to follow its orders blindly or develop their own informal, non-standard methods. If there are definite special requirements of the system, such as multiple warehousing, spares/service part planning, or lot control and traceability, they must be specified so that appropriate software can be purchased or developed that will adequately meet the user requirements.

Obtain vendor back-up

The system selected, besides meeting the specifications, should be backed up with complete user documentation and vendor support for education and training. The vendor should be able to provide continuing support, especially in problem solving (debugging) during commissioning and implementation and for future enhancements.

Put people first

Overall, the software and computers are not as important as people, plans and procedures.

Software module installation

There may be certain fundamental areas in the computer systems portion of the plan which must be taken into consideration when preparing the detailed plan. For example, plans and costs for integrating and/or modifying any existing systems developed by or for the company, that are to be retained in the new system architecture.

Sequence of modules
Major functions of MRP software
1. Database function
2. Calculating and planning functions
3. Execution functions

Conversion
1. Cut-off/cut over
2. Parallel operation
3. Pilot product conversion
4. Progressive by module conversion
5. Conference room pilot

There is no single best sequence of installing an MRP2 system. The project team must select the route that best matches operating conditions and the company's needs. Software for MRP is generally delivered as a set of modules. The logic of how these modules work together may affect the company's readiness to undertake each part. The planned sequence of installation is important because it will determine the priorities and schedules for

Preparation procedures
Elements of the plan for education and training
Testing of software, systems and conversion

The planned sequence is incorporated into the detailed project plan.

Three major functions of MRP2 software

When deciding the most practical way to install a system, think of the three major functions of software:

1. Data and file handling
2. Calculation and planning
3. Execution and monitoring

These functions may appear in one module or several modules, according to the system. Prepare for each of these functions (Figures 13.12 and 13.13).

Database

Database software modules hold basic data. They can be started first because they do not depend on the operation of the rest of the system. Several database functions can be worked on at the same time. This is useful because the effort to prepare accurate data for the system can be tedious. In each case, the initial

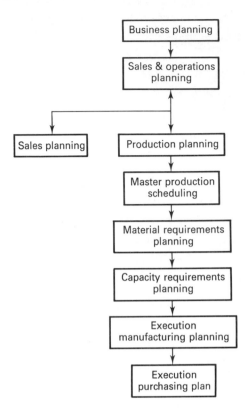

Figure 13.12 Business planning and control by level and activity.

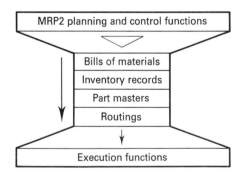

Figure 13.13 Relationship of MRP databases, planning and execution functions.

step is full understanding of each element in each record and how to use it in the total system.

Bills of material

Loading can proceed as soon as the bills have been audited and passed for accuracy. Direct dumping of old BOMs into the new system is not recommended unless they are known to be over 95% accurate. Once the new BOM system is functioning, it must be carefully maintained through formal engineering change procedures. This database is now available to the entire user community as required. In many cases, only currently active bills need to be loaded at the start.

Item master file

The item master file contains a large amount of essential information that the MRP1 and MRP2 functions require about each item. Data collection, auditing and entry takes a long time when starting from scratch. Decisions need to be made about product and part lead times, MRP codes, time fences, lot-sizing rules and modifiers such as multiples, ABC classifications, scrap factors, shrinkage factors, commodity codes, product codes, planner codes, and so on. The project team, with key users, should define the meaning of every element then assign responsibility for their collection, verification and entry.

Routings

Accuracy verification and data entry are required before job operations can be scheduled for production control and capacity requirements planning. Before routing data is entered, procedures, for entering and maintaining the data should be established. Routing data includes

- Cost centres
- Resources
- Jobs per resource
- Job codes
- Standard cost rate per resource

Costs

The manufacturing cost elements of material, labour, and burden or overhead rates must be defined before they can be entered. These may be manually entered or automatically calculated using other parts of the system, such as using the BOM system for cost build-up of parts and assemblies, or using routing data for standard labour costs.

Calculating and planning functions

Beyond the installation of module software into the computer, very little data entry is required to start calculating and planning functions. However, the presence of accurate database information is a prerequisite to their operation. Further extensive testing of these modules should be undertaken before they are used. This testing will not only prove the dependability of results and systems but will train people in their use and interpretation.

Execution functions

Execution functions record the day-to-day activity of the system. Inventory transactions of issue and receipt, etc., are used to update the perpetual inventory records. The opening, releasing and closing of work orders and purchase orders update the open order balances used by MRP. The order entry functions, if included in the project, update the master planning data as well as the inventory files. Figure 13.9 shows a project planner for MRP *installation* of hardware and software (not *implementation*). This chart indicates data input, conversion tasks and testing periods, underlying the higher level, modular implementation chart.

Changeover to the new system

When the project team is satisfied that procedures and training for operating the system, or a significant part of it, are completely ready for dependable operation, it can authorise the changeover from the old system to the new (Figure 13.14). There are several ways to proceed with conversion, some relatively low risk, and some very high risk.

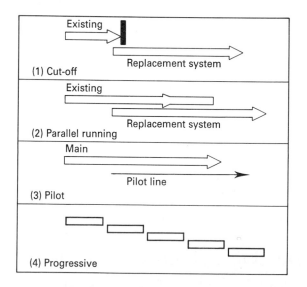

Figure 13.14 Changeover methods.

Cut-off

Cut-off or cold turkey simply involves turning off the old system and turning on the new. It assumes that all new system testing has been completed perfectly, training is thorough and complete, and the system can be relied upon. There is a great risk of the new system failing. After failures, it is often difficult to recover the previous system because the old data was not maintained.

Parallel running

Parallel running maintains the old system while the new system is started and tested. This is a low risk method, and the new system outputs can be thoroughly compared to the old. However, it involves several complications regarding the computer system and a massive workload on people across the company. All transactions must be duplicated into each system, and two complete sets of outputs must be reviewed and cross-referenced. This conversion is an expensive and time-consuming but it is fairly safe. There is a danger of errors due to incorrect input and human mistakes, rather than system malfunction or inadequacies.

Pilot product

The pilot approach installs a system on only one product (or product line), operating it in parallel for testing and training, and then proceeds to expand the installation to other product lines. If there is a suitable pilot product, typical of the rest of the enterprise, for a short time it can be run separately, without a high proportion of parts and subassemblies that are being planned by the existing system. Otherwise, the problems are similar to parallel running.

Progressive

Progressive installation is one module at a time. Successfully add a module then move on the next one. Module-by-module conversion is very common; it can incorporate parallel running or pilot line approaches; and it spreads out the workload of preparing for conversion. Effort can be concentrated on the module under conversion. Each module is tested in isolation and in relation to other functioning modules. At the end of each test, conversion is signed off by the project team before moving to the next module.

Conference room pilot

Table-top simulation of a new system is often carried out part-time by future users in a conference room setting. (Figure 13.15). A full set of test data is created, all transactions are processed, their outputs acted upon and the resulting data fed back into the system. All activity stays within the test system

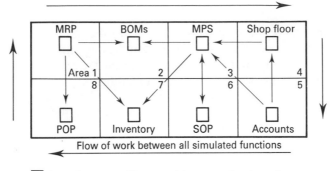

Figure 13.15 Conference room pilot around the table to work through all aspects of system operation and procedures. Not all functions are shown.

and results can be monitored to see if the system is working properly. This is an excellent way to test procedures and to train personnel in their use. It greatly increases confidence when going live and can be used with any other change-over techniques, well before software installation.

■ Detailed implementation

The creation of a detailed plan is the responsibility of your own project team, but a standard plan is available from Oliver Wight or the BRIM group in the UK and the US. This implementation plan is a generalised framework applicable to nearly any company. Its primary uses are

- To provide a clear statement of priorities.
- To separate the vital and trivial and to keep them in perspective.
- To provide a road-map for implementation.

The implementation plan outlines the basic functional areas before they are broken down into specific actions and milestones. This provides a very practical plan organised to constantly focus attention on items that have the greatest impact. The *people* part of an MRP/ERP system is fully 80% of the system. The system will only work when people understand what it is, how it works and what are their responsibilities. For this reason, education and training are listed at the front of the implementation plan. The computer software and programming effort is less likely to undermine the success of an MRP system, so this topic is covered later in the plan. The implementation plan also provides a detailed schedule of events that have to be accomplished in order to implement the system. *The most effective way to use the plan is to tailor it to each company and then use it as the agenda for management reviews of implementation progress.*

■ Practicality

An implementation plan is not a theoretical exercise. Since the first Oliver Wight plan was developed, it has been used successfully by a number of companies. Would these companies have been successful without it? That is impossible to say, but the plan does work, it is practical, and those who have used it swear by it.

■ Implementation plans help current users

The Oliver Wight implementation plan is also meant for companies using an MRP system. Many companies have MRP technology but they are not using it well. A reimplementation plan can help these companies, as the tasks for improving an MRP2 system are the same as those to implement it correctly. Those that have already been done can be deleted from the plan. 'MRP: Making it Work', an Oliver Wight video, shows how to use the plan to implement MRP2.

■ Performance measurements

For any company working toward class A, world class performance with its MRP2 project, the implementation and maintenance of a formal operating performance measurement system is essential. Performance measurement on each main operation sets up important milestones and guides continuous improvement under total quality (Figure 13.16).

Figure 13.16 represents the three levels of management planning and execution and establishes the responsibilities, accountabilities and measurements required to chart performance for each function. Performance

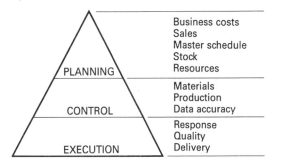

Figure 13.16 Management planning, control and execution.

439

measurement establishes a formal process which the steering committee can use to communicate objectives and performance expectations throughout the operation (for class A, 95% or better performance in each functional area).

The first step towards a performance measurement system is the establishment of performance objectives, reviewed regularly to monitor performance. They should be clearly defined, easily understood and measurable, with responsibility for each assigned to an employee or group in the related functional area. The next step is to measure current, pre-MRP2 performance against these objectives to establish a baseline.

Baseline data for measuring performance

An important step that is frequently overlooked is the planning for collecting base reference data at the start of the project. It is extremely important to identify the existing initial conditions, as only in this way can progress in achieving the objectives be measured, or time and effort justified. These baseline figures become the performance measures of the overall system as it progresses. Fifteen or twenty items can be defined then measured and reported at frequent intervals. This data and subsequent performance tracking are essential to secure the continuing commitment of management and all users of the project. Here are a few basic measures as examples:

1. Inventory
   - Inventory value (in raw stock, WIP, finished goods)
   - Inventory turns (in total, by category, by ABC class)
   - Inventory record accuracy (cycle count results)

2. Work/job orders
   - Percentage of orders completed on time
   - Accuracy of open order balances
   - Number of open orders
   - Number, or the standard hours, of past due orders
   - Percentage of orders rejected totally or partially
   - Kits to be picked and kits picked on time

3. Purchase orders
   - Percentage of orders (or items) received complete and on time
   - Percentage of items rejected
   - Accuracy of open order quantities and due dates
   - Percentage of orders placed in three days or less
   - Price variances against standard costs

4. Customer orders
   - Percentage of orders, or dollar value, shipped complete and on time
   - Percentage shipped one week late, two weeks late, etc.

5. Backlogs
    - Number of hours/days of receipts waiting to be received
    - Number of hours/days of receipts in inspection
    - Value of material in material review awaiting disposition

6. Exceptions
    - Percentage of orders past due
    - Number and percentage of orders to be rescheduled or cancelled

These are only suggestions; others may be required. The aim is to find items that are indicative of how the total business, production and planning system is working or how the company is working with the system. It is very important that brief procedures for calculating these measures be written, defining clearly what is to be counted, how the data is to be recorded, and how the calculation is to be done. For example, if the measure were to be percentage of orders shipped complete on time, is it the number of orders, or is it their value, or both? Does 'on time' mean the customer's required by date, or our promise? Writing the procedures helps to define what is being measured, and prevents subtle changes that can reduce its relevance over time. *Also to be agreed are actions and responsibilities that focus on the improvements required.*

Once the objectives have been determined and a good set of current operating performance measurements are in place, progress checks must be applied regularly to monitor performance improvements and to identify problem areas. A report will ideally contain both business and shop-floor measurements. A report card may help to formalise this process (Figure 13.17).

Definitions of performance measures

Business plan

The business plan lays out what, when and how much the company needs in markets, products and profits to meet its overall business objectives. These objectives are stated in value or percentage value of income, investment or rate of return sought for each month and year in the plan. A key measurement for business is return on investment (ROI) which calculates the income earned from the investment required to support product and market opportunities. Business plan performance is measured as the percentage that the actual ROI constitutes of the planned ROI.

Sales plan

The sales plan states what, when and how much product is required to meet anticipated customer demand. These objectives should be stated in cash value and units by product line for each month and year in the plan. Sales plan

Functional area	Responsibility	Performance objective	Performance measurement	Class A performance
Business plan	General Manager	Return on investment	Actual ROI Planned ROI	95%
Sales plan	Sales dept.	Sales performance	Actual sales Planned sales	95%
Production plan	Manufacturing	Production performance	Actual prod Planned prod	95%
Master schedule	Manufacturing	MPS performance	Actual MPS Planned MPS	95%
Materials plan	Materials dept.	Schedule reliability	Orders released on time	95%
Capacity plan	Manufacturing	Capacity performance	Hours produced Planned	95%
Bills of material	Engineering depts	BOM accuracy	Accurate bills Total bills	99%
Inventory control	Materials dept.	Inventory accuracy	Parts correct Total parts	99%
Operation cards	Manufacturing or Production engineering	Routing accuracy	Accurate op. cards Total op. cards	99%
Purchasing	Purchasing dept.	Schedule accuracy	Parts delivered Parts due	95%
Shop floor control	Manufacturing	Schedule accuracy	Parts completed Parts due	95%
Schedule performance	General manager	Delivery performance	Units delivered Units promised	95%

Figure 13.17 Performance measurement report card.

performance measures actual sales as a percentage of planned sales by month. There are three aspects of class A sales plan performance; meeting the sales **dollar** plan within a +/−5% tolerance, meeting the sales **unit** plan within a +/−10% tolerance, and meeting the sales **mix** plan within +/−15% for 85% performance.

Production plan

The production plan says what, when and how much is required in production rates and levels of output to maintain the desired level of finished inventory or backlog to meet the sales plan and provide satisfactory customer service. These levels should be stated in units of production by product line for each month and year in the plan. Production plan performance represents actual production as a percentage of planned production.

Master production schedule

The master production schedule (MPS) bridges sales and manufacturing by detailing what, when and how much of each product, model, feature, option or product mix level must be scheduled for production to meet the sales plan. Schedule performance measures the actual production as a percentage of the planned MPS by model, feature, option or product mix.

Materials planning

Materials planning determines the schedules of what, when and how many parts are required to produce the product and to maintain part priorities required for the MPS. The key measurement of materials planning performance is schedule release reliability, which indicates whether orders are released on time to meet the MPS. Performance in this operating area is expressed as the number of orders released on time as a percentage of total orders.

Capacity planning

Capacity planning specifies what, when and how much capacity in labour and machine resource is required to produce the MPS. It should be stated in standard hours required by departments and work centres for each week, for each month and for the year. The measurement of capacity planning performance reflects the number of hours produced as a percentage of the hours planned by work centre and department.

Bills of material (BOMs)

Bills of material specify what, when and how many parts and materials are required to produce the product, as well as the assembly or process relationships. BOM accuracy is the key measurement, indicating whether the BOM really represents the product as it is being produced. BOM performance calculates the number of bills 100% accurate as a percentage of the total number of bills. Even if only one part on the bill is inaccurate, the bill does not reflect reality within the plant. A procedure for BOM accuracy measurement is in Appendix D. In many make-to-order operations, it can be valuable to measure the percentage of jobsheets or complete BOMs on time to the MRP deadlines, indicating how well the order processing and definition/interpretation activities are being performed.

Inventory/material control

Inventory/material control states what, when and how much inventory is on hand to produce the product. Performance is commonly measured by comparing the inventory record to the physical inventory, calculated as the number of parts correct (where the physical count equals the stores record value within an acceptable variance as a percentage of the total number of records counted. A further measure is the number and duration of material shortages to production, calculated either as a percentage against number of successful issues, or as a percentage against total active stores items.

Operation or job cards

Operation or job cards specify the nature of the operations and the number of standard hours required to produce the product. They specify the operations and sequences, the machine or work centre, tooling, set-up and run hours for each listed operation. Accuracy measurements indicate whether the steps represent the operations and sequences as they actually occur in the plant. Accuracy is calculated by taking the operations in agreement with the job cards as a percentage of the total number of operations specified. The level of change in the schedule can be indicated by measuring the number or percentage of amended and/or additional job cards per period.

Shop-floor control

Shop-floor control specifies what, when and how much labour and material is required on the shop-floor to deliver the manufactured parts on or before the due date to meet the MPS. The key measurement of shop-floor control is production schedule performance. The measurement shows whether the parts are being completed on time in the shop. It calculates the number of manufactured parts completed as a percentage of total parts due, and it should be stated by work centre, schedule performance or assembly line.

Purchasing plan

The purchasing plan details what, when and how much purchased material is required to meet the MPS. The key measurement of purchasing is schedule performance, which indicates whether your vendors are delivering the parts on time. Schedule performance reflects number of purchased parts delivered as a percentage of the purchased parts due. Percentage of bought-out parts rejected for nonconformance should also be measured, tied in with delivery performance to obtain a *supplier performance rating*.

Delivery performance

Delivery performance defines what, when and how many units to build, ship and deliver on time to meet the sales plan. Delivery schedule performance measurement tracks whether the product was delivered to the customer when it was promised, and reflects the number of units delivered as a percentage of the number of units promised (Figure 13.18 and 13.19).

Number of cases delivered and reported site shortages

Wk	Confirmed Short	Spec change	Shorts (Damaged)	Disputed Short	Total all shorts	Cases Delivered	Average Shorts per Case
34	6	2	16	10	34	80	0.43
35	24	4	6	4	38	134	0.28
36	2	0	2	6	10	84	0.12
37	30	0	9	27	66	141	0.47
38	14	0	6	10	30	158	0.19
39	0	0	2	0	2	69	0.03
40	4	0	3	5	12	159	0.08
41	6	0	1	0	7	49	0.14
42	5	0	6	7	18	102	0.18
43	11	0	1	3	15	122	0.12
44	6	0	5	5	16	76	0.21
45	1	0	1	4	6	64	0.09
46	3	0	4	0	7	30	0.23
47	1	0	2	2	5	118	0.04
48	0	0	0	0	0	16	0.00
49							

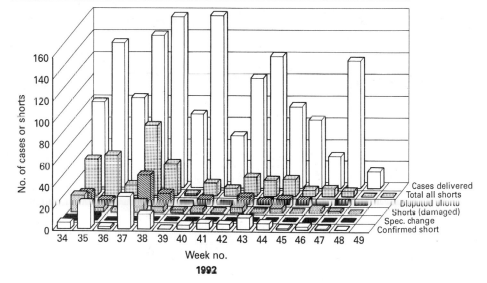

Figure 13.18 Analysis of complete delivery performance.

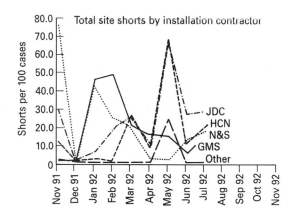

Figure 13.19 Analysis of site installation shortages.

Function Area	Measure	Previous year Aver.	Dec ('91)	Jan ('92)	Feb	Mar	Apr	May	Jun	Jul	Aug	Sept	Oct	Nov	Cum. Y.T.D.	Comment
Planning the Business	Business Costs v Plan	79.6	80.7	100	100	89.2	99.3	95.8	94.5	94.9	100	100	100	100	92.4	Costs reduced in line with sales
	Sales v Plan	83.1	44.2	99.4	81.3	87.0	68.2	73.5	75.0	81.2	100	100	96.1	100	78.4	Sales plan revised
Planning Manufacturing	Master Schedule v Plan	72.7	99.7	88.6	87.8	70.8	78.0	92.1	76.1	86.8	85.6	97.2	83.7	88.8	86.3	Scheduled in line with sales
	Materials v Plan (Stock turns)	77.0	48.7	70.6	72.9	76.0	78.0	79.0	82.0	85.0	89.0	73.6	74.6	76.0	75.4	High inventory to be reduced
Accuracy of Data Records	Bills of Material accuracy	60.0	76.0	74.0	77.0	79.0	82.0	85.0	90.0	87.0	91.0	93.0	92.0	93.0	84.9	Varies with random sample
	Inventory Clcle Count accuracy	77.7	90.6	91.0	93.0	85.4	84.0	88.0	82.6	87.3	95.0	88.7	93.6	83.9	88.6	On improving trend again
Achieving The Manufacturing Plan	Bought in Short per month	87.0	87.2	90.8	93.0	97.0	98.1	98.9	98.8	97.8	99.2	98.7	98.7	98.4	96.4	Continuing good
	Job Lines on time	88.1	85.3	56.2	60.0	87.9	59.0	68.2	86.7	75.0	68.0	75.8	74.1	77.0	72.8	Fluctuating with sales volume
	Production v MP Schedule	91.0	94.8	100	98.6	100	88.2	97.8	98.1	73.4	98.5	100	99.5	100	93.6	Shop keeping up with schedule
	a Plant Total (Amb'dor % Complete)	95.0	94.6	84.2	85.5	89.2	89.0	92.6	94.8	88.5	88.5	88.4	95.0	96.4	90.5	Improving trend
Meeting Quality Plan	Site Shorts	75.9	95.2	75.1	70.2	78.2	81.0	79.0	83.0	78.6	81.0	78.3	86.6	92.1	81.5	Worse than ex-factory
	a Orders Shipped Complete (Cases)	86.0	100.0	100.0	99.1	98.4	100.0	97.9	100.0	97.5	96.7	98.1	100.0	100.0	99.0	At leaving factory
	a On Time Delivery	89.0	98.4	100.0	98.9	100.0	97.0	100.0	99.3	99.2	98.1	99.0	99.4	99.0	99.0	Continues very good
Total	OVERALL PERFORMANCE	81.7	84.3	84.5	84.8	86.5	84.8	88.3	89.3	87.1	90.1	89.1	91.1	99.5	87.5	
CLASS	90+(A),85+(B),80+(C),70+(D).	C	C	C	C	B	C	B	B	B	A	B	A	A	B	

Explanation

	Calculation
Act. costs/planned costs × 100	£1871/1979 × 100 = 94.54%
Act. sales/planned sales × 100	£1807/2833 × 100 = 63.78%
Sched./prod. plan value × 100	685/900 × 100 = 76.11%
Act. stk. turns planned turns × 100	2.48/3.90 × 100 = 63.58%
As per procedure for BOM accuracy	9/10 × 100 = 90.00%
As per procedure for cycle counts	518/627 × 100 = 82.62%
Stores issues – shorts/issues × 100	6344–77/6344 × 100 = 98.78%
Total job lines – late/tot. × 100	30–4/30 × 100 = 86.66%
Act. cases produced/MPS × 100	672/685 × 100 = 98.10%
As USA 1	100 – (17/324 × 100) = 94.75%
Case on Site with Shortages	330 – 56/330 × 100 = 83.03%
As USA 2	100 – (0/346 × 100) = 100.00%
As USA 3	100 – (1/140 × 100) = 99.29%

Figure 13.20 A performance report used to track progress. Note the adapted measurements and percentage breakpoints to suit the business.

Other performance measures may be included where appropriate, particularly to measure interdepartmental (customer/supplier) achievement of objectives. For example, we may want to measure the percentage on-time and completed sales order details from estimating to order implementation (their customer function) (Figure 13.20). Where make-to-order production is required, shipping complete products on time may be a weak area, and should be tracked to see the effect of improved procedures and systems.

Weighting of performance measurements

Different viewpoints exist as to the relative weighting of performance measures used on a company-wide report; there is no rigid standard. But remember the objective is to track business performance as well as key accuracy levels in areas such as BOMs and inventory. The author favours unweighted values, as shown in Figure 13.20. However, the method used to measure and derive a percentage performance of *every* item on the report should be discussed and agreed with the management and project teams before use.

■ Summary

More and more people are looking for information on the implementation and operation of MRP2 systems. Convinced of its value, they desire a proven strategy for implementation. By describing failures as well as successes, I hope to have given the reader an understanding of what is required to track progress and ensure success.

Resource planning is essential for improving business performance. By successfully creating and running ERP, you have established a firm foundation to become world class.

■ Further reading

A. T. Kearney Management Consultants (1990) *Leaders and Laggers in Logistics*.
Bititci, U. (1993) 'Measuring your way to profit'. Paper presented at FAIM 1993.
Burbridge, R. N. (ed.) (1988) *Perspectives on Project Management*, Peter Peregrinus.
Landvater, D. V. and Gray, C. D. (1995) *MRP2 Standard System Handbook*, Oliver Wight.
Oliver Wight Co. (1981) *MRP2: Unlocking America's Productivity Potential*.

'Making it Work', a video about MRP, is available from Oliver Wight Video Productions.

Appendix A

Basic Business Checklist

and Key Ratios

■ General questions for each function

Objectives

- State the department or function objectives, if any.
- Who defined them?
- What factors determine these objectives?
- Are they consistent with the overall objectives of the business?
- Have they been formalised and communicated?
- Are they reviewed and up dated? By whom?
- Are the current objectives achievable?

Policies

What policies govern management of the function?
- Have such policies been stated and communicated?
- Is the function regarded as being subordinate to other activities?
- Who co-ordinates, controls and reviews the policies and objectives?
- Describe the role(s) of the function in the firm.
- Are the policies consistent with the overall policies of the business?

Organising

- Who has overall responsibility for the function or department?
- Describe the organisation and produce, if possible, an organisation chart.
- If job descriptions exist produce them. Do they match current duties?
- Appraise internal/external communications. Is there effective communication of functional plans to other functions?

- Interpersonal clarity and understanding of roles?
- Are the organisation and responsibilities acceptable to other staff?
- When was the organisation reviewed last?
- What changes resulted?

Managing/controlling

- Quality of general communication?
- Selection, compensation and motivation of staff?
- Plans and status of management development programme – if it exists.
- List key personnel, their supporting staff and identify their qualifications. Evaluate calibre of staff (number of graduates, members of professional bodies, etc.).
- Check for control procedures in relation to key result areas – are formal procedures in place?
- Identify process and speed with which corrective action is taken when problems are encountered.
- Is there proactive, or only reactive approach to problems and improvements?
  Describe the lines of communication between the function and the rest of the business.
  - Are there any standards of performance?
  - Appraise their appropriateness.
  - To what extent does the department feel that its work is hampered by lack of facilities?
  Appraise the nature of the control exercised over departmental expenditure.
  - Who is responsible for controlling the departmental effort?
  - What sort of information is used for controlling expenditure and costs? Comment on the quality, frequency and appropriateness of such information.
  - How capable are the staff responsible for controlling expenditure?

■ General management

- Who are the main stakeholders in the business – owners, shareholders, senior directors?
- List them in order of importance and contribution as perceived by management.
- General direction of the business – is there well-understood and agreed direction for business direction at a senior level?
- List the corporate objectives as perceived by management.

- How are objectives measured, how often, and by whom?
- Are these formally stated? Are they generally known within the company?
- List and describe external factors that can significantly affect achieving these objectives (e.g. market change, increased competition, etc.)
- Are there plans in place to meet these changes?
- Is there a set of general performance measures in place? Are they the traditional basic management accounting and production efficiency measures, or do other measures exist? (e.g. performance table as in Chapter 13)
- Is a historical record of any performance measures available? Can any trends be measured?
- Does the company regularly act on these measures to secure continuous improvement?
- Are cross-functional teams used to tackle particular problems?
- Describe how the planning of business objectives vs. performance aims is carried out.
- Succession plans, planning and preparing suitable people to replace key posts in the company, are highly desirable. Are plans in place for this, and are they reviewed at regular intervals?
- Is there a structured programme for management development in place and operating?
- How much autonomy are employees allowed to define on-the-job behaviour?
- Are formal procedures in place in any or all functions? Are they used?

Management information

- Determine the quality and frequency of forecasts
- How far out into the future is forecasting done?
- What information sources are used?
- Are the forecasts used in running the business?
- To whom are the forecasts communicated?
- Is the accuracy checked or compared? How often?

Business control

- Is an integrated management information system used?
- How fully is the information used to run the business?
- Describe the system of control and the purpose of information produced.
- Investigate quality, frequency and appropriateness of existing procedures including budgets.
- Gauge competence of personnel responsible.
- How much effort is needed to collect information?

- What are results compared with, and how?
- Are reasons for variations examined on a regular basis?

Organisational issues

Describe the present organisational structure
- Is an organisation chart available? Obtain organisational charts and examine.
- Is there a reasonably flat control structure, or are there (too) many layers and few direct reports?
- Levels of responsibility/accountability?
- Does the organisation conform to the current needs of the business?
- Is centralisation or decentralisation planned or required?

Chief executive/managing director

- His or her powers and role in the business?
- Relationship with others?
- Esteem in the company?
- Clarity of responsibilities?
- How is information communicated downwards to lower grades? How effective is it?

Communications

Between which points in the country and within what businesses/levels are communications required?
Explore the following:
- State any communications objectives that exist.
- Are they understood?
- Who is responsible for planning and designing communications systems?
- Effectiveness of communication? – appraise and describe.

Planning

Is long-range planning a formal process? If yes:
- Over what period?
- Formality and quality of the process?
- Frequency and regularity?
- Whose responsibility is it to run and act on the planning sessions?

- To whom are the results communicated? Are actions defined?
- Appraise quality of implementation methods and procedures.

Management services

This is an attempt at appraising the quality of services which exist to help management to perform its job. Under this heading the nature, aim and effectiveness of each service must be identified. In each case the relative cost of using outside subcontract or consulting specialists should be considered. Likely services:

- Organisation and methods (O&M).
- Work study
- Operational research
- Public relations
- Law
- Taxation and insurance
- Patent, copyright and registered trademark
- Logistics/distribution/transport advisor
- Packaging advisor
- Health and safety/COSHH
- Quality Assurance – BS 5750, ISO 9000
- Computing / MIS contract services

■ Marketing

General

- Which sector of industry does it fit in?
- Identify market segments and quantify relationships.
- List products and quantify their overall performance over five years.
- Trends of shift by segment and/or by product may thus be identified.
- Competition – size, number and type (brand or function).
- Relative strength of competitors (relate to previous example of company comparison).
- General facts and data about the industry and its development, including the environment, technologies, growth, specific problems.

Marketing orientation compared with similar firms

- Market share
- Market penetration

- Growth
- Level of competitiveness in market, e.g. very high, high, medium, low, very low
- Stability of market, e.g. very high, high, medium, low, very low
- Frequency of new products, e.g. very high, high, medium, low, very low
- Profitability by product/market
- Profit/return on invested capital
- Profit margin on sales
- Sales vs. capital employed (turns)
- Sales vs. fixed assets (turns)
- Sales vs. stock value (turns)
- Sales per employee
- Production value/person
- Profit per employee
- Others

Information

- Is forecasting a formal, regular process?
- Is a proprietary forecasting system or package used?
- Who uses information?
- Frequency of review?
- How far ahead does the process look?
- Levels of accuracy – are reviews carried out, and how?
- Scope and quality of market/marketing research?
- Control procedures for measuring effectiveness?

Planning

- Frequency of planning and review cycle?
- Are they based on facts/forecasts or mechanical extrapolation?
- By whom?
- To whom are they communicated?
- How dynamic is the planning process?
- How reasonable and achievable are the plans?
- How effective have plans been in past years? How are they rated?
- Which resources are covered by plans?

Product

- Does the firm have product objectives?
- General assessment of product quality and standing in market.

- Evaluation of comparable competitive products.
- Life-cycle performance (sales, profit, investment recovery vs. anticipated life per product/market). Is life-cycling concept used? Number of years/months?
- Product development and introduction?
- Number of different and active products and variants; trends?
- Quality of innovation, R&D?
- Frequency of new product ideas?
- Formal screening procedures for new products including:
    Product evaluation in 'New Model Introduction' procedure
    Business/market analysis
    Controlled experimentation/prototyping
    Commercialisation
- Rate/ratio of successful launches?
- Product design and styling?
- Quality of service?

Pricing

- Does the firm have a pricing strategy and objectives? If yes, what are they?
- Price elasticity/flexibility and rules for control?
- Identify the firm's behaviour in the face of competitive pricing practices

Marketing/Promotion

- Expenditure over five years compared to sales by product?
- Does the company operate an advertising by objectives approach?
- Methods for determining promotion budgets
- Is effectiveness measured? If so, how? Is measurement used to feed back and amend strategy?
- Quality of media planning?
- Is an advertising agency used? If so, its reputation?
- Quality of information about competitive promotional strategies and responsiveness?
- Costs of promoting each product? Total and per unit costs?
- Quality of sales aids used?

Selling

Sales strategy and objectives

- Sales objectives, now and over a five year period. Identify changes

- Who is responsible for such objectives?
- Are objectives consistent with marketing objectives and other ingredients of the marketing mix?
- Are they accepted by everyone?
- Are they based on sound measurement and forecasting techniques?
- How accurate have previous sales forecasts been? Are regular reviews carried out? If so, who is involved?
- Identify most important factor that affects selling results.

Sales planning

- To what extent are selling activities planned?
- Period? Frequency? Routes and priorities?
- Are plans consistent with other functional plans and resources?
- To whom are they communicated?
- Have they been achieved in the past? If not, has management responded to deviations?
- Has management identified factors that affect selling results?

The organisation of selling

- Appraise quality of management and sales force in measurable norms.
- Who is responsible for determining sales objectives, targets and standards of performance?
- Is the organisation capable of achieving its objectives?
- Consider the quality of:
      Recruitment
      Training of sales staff
      Development and promotion
- Motivation
      Performance related pay/bonus scheme? If used, does it effectively reward, and are targets appropriate?

Controlling sales force

- How is performance measured? (e.g. sales quotas, selling territories, number of calls, sales volume, account development, profitability of sales)
- Average cost per call for each salesperson?
- What steps are taken to improve performance?
- Quality of information, frequency and appropriateness? Overall appraise the quality of sales effectiveness control.
- What are selling expenses as percentage of sales?
      overall
      by market
      by product/group

per order (average)
per salesperson

Customers

- How many customers are there?
- How many account for 80% of sales (Lorenz curve or Pareto)
- Are customers credit rated?
    Appraise credit control procedures
    Describe bad debt record (use days to pay records)

Distribution

Planning

- Describe the firm's distribution plans.
- Are they dynamic and effective? How are they measured?

Facilities

- Describe the firm's physical distribution facilities and try to gauge the effectiveness with which they are utilised:
    a. warehousing
    b. packing
    c. despatching
    d. transport
    e. delivery
    f. service

In relation to each facility consider:
- Level of utilisation
- Comparability of own service with cost of external contractors
- Competitive practices (Inter-firm analysis)
- Level of service offered and expected by customers (differences often exist)
- Cost analysis by product and market.

Channels of distribution

- Identify available channels
- Describe channels actually chosen; try to explore reasons for such a choice
- Gauge performance of middlemen against:
    a. their own past performance
    b. each other

This should identify good and poor performers in relation to each product and each market.

■ Are existing channels effective enough to achieve firm's objectives?

Controlling

■ Carry out and examine distribution costs analysis.

Business performance ratios

There are many different ratios and measures for business performance, in each sector of industry and commerce. Selecting and interpreting those relevant to a particular business depends on several factors, including business type, strategies and method of operation. Compare over a period and between companies.

$$\text{Ratio } 1 = \frac{\text{Marketing contribution}}{\text{Marketing assets}}$$

Analysed by product
Marketing contribution = sales − (marketing costs + variable manufacturing costs)

Marketing assets = finished goods stock − debtors − distribution facilities

$$\text{Ratio } 2 = \frac{\text{Marketing costs}}{\text{Sales}}$$

$$\text{R3} = \frac{\text{Sales}}{\text{Marketing assets}}$$

$$\text{R4} = \frac{\text{Warehouse costs}}{\text{Sales}}$$

$$\text{R5} = \frac{\text{Distribution costs}}{\text{Sales}} = \frac{\text{Home dist. costs}}{\text{Home sales}} + \frac{\text{Export dist. costs}}{\text{Export sales}}$$

$$\text{R6} = \frac{\text{Advertising costs}}{\text{Sales}}$$

$$\text{R7} = \frac{\text{Selling costs}}{\text{Sales}} = \frac{\text{Home selling costs}}{\text{Home sales}} + \frac{\text{Export selling costs}}{\text{Export sales}}$$

$$\text{R8} = \frac{\text{Discounts}}{\text{Sales}}$$

$$R9 = \frac{\text{Sales office costs}}{\text{Sales}}$$

$$R10 = \frac{\text{Sales}}{\text{No. of orders or invoices}}$$

$$R11 = \frac{\text{Sales office sales}}{\text{No. of orders or invoices}}$$

$$R12 = \frac{\text{Finished good stock}}{\text{Sales}}$$

$$R13 = \frac{\text{Debtors}}{\text{Sales}}$$

$$R13 = \frac{\text{Debtors}}{\text{Average daily sales}}$$

There are a number of useful ratios (see: *How to Use Management Ratios* by C. A. Westwick). They must be used over a reasonable period (at least five years) to show trends. If used in an interfirm or interdivision comparison, ensure that data used has been assembled and compared in a valid and comparable manner. Before using any ratio try to gauge its usefulness and appropriateness to the company, and the purpose of the review.

■ Production

General

Production resources

- List processes and technologies and productive resources allocated to each.
- Appraise interfirm status in relation to above (See interfirm ratios later).
- Plant utilisation and general productivity.

Production planning

- Purpose?
- Period and frequency?
- Consistency with objectives?
- Are they based on facts/forecasts of demand?
- Who is responsible for these plans?

- To whom are they communicated?
- Degree of change in the planning process?
- How reasonable and achievable are these plans?
- How effective have plans been in past years?
- What are the scarce resources or other limiting factors on production effectiveness?
- Have they been identified for planning purposes?

Standards

- Describe standards used.
- Procedures for reviewing and revising standards.

Production schedules

- Describe production schedules and procedures.
- Is MRP1 or MRP2 used?
- If so, are measures of performance used?
- Is master scheduling appropriate and/or used?
- Appraise their quality and effectiveness in relation to marketing facts.

Staffing

- Describe staff planning to meet future requirements.
- Is there a programme in place for training and progression?
- Have staff shortages occurred.
- List staff qualifications appropriate to their jobs.

Managing/controlling

Plans and decisions

- How are production decisions communicated? Appraise staffing policies.
- How are production plans put into effect?
- How are production activities co-ordinated with each other and with all external activities?

Standards of performance

- Who determines them?
- Frequency of returns
- Appropriateness (efficiency, throughput, etc.)
- Compared with what?
- Are reasons for variations examined?

Output levels

- What were the highest levels of output achieved?
- Increase (or decrease) in productivity in the last five years?
- Appraise quality of production control procedures.

Materials

Specify materials used

- What objectives determine use of materials?
- Are they accepted by everyone?
- Policies affecting use of materials?

Information

- Frequency and accuracy of material forecasts?
- Who uses the forecasts?
- Are materials forecasts co-ordinated with other forecasts?

Planning

- To what extent are materials planned, generally or specifically?
- Is MRP1 used? How effectively?
- What period ahead do plans cover?
- Appraise the quality of these plans.

Managing

- How are materials plans put into effect?
- What corrective action is taken to change a materials situation where a change is indicated by adverse results.?

Purchasing

- Is purchasing considered to be a profit-earning activity?
- Is there a purchasing plan?
- Have economic lots been decided for all major supplies? Are vendor schedules used?
- Vendor engineering and interfacing?
- Describe purchasing organisation, is it centralised or decentralised?
- Quality of personnel?
- Service to production department?
- Effectiveness of records and paperwork?
- Control procedures?

- Cost of purchase orders?
     stock-turn measurement?
     is 80% of cost in 20% of items?
     total cash discount obtained?
     ratio of such discounts to value of purchase over a period?
     level of activity, number of orders placed, number of invoices handled, number of orders late, number of sales staff seen?
- Investigate procedures for evaluating suppliers and ensuring that alternative sources of critical items exist.
- Are approved suppliers' lists in existence and used? Describe procedures regarding above.
- Plans for contingencies (e.g. fire or damage in suppliers' plant).

Record of suppliers' performance

- Quality control of supplies?
- List suppliers and show breakdown figures of values supplied over a five-year period.
- Indicate credit arrangements offered by suppliers.
- Describe supplier record of past performance in the field of creative product development.
- Describe receiving procedures.
- Is the procurement of materials co-ordinated with requirements?

Storing

- Describe stock policy and procedures.
- Who is responsible for stock control?
- Identify MRP or reorder controls used.
- What level of protection against stock-out is in use?
- Describe stock records and appraise their quality.
- How many items in stock? Calculate $/£ value of total individual items, provide information on rate of stock turnover.
- Is cycle counting employed? Check accuracy.

Maintaining quality

- Yield on materials used?
- Percentage and cost of scrap?

Maintaining economy

- Procedures for cost reduction?
- Evaluate analysis procedures and appraise their quality.
- What is the percentage cost of materials to finished goods for the last five years?

Controlling

- Appraise nature of materials control and supportive procedures
- Who is responsible?
- Describe the purpose, frequency and appropriateness of such control procedures?

Ratios

$$R1 = \frac{\text{Production contribution}}{\text{Production assets}}$$

$$R2 = \frac{\text{Production costs}}{\text{Sales value of production}}$$

$$R3 = \frac{\text{Direct materials cost}}{\text{Sales value of production}}$$

c.g. Company 1 = 42%
Company 2 = 50%

$$R4 = \frac{\text{Direct labour cost}}{\text{Sales value of production}}$$

e.g. Company 1 = 13%, Company 2 = 14%

$$R5 = \frac{\text{Production overheads}}{\text{Sales value of production}}$$

The following ratios are used as indicators of progress, and are marked where a rise or fall in the resulting ratio is a cause for concern

RM Performance

$$R6 = \frac{\text{Raw materials stock}}{\text{Average daily purchases}}$$ Rising

Shop-floor efficiency

$$R7 = \frac{\text{Work in process}}{\substack{\text{Average daily value of issues to production} \\ \text{and products completed}}}$$ Rising

Finished stock

$$R8 = \frac{\text{Finished goods stock}}{\text{Average daily value of production completed}}$$ Rising

Utilisation of premises

$$R9 = \frac{\text{Sales value of production}}{\text{Area of factory premises}}$$ Falling

Utilisation of plant and machinery

$$R10 = \frac{\text{Value of plant}}{\text{Sales value of production}}$$

Rising

Average age of plant

$$R11 = \frac{\text{Plant at depreciated value}}{\text{Plant at undepreciated value}}$$

■ Research and development

Planning

Forecasts

- Are forecasts made of long-, medium- and short-term R&D requirements?
- Appraise the accuracy of past forecast.

R&D plans

- Is the planning process reviewed regularly?
- How are research proposals appraised?

Managing/controlling

- List the plant and equipment that the R&D department possess
- To what extent does the R&D department feel that its work is hampered by lack of facilities?
- Describe the use that is made of other research laboratories
  Appraise the record of the department's originality/success/failure
  Calculate annual R&D expenditure over a five year period. What percentage is this of the Company's turnover?

Design

- What are the objectives of the business in relation to design?
- Where does design fit into the firm's activities?
- What are the policies which govern design management (e.g. function, cheapness, technical superiority, appearance)?
- Describe the way design requirements are planned and implemented.
- Appraise the quality of the firm's design record.
- When were the designs of the company's products last reviewed?
- What is the relationship of product life-cycle to design planning?

What are the following:
- Total cost of design efforts
- Costs of internal and external effort
- Ratio of design costs to the overall business turnover
- Ratio of design to production costs

Ratios

$$R1 = \frac{\text{Expenditure on R\&D}}{\text{Annual sales}}$$

$$R2 = \frac{\text{Expenditure on R\&D}}{\text{Total no. of employees}}$$

$$R3 = \frac{\text{No. employed in R\&D}}{\text{Total employees}}$$

$$R4 = \frac{\text{Salaries on R\&D employees}}{\text{Total expenditure on R\&D}}$$

$$R5 = \frac{\text{No. qualified scientists employed on R\&D}}{\text{Total qualified scientists in firm}}$$

■ Personnel

General

Planning

- Forecasting personnel requirements
- Accuracy of past forecasts?
- Staff budgets?
- Is the planning process reviewed regularly?
- How effective is personnel planning?
- Management of change?

Control

- Cost of running the function and trends?
- Do standards of performance exist in relation to personnel activities?
- Describe information procedures and appraise their quality/frequency.

Specific areas

Management development

- Policy which governs management development?
- Quality of planning?
- Budgets and their control?
- Programmes (in-house and external) and their appropriateness?
- Induction training?

Selection and recruitment

- Policies?
- Quality of procedures and records?
- Cost-effectiveness of advertising over a period?
- Is patterned interviewing used?
- Tests and their effectiveness?
- Turnover of new recruits?

Employment conditions

- Hours of work? Compare with other firms in locality and industry.
- Hours lost through absence?
- Percentage of overtime to normal time at present?
- Cases for overtime?

Remunerating

- Describe the wage and salary policy.
- Are annual basic salaries and wages guaranteed?
- Are incentives used?
- Describe fringe benefits including holidays.
- Describe the way remuneration is planned for the future.
- Assess your perception of the level of personnel satisfaction with the system.
- Is job evaluation being used to determine basic rates of pay and salary?
- If yes, describe and appraise.
- Are merit rating techniques used?
- How are salary increases determined?
- Cost of wage, salary and fringe benefits administration?
- Describe the way remuneration practices are controlled, in particular:
  - Are levels of total remuneration of individual managers examined?
  - Are fringe benefits consistently applied?
  - When was the wage and salary structure last overhauled
  - What proportion of total wages is represented by fringe benefits?

Promoting

- Is promotion policy specified and known to all employees?
- Succession plans?
- Records of promotion plans?

Leaving

- What is the level of performance of employees who leave?
- Are their reasons for leaving analysed?
- Are exiting interviews carried out?
- Analyse labour turnover by department, sex and cause.
- Obtain labour turnover figures, in terms of cost and numbers. Look at trend over a period.

Industrial relations

- General policies?
- Who is responsible for negotiations?
- Trade unions?
- Are supervisors and managers trained to interpret trade union agreements? And to negotiate?
- Frequency of formal meetings with trade unions?
- Proportion of time spent on negotiations?

Suggestion scheme

- Usefulness of scheme?
- Frequency of suggestions?

Morale

- Assess morale at the moment.

Ratios

$$R1 = \frac{\text{Personnel costs}}{\text{Average no. of employees}}$$

1. Recruitment and replacement
2. Training and retraining
3. Other personnel dept. costs
4. Wage increases in excess of national agreement.
5. Production lost through poor industrial relations.

Recruitment

$$R2 = \frac{\text{Recruiting costs}}{\text{Recruits retrained}}$$

Training

$$R3 = \frac{\text{Training costs}}{\text{Average no. of employees}}$$

Industrial relations

$$R4 = \frac{\text{Cost of poor IR}}{\text{Average no. of employees}}$$

$$R5 = \frac{\text{No. of employee days lost through strikes}}{\text{No. of employee days worked}}$$

$$R6 = \text{Wastage} = \frac{\text{No. of leavers}}{\text{Average no. of employees}}$$

$$R7 = \text{Stability} = \frac{\text{No. of employees with 12 months service now}}{\text{Total employed a year ago}}$$

$$R8 = \text{Skill conversion} = \frac{\text{No. of employees with 12 months service now}}{\text{Total employed now}}$$

Appendix B
IDEF Analysis Example

■ IDEF analysis of information flow and activity

First developed and used by the US Air Force to describe their large and complex operational systems, IDEF (Information DEFinition) contains several levels of diagram to deal with top-to-bottom coverage of system operation and information flow. At the top level – IDEF 0 – the diagrams are used to describe high level company operations. For example, the IDEF 0 node shows a basic business operation. Next level down, the business is broken into the main functional groups:

Commercial activities	1
Design	2
Planning and control	3
Production	4
Delivery and installation	5

All blocks describe activities, rather than titles or functions. These blocks are constructed showing main input and output information, together with resources and constraints on the activity.

Once the Level 0 charts are defined, each activity area of the IDEF 0 is exploded down to the next level. In this example, Commercial activities (1) could be the subject of up to five activity blocks, which are then in turn exploded down to another level. This process continues until a level of decomposition is reached that adequately describes the aspects of business and operational activity and information desired.

Every listed activity and input, output etc should be consistant between each level, with no untraceable appearance of new elements. Written text is also used to further describe each chart and operations. The IDEF numbering system allows easy referencing of each level and area of the whole IDEF study.

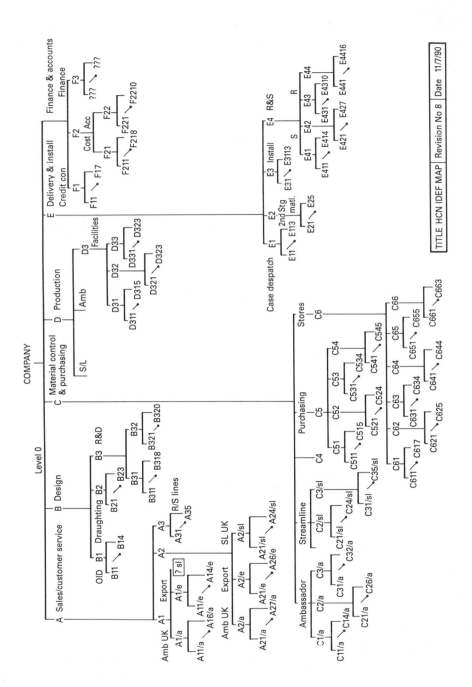

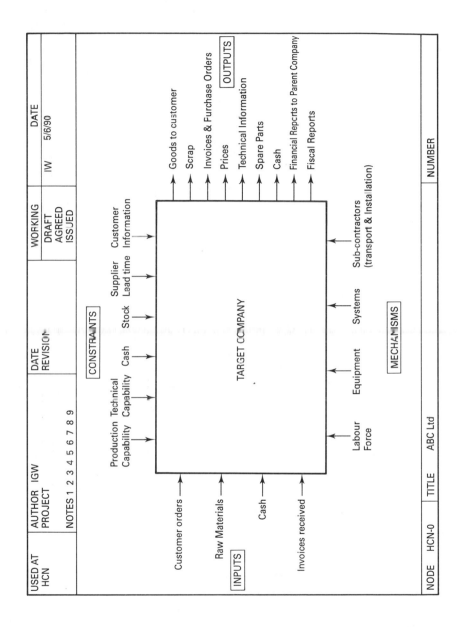

| USED AT HCN | AUTHOR IGW PROJECT | DATE REVISION | WORKING DRAFT AGREED ISSUED | DATE IW 5/6/90 |
| NOTES 1 2 3 4 5 6 7 8 9 |

INPUTS
- Customer orders
- Raw Materials
- Cash
- Invoices received

CONSTRAINTS
- Production Capability
- Technical Capability
- Cash
- Stock
- Supplier Lead time
- Customer Information

TARGET COMPANY

MECHANISMS
- Labour Force
- Equipment
- Systems
- Sub-contractors (transport & Installation)

OUTPUTS
- Goods to customer
- Scrap
- Invoices & Purchase Orders
- Prices
- Technical Information
- Spare Parts
- Cash
- Financial Reports to Parent Company
- Fiscal Reports

| NODE HCN-0 | TITLE ABC Ltd | NUMBER |

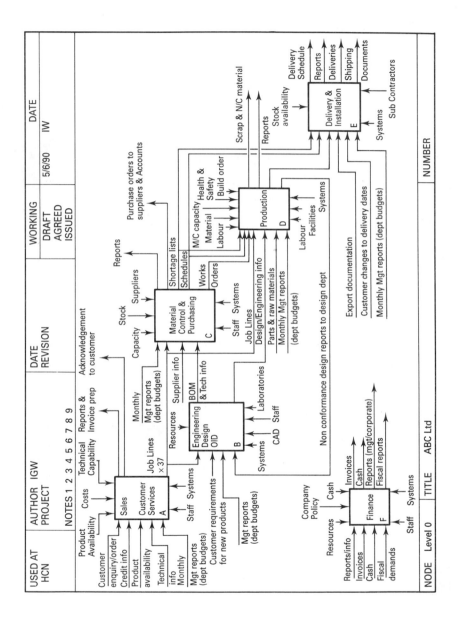

■ Business review study

Level O

A. Sales and estimating

Sales receives orders by telephone, fax, letter and construct an outline requirement order (job line). It negotiates with material control regarding delivery dates and acknowledges this date back to the customer. Job lines are created, credit checks are made with accounts, the job lines issued to other departments and invoices are prepared.

B. Design and order interpretation (OID)

OID will receive job lines from sales to construct a bill of manufacture for each case ordered and supply technical information to material control allowing material requirements to be assembled for the orders.

Design also receives customer requirements for new product design and produces design drawings and bills of materials (BOMs) for production departments. It also produces CNC programs for the Amada press shop equipment.

C. Material control/purchasing

Material control works with estimating on delivery dates and produces a master schedule and this schedule information is passed to purchasing for the generation of purchase orders. Material stock is also controlled by the department, and works orders and shortage lists are prepared and issued to production.

D. Production

Production works to schedules received from material control, producing parts and subassemblies as well as complete case units. It uses drawings supplied by design to produce non-standard assemblies. Finished cases are passed to dispatch via quality control.

E. Despatch and installation

Despatch receives, stores and ships the finished case units and has some customer input to actual delivery dates, preparing a delivery program. On leaving the warehouse despatch notes are generated. Installation co-ordinates with subcontractors to arrange for the installation of cases in customer sites.

F. Finance

The finance department handles all accounting functions and is divided into three main areas: cost accounts, credit control and management accounts.

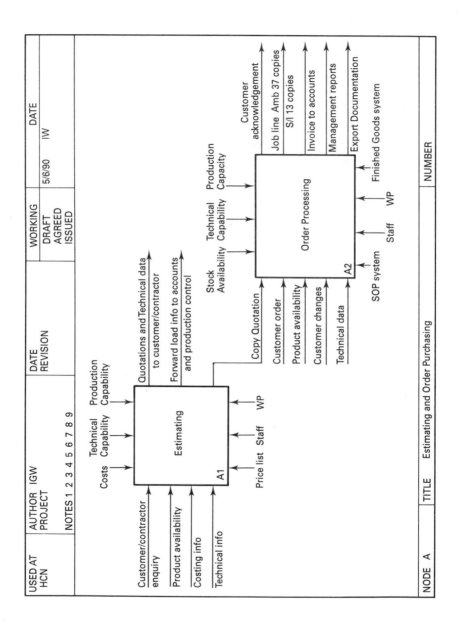

Level A

A1. Estimating

The estimating function receives requests from customers or contractors and will produce and return quotations or product availability. Internally it receives requests for costing information and general order enquiries. It also passes the forward load information to accounts and production control.

A2. Order processing

The order processing function uses the accepted quotations to generate various stages of customer orders: P, provisional; B, booked; and F, firm. It produces customer acknowledgements, job lines (30+ copies), invoices for accounts, management reports and export documentation. It uses the current SOP system, finished goods system and word processing.

Level A1: Export sales

A11./E   Receive enquiry

Enquiry received and logged in manual book by Anne or Liz. Enquiries are often faxes or letters. May request price or delivery information from other departments. Price and delivery data may be appended to enquiry or verbally communicated.

A12./E   Determine price

Determining the price may require the use of a price list but when options are not standard then contact is made with design for specific technical information and/or part numbers. For a one-off order, a cost breakdown is obtained from accounts (David). The estimating department is used as a back-up if export do not have all relevant information.

A13./E   Determine delivery date

Dates are determined by verbal discussions with production control. Enquiry and negotiation is logged if necessary. If a delivery date is acceptable this is logged by production control once a date is confirmed, possibly the same day.

A14./E   Prepare official quote

A quote number is allocated and the word processing department prepare an official quote on headed paper. It contains customer and date information and is sent to the customer.

Appendix B

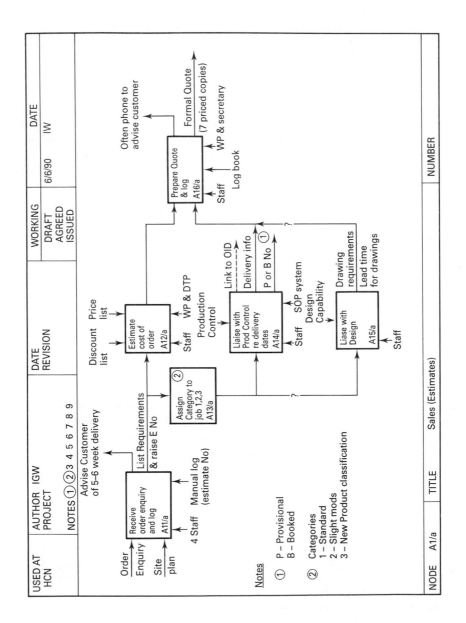

476

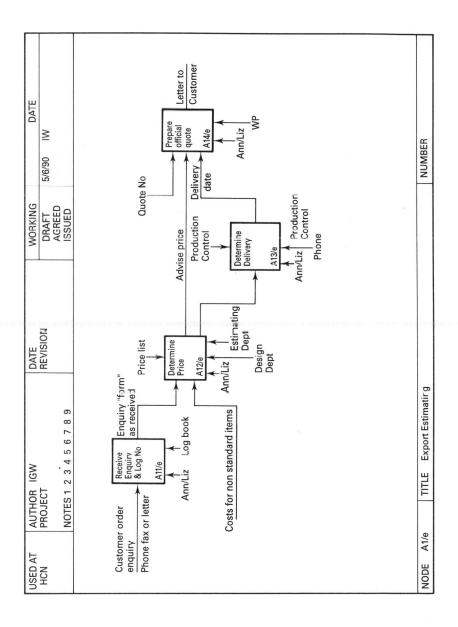

477

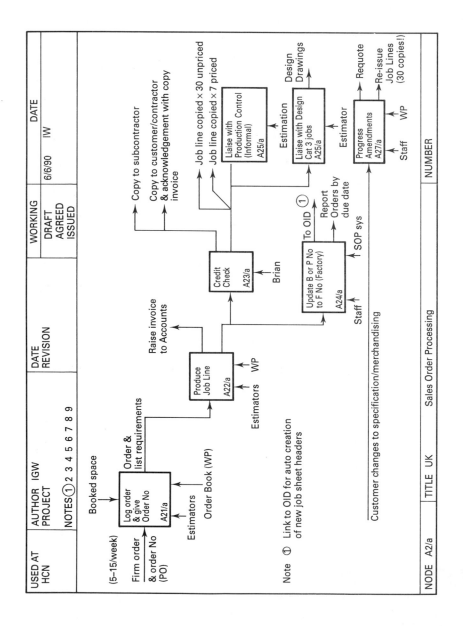

Level A2: order processing

A21./A Log order and give order

When firm orders and customer order numbers are received the order is logged into a book and given a company order number.

A22./A Produce job line

The order is issued with requirements information to produce a job line using word processing. Five to fifteen orders are received per week normally five weeks prior to delivery date.

A23./A Credit check

Credit control (Brian) must check and physically stamp all job lines prior to issue. When the job lines are returned they are copied and distributed as follows:

- 30 unpriced copies
- 7 priced copies (see circulation list)
- Copy to customer/contractor + order acknowledgement + invoice copy
- If required copy to S/C for electrics

A24./A Change booked and provisional orders to confirmed

The order in the computer-based SOP system has its prefix changed from B or P to F. The SOP system then generates a report of orders by due date.

A25./A Liaise with production control

Informal liaison with production control occurs regarding the job line interpretation.

A26./A Liaise with design on category 3 jobs

Liaison with design and OID to (a) produce new or amended drawings as required and (b) generate BOMs for the job. Liaison only occurs when the job is confirmed, F prefix, which is normally five weeks before the due date. But design has a four-week lead time on drawing.

A27./A Progress amendments

Amendments to specification/date from customer are processed (**without a further credit check**) and job lines are reissued and copied. The old job lines are then withdrawn and requotes are sent to customers.

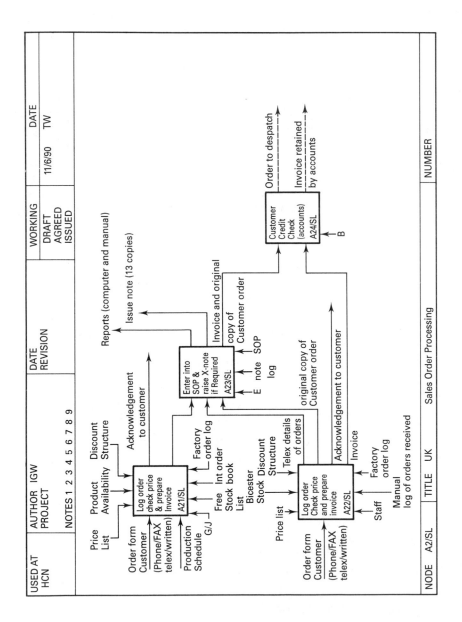

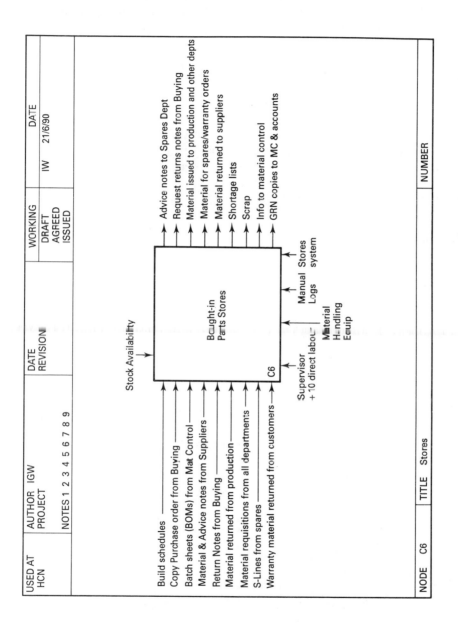

Notes

Cost of amendment is not governed by a formal policy. The sales function attempts to pass the cost onto the customer but is rarely successful; it tries to resist amendments once the order is in the shop.

Areas of improvement

■ Introduce a formal change control mechanism with calculation of the true cost of change.
■ Reduce the number of copies of job lines ASAP by use of an OID system.
■ Rationalisation of sales and OID functions by integration.
■ Earlier involvement of design in drawing needs.
■ Improve accuracy of amendment costing.
■ Increase technical knowledge of sales function.

PRODUCTION CONTROL	APP'X NOTE	COMMENTS	ACTIVITY VALUE £K	ACTIVITY VALUE £K	DIFF £K
1. PRE-SHOP CONTROL					
– No rough capacity planning aid (tenders)	16	Integrated capacity planning			
– Inadequate Mfg programme ('safe' leadtimes, not efficiently linked to shops)					
– No formal update of standards with actuals (at all levels)		Comprehensive master scheduling			
– Mfg programme produced on stand-alone PC system	17		1		1
– Logsheets not generated automatically		Shop floor data collection			
– No single, comprehensive, on-line production control	18		7		7
system (CMF, spares, shop scheduling, stock control)	40	Integrated mfg programme at all levels			
– No fast, accurate feedback on progress/shop status (weekly review meeting)	18	Single database			
– Inadequate capacity planning	19	P control	7	2	5
– Progress in pre-shop	20	System	59	34	25

– Take on business with little knowledge of impact
– Loss of competitive advantage
– Capital employed unnecessarily high
– Lack of control leading to loss of priorities (real)
– Not realising potential for increased throughput
– Not using best possible data for planning (capacity, process etc)
– Unnecessary effort/wasted time
– Duplicate data; reprocessing same data; high date maintenance (out of date/corrupt data)
– Batch systems give rise to lengthy response times (PC takes 1 month to process new drawings – batch system)

Action quick fixes:

From the analysis, opportunities for appropriate quick fixes will emerge.

Based upon return these will be actioned to provide immediate benefit to the company.

Appendix C

MRP2 System Specification,

Invitation to Tender, and

Requirements Comparison Table

■ Company background

The company manufactures commercial refrigerators for the retail food trade. Products are manufactured on a single site in Scotland. The company is the leading manufacturer in the UK, with two main product lines for remote (supermarket) cases and for integral (small shop) cases. The business is market-driven, requiring to build and deliver cabinets on short lead time, modified to customer specification. Turnover in 1991 was £25 million, rising to £28 million in 1992.

■ System specification

1. Hardware

The hardware and application software will be installed at premises in Glasgow, Scotland. Suppliers should use the information on processing volumes to arrive at a suitably sized computer configuration. However, it is felt that the minimum initial configuration will be for a system that supports the following peripherals:

100 terminals
40 printers
25 bar-code terminals

It is essential that the proposed hardware is capable of expansion to cope with extra workload which in turn will mean the addition of further memory, disks, terminals and printers. We envisage that any proposed system should be capable of expansion to 130 terminals with no detrimental effects on system performance.

2. Procurement and implementation schedule

The expected timetable for the selection of a supplier and start of implementation is as follows:

Draft tenders to be received	14 November 1990
Shortlisting of top two suppliers	14 December 1990
Confirmation of supplier	15 February 1991
Agreement of project plan and contracts	March 1991
Equipment delivery and commissioning	April 1991
Client acceptance of first phase of software implementation	May 1991
Phase 1 fully operational including all software	End August 1991

Suppliers are asked to commit to this schedule and may suggest any changes they consider reasonable.

3. Selection criteria

The criteria for selection which will be applied are as follows:

Vendor ability to meet the requirements
> The quality of the proposed solution in terms of its ability to meet the specified needs of the client and in particular to provide:
>> A full turnkey solution.
>> Full integration between modules.
>> Real-time update of information.
>> Multi-user access to all application areas.
>> Full enquiry facilities.
>> Full audit facilities.
>> Ad hoc report generator.
>> Links to existing order entry/OID systems and database systems (written in Foxpro).
>> DOS and Novell network platform.
>> Effective system security.
>> User-friendly application software.
>> The flexibility of the software and hardware, including availability of source code.
>> The level of system management required.

Vendor experience and credibility
> The credibility of the packaged software and hardware proposed.
> The experience of the supplier in implementing systems of a similar type.
> The ability to meet the required timescales.
> Financial viability of the supplier(s).

485

Vendor support and maintenance
> The level of continuing support and maintenance, including user documentation and on-line help facilities.

Vendor costs
> The initial capital outlay.
> Annual charges.

4. Future strategy

The senior management believe that it is now appropriate and important to consider the application of a fully integrated business information system which will provide the following:

> Accurate and timely information on sales, purchasing, production control, stock and financial performance.
> Efficient and accurate sales order processing.
> Efficient and cost-effective manufacturing.

5. Status of manual systems

The client is in the process of developing improved procedures for the various business functions and introducing manual systems to control the operating activities in these functions. The capacity of the manual and computer-based systems is currently being exceeded due to the increase in the number of transactions that are required by present and future production levels.

5.1. Estimating and order entry

Estimating and order entry is a key area requiring integration with the proposed systems. In-house systems have been developed to perform order interpretation against confirmed jobs, using bills of material to produce fully defined jobsheets for MRP processing. It is anticipated that any incoming systems will be required to interface fully with these bespoke applications.

5.2. Current status of computer systems

There are currently several 'in-house' computer applications, based on the Foxbase/Foxpro 4GL. These address discrete areas of operation, but are not integrated. All applications run on a Novell local area network, with approximately 80 nodes. The network has recently been upgraded and reconfigured to provide redundancy in the event of cable failure.

Operating system	DOS Novell 2.1
File servers	386/286 servers
Terminals	70 workstations/PCs
Printers	60
Hard disk	1.2 Gbyte
RAM	20 Mbyte
Tape Streamer	60 Mbyte

The current disk space is approx 90% used and may also require expansion in the near future.

■ Systems requirements

1. Functional structure

It is thought best to divide the company structure into business functions and functional activities and discuss each area and its requirements separately.

Business function	Functional activity
Estimating and order entry	Tendering, new products (sales order processing), order interpretation, forecasting
Planning	Production and stock control, scheduling, material requirements planning, master scheduling, purchase order processing
Production	Capacity planning, work in process tracking, shop-floor data collection, maintenance control
Finance (existing system)	Nominal, sales & purchase ledgers, cost accounting, payroll & management reporting

In each activity the key criteria for any new system have been specified, along with the main inputs, outputs and transaction volumes. In addition, the minimum requirements for all modules of any computer system selected should include the following:

Ease of use to accommodate the varying skill levels within the company.
Ease of administration of the system to reduce the internal resource requirements in managing the system.
Flexibility to allow the system to expand and grow with the business.
Reliability to ensure continuity of operation.
On-line access to reduce the amount of paperwork and permit timely availability of information.

487

Fast response to ensure productivity and credibility with users.
Full integration between modules.
Full enquiry facilities.
Full audit facilities.
Ad hoc report generator.
Links to existing OID and BOM systems.
Real-time update of information.
Multi-user access to all application areas.
User-friendly application software.
Effective security system.
Modem/remote data entry facility, and links to spreadsheet and database packages.

2. Order entry and interpretation

Order entry and interpretation is responsible for producing tenders and jobsheet headers, which effectively are the result of every sales order, so it should maintain the control by means of an SOP module integrated to an existing OID system.

It is our intention to implement a system which will make all necessary finished stock, production and customer information available at the order entry stage. This information must be on-line and real-time. Sales orders will be entered directly into the system. It is intended that by providing sufficient and accurate data at the time of order entry, a sales order can be rapidly created, with the appropriate information made available to the planning and finance functions. The module should provide the following facilities:

Direct order entry customer database
    Customer code
    Company name
    Address
    Telephone/fax/telex
    Contact
    Payment terms
    Pricing
    Credit rating and status
Ability to create
    Quotation
    Sales acknowledgement
    Internal sales order
    Dispatch note
    Address labels for outgoing goods
    Invoice
Sales order record
    Sort facilities

Multicurrency facility
Stock availability and allocation (flagging)
Scheduled production availability
Order splitting for production and dispatch
Generation of invoices from outside SOP
Addition of carriage and picking details
Automatic functions
    Pricing from customer code
    Discount by job, drawing, part number and/or customer
    Override pricing
    Real-time update of stock allocation records
Sales order forecasting
    Forecast from previous sales and manual input
    Manual/automated forecasting
    Product consolidation
Key data for sales orders
    Customer account number
    Customer name
    Customer delivery addresses
    Delivery date requested
    Quantity
    Drawing/part number
    Customer order number
    Job number
    Delivery date given
    Special requirements
    Order progress flag
    Sales area code and sub-area
    Sales person

Most of the information generated by the system will be required for on-line enquiry only, however, there is a need for any data captured during the sales order process to be available for enquiry and report generation. Reports/enquiries required include

Outstanding orders
    by customer
    by date, by part and in total
    by job, drawing and part number
    by sales area
Overdue orders
    by customer
    by date, by part and in total
    by job, drawing or part number
    by area

Scheduled dispatches
    by customer
    by date, by part and in total
    by job, drawing or part number
    by area
Price lists
    complete print out
    selective print by job, drawing or part number
    complete range of prices by job, drawing or part number
    price per job, drawing or part number referred to a common currency
        for comparison
    the ability to manipulate price lists by factoring prices from base price
        in one price list and by factoring prices between price lists
Historical sales analysis
    by customer
    by date, by part and in total
    by job, drawing or part number
    by area/salesperson
Exception reports
    all orders and invoices with non-standard prices
    all orders and invoices with non-standard discounts
    all orders and invoices which have a gross margin less than a preset
    minimum value
Delivery performance
    delivery time vs. requested delivery
    acknowledgement time vs. requested delivery
    acknowledgement time vs. actual delivery time
    acknowledgement time vs. order receipt
All by customer, job, drawing or part number areas as a percentage of period parameters, i.e. weekly over a range of weeks.

Data transaction volumes and estimated workload have been estimated as follows:

	Current average	Anticipated maximum
No. of customers	200	500
No. of carriers used	24	30
Sales orders outstanding	100	200
Item pages per order (pages)	2–3	
Invoices per week sent out	30	50
Item lines per invoice	10	100

The sales order processing module should provide real-time integration with the sales ledger, stock control and scheduling areas.

3. Production

The production function has distinct but interrelated activities which will benefit from the implementation of a suitable computer system. This will be achieved by improved collection and processing of information regarding production, inventory and order progress.

4. Production and material control and material requirements planning

Production and material control and material requirements planning should provide facilities for works order processing and production scheduling. It requires full interfacing to the BOM/specification system to allow the material requirements planning of purchased and manufactured parts. The MRP software should carry out the following standard MRP functions:

- Gross requirements planning of all BOM parts, from jobsheet system.
- Netting of the requirements against the existing and planned stock (for purchased parts), and allocating parts as required in the stock files.
- Use of the part parameters set in the stock control system (e.g. lead time, reorder policy) to calculate order quantity and release information.
- Generating a file of recommended purchase requisitions replanned/ expedited/cancelled requisitions.
- Generating exception reports covering the above, plus jobs to be rescheduled and part shortage forecasts.
- Facility to back jobs out of a schedule and de-allocate associated parts.
- Master scheduling facilities for case production planning, with user-definable time fences for firm and forecast schedules.
- Ability to run MRP from the master schedule or individually entered jobs.
- Ability to construct and test several master schedules and MRP runs for comparison.
- Facility for capacity planning of a schedule using BOM job times and workcentre capacity data.
- Ability to add repair and modification work to planned and existing cases, for both purchased and manufactured parts.

The company requires a system for schedule generation of work orders which will enable the creation of a schedule. The scheduling system should generate weekly (and potentially daily) production schedules, which would be split by production area and give job details, estimated production time and capacity utilisation. Each schedule should be available for authorisation prior to the automatic generation of the corresponding works orders. The works order processing software should also generate the following reports:

Outstanding works orders
    by job/drawing number
    by date parameters
Overdue orders
    by job/drawing number
    by date parameters
Scheduled vs. actual completion date
Printing of batch cards for defined production areas

Current and expected transaction volumes have been estimated as

	Current average	Anticipated maximum
No. of works orders per year	5,000	10,000
No. of works orders outstanding	1,500	2,500
Number of cost centres	10	20
Number of direct employees	70	130
Characters in drawing/part no.	—	12
Drawing/part no. in use	15,000	30,000

The production control module must have BOM/specification facilities, with links to the estimating stock control and shop-floor data collection modules.

5. Stock control

Stock will be stored by job or part number, and should move on a FIFO basis. It is intended to develop the stock control system to provide effective materials planning and control information. Information should be available on the current stock allocated against orders, available and forecast stock levels, etc. A main function of the stock control system will be to provide information for MRP processing, including safety stock levels, reorder levels and quantities, and ordering policies. The objective is to provide a high level of control over component parts, allowing us to reduce inventory holding levels. The stock control module should provide the following facilities:

Stock adjustments
    Receipts
    Issues
    Returns
Stock allocation
    Available stock + allocated stock = total stock
Multiple stock locations
Printing of bar-code labels and packing labels
Bar-code data entry for stages of manufacture
Goods movement

It is essential that the stock control system should cater for the following data which is required for finished products and for raw materials and consumables:

Part and drawing number
Description
ABC categorisation
Physical stock
Allocated stock
Available stock
Material on works order
Lead time
Minimum and maximum stock levels
Reorder quantities
Cost
Usage history
Possible substitute products
Stock location
Stock returned from customers

The following facilities are required:

- facility for second vendor
- show units per each vendor to date (we intend to divide sourcing between at least two suppliers)
- activate ABC classification, based on set extended values
- for user to force a classification outside ext. value
- provide extended value field, used for the ABC classification
- for user to redefine cut-off points for ABC (sterling)
- for user to define the review periods for each class/item: n days, weeks, months
- for part to be controlled by specified methods: JIT/to order, fixed period (period review), fixed quantity, reorder point (ROP) control
- the above control methods to default to the ABC classes as follows:
    - A   JIT/to order, fixed period
    - B   fixed period or quantity, ROP
    - C   fixed period or quantity, ROP
- user override to be allowed; for fixed period, the reorder level to be recommended by calculation and consideration of lead time, forecast demand, period length, and safety stock required
- default review periods to be provided and shown:
    - A   weekly
    - B   monthly
    - C   4 monthly

    user override of these to be allowed
- standard deviation of the 12, 6, 3 month average usages to be shown alongside, also (2.5) standard deviation (s.d.) value as maximum safety stock recommended. $n \times$ s.d., with n being user-definable

- desirable for lead time to contain trail of actuals, being used to prepare average and s.d. of lead time
- reorder level to be enabled for the appropriate control methods; a recommended ROL to be displayed, calculated as

(lead time × average usage in l.t.) + safety stock

- if standard fixed time periods (period review) are the selected control method, reorder quantity to be calculated based on expected usage over the time period, with facility to include price breaks and economic order quantity options
- when period review method used, facility to check usage against planned and generate an exception report if outside planned limits
- field for maximum stock level to be carried, recommended as ROQ + safety stock
- forecast facility, to create a calculated and/or manual forecast by part, based on historical data with smoothing, plus additional +/+ for other known or expected factors. This facility must compare actual demand against the above forecast, and using the calculated error, update the forecast to reflect the error; it should then revise the order quantities or periods used, as appropriate.
- settable flag for part to be MRP or non-MRP (related to following section on the BOM schedule explosion system)
- cycle-counting facilities based on ABC classification

5.1. Reports/exception reports

1. Automatic printing of cycle count paperwork daily, for the required groups of ABC items. User-defined total number of counts/day.
2. Facility to update/reconcile counts with planned stock. Reason fields and authorisation field/code. Also to allow updating from unplanned stock checks.
3. Facility to disable cycle counting on a specific part number.
4. Recorder reports for B and C class items w.r.t method of control.
5. Exception reports for usages of any part outside planned usage/replenishment.

Material should be accessible on the system by job, or part number. Information should be updated in real-time to give accurate stock availability to the sales and scheduling functions. Major print processing should be done overnight.

There is a requirement for the following stock reports to be available. Some of these are required regularly and some on an *ad hoc* basis:

Stock list giving actual stock allocated to each stock order
Minimum and maximum stock level
Detailed stock list

Stock valuation
Stock movements report
Stock amendments report
Master file changes report
Stock-taking report
Stock control ratios

The materials planning department will make day-to-day *ad hoc* enquiries. The system should be able to give immediate replies to various onscreen and print enquiries, For example:

Excessive stocks report by specified parameter
Low stocks report by specified parameter
Slow-moving stock by specified parameter
Negative stock report
Possible substitute products
Sales orders split over periods of delivery date combined with stock orders due over periods of delivery date
Stock master file scan facility on each enquiry made
Answers to 'what if' sales orders

The current volumes are expected to increase as the company grows. Expected volumes are

	Current Average	Anticipated maximum
Characters in drawing/part no.	—	12
Characters in description	—	50
Number of stock locations	4	8
Number of stock bins	3,000	4,000
Number of stock movements per week	500	1,000
Number of months movement history	—	18

The stock system should be able to interface directly with the order interpretation, MRP, purchasing, manufacturing control systems and shop-floor data collection (bar-code) system.

6. Shop-floor data collection

It is our intention to utilise a system that will minimise the amount of manual data entry in the production and warehouse areas. We envisage a shop-floor data collection (SFDC) system based on bar-code readers and alpha numeric terminals. The system should provide for automatic entry of job data using, for example, bar-code tickets generated by the shop-floor documentation system. Stations would be used to read in data as job was (i) loaded, (ii) fabricated and assembled, (iii) tested and (iv) dispatched. This facility must provide real-time integration to production control and stock control modules. Also to cost accounting/management reporting to the finance area, to allow cost analyses and reporting facilities for management purposes. This module should provide the following facilities.

Material control.
Direct and automatic job data entry.
Bar-coding system for all material and jobs.
Alpha-numeric terminals and printers in production and warehouse areas.

Data capture at:

Goods-in, stores issue, press shops, wood mill, fabrication, welding, foaming, wiring, engineering, test, warehouse in, dispatch.
Hand-held units for stock-checking.
Data validity checking.
Primary data entry point to stock control system.
VDU access to current/future production schedules for overseers.
Production reporting and on-line VDU access.
Sign-off from production schedule using VDU input.
Extensive links to cost accounting/Management reporting.
Ability to create and/or issue bar-coding labels.
Reprint facility.

Key input data for SFDC includes

Job number
Drawing/part number
Date/time
Work centre code
Operator code for information
Sign-off and production schedule (via VDU) to shop-floor

Reports on the production performance of each work centre would be required, together with financial equivalents for the production period. This to be by operator, by job or by case. Examples include

Production output
Date scheduled vs. date completed
Production vs. capacity (%)
Standard hours
Production time vs. standard time
Non-productive time
Quality control failures/returns/scrap
Materials usage
Cost of production for fabrication, wiring, test

The system would require to run in real-time to provide up-to-date process times and production data. Many of the above outputs may be from the financial systems, derived from data received from SFDC. Communication between SFDC and costing should be automatic, but not necessarily on-line. Relevant data volumes are given in the sections on stock control and production control.

7. Purchasing

The proposed purchasing system is required to give effective support in order progressing and reordering for bought-in stock, and non-stock items. The MRP system should give purchase order suggestions based on the current stock/order position and the lead time required. This should be able to be modified and manipulated before converting to a firm order. The system should then prompt for order acknowledgements, order confirmation, etc., according to the expected delivery date. The system should provide historical data on the purchase of stock items and their movements.

Inputs will be both computer generated and manually input. Purchase orders will also be generated internally and input manually. Order progress will be maintained daily on information received. Goods received (GRN) data will allow automatic GRN matching to purchase orders.

Processing can be done overnight. A delay of one day should be acceptable for the user needs. This would cover items on order, due dates, etc. Processing would be required during the day for order raising, modification and printing. This area shares much of the same data as stock control with some additional items.

8. Purchase order processing

Purchase order processing, should also provide the following facilities:

- Electronic purchase requisitions from approved MRP output
- Direct order entry
- On-line verification of nominal ledger code when input at purchase order stage
- Password access by buyer to part files regarding price details, updating, etc.
- Field to calculate and enter economic order quantity and minimum recommended quantity
- Buyer field per part
- Fields for price change/date effective
- Obsolete stock parameters
- Flag for goods in inspection, with memo field of test to be carried out
  Supplier database
    Supplier code
    Company name

            Address
            Telephone, telex, fax
            Contact
            VAT number
        Ability to create
            Purchase order
            Goods received note
        Purchase order record
        Goods received note (GRN) matching process
        Multicurrency facility (desirable)
        Vendor rating and memo field

Key data for purchase orders includes

    Order date
    Order quantity
    Reason for purchase
    Expected due date
    Date acknowledged
    Date of delivery
    Date delivery confirmed
    Order number
    All above for each part in order may be part delivered.
    Supplier
    Supplier code
    Goods received number
    Invoice number

The control of incoming goods requires regular monitoring of the stock/ purchasing position. It is expected that the stock order requirement and purchase order processing would be done over a weekly cycle, or more frequently. This may require running the main reports on order progress and purchases required overnight. Other reports would be ad hoc during the working day.

The volumes are to some extent the same as the stock control area. Purchase orders are done on a monthly basis for certain high use items and daily for spot orders.

	Average	Maximum
Number of orders per month	600	1,000
Number of orders outstanding	150	300
Number of lines per purchase order	5	50
Invoices received per week	120	250
Number of items per invoice	5	50

The main control requirements on such a system are to match the goods received with purchase orders. It will also be necessary to monitor the value of orders placed and deliveries due over each period for accounting control. The purchasing area should provide on-line enquiry facilities to meet day-to-day enquiries. Reports and enquiries required include

Order status by order number
Deliveries due per product per product group
Answers to 'what if?' sales figures, i.e. possible shortages given expected delivery dates
All orders are acknowledged
All orders with no delivery date
All orders that are at a certain progress stage
Purchase order under which a drawing/part number was last received
Outstanding orders
Overdue orders
Receipts due
Supplier analysis

The purchase order system should provide real-time integration with the purchase ledger and stock control functions.

The financial function is to be served by integrated modules for sales, nominal and purchase ledgers. There will be integration to the sales function and the operations functions. Comprehensive management reporting facilities are required:

- There must be on-line, real-time updating of information across the ledger systems, with the nominal ledger being updated immediately as a result of entries in the purchase or sales ledgers.
- There is a requirement for the facility of 'forward' postings before period-end close of all ledgers and also periodic 'backwards' postings for the current and previous financial years, in the nominal ledger only, plus the ability to call for trial balance in any period(s).
- Maximum flexibility in account code structure is desirable, with the ability to analyse the ledger, and produce reports designed by the user for a free-format code. Additional attribute field(s) to be held against each account code (e.g. as cost centre or for reporting).

9. Ledgers

9.1. Nominal ledger

The nominal ledger (NL), should be fully integrated with the sales and purchase ledgers, stock, payroll and the cost accounting and management reporting module. Other input will be via journals. The module should provide the following facilities:

On-line, real-time data updating from other ledgers.
On-line input and comprehensive enquiry facilities.
Screen formats at user discretion.
Detail of subsidiary ledgers, e.g. individual sales and purchase invoices (to allow enquiries without leaving the nominal ledger).
Detailed on-line transaction history of nominal accounts for specified selected periods of time.
Bank reconciliation function.
All transactions retained on-line to end of annual accounts preparation (12 months after year end).
User-definable account code analysis.
A minimum of one budget version against each account code.
Facility for 12× monthly budget.
User-definable budget phasing.
Budget comparisons.
Trial balance.
Profit and loss reports and balance sheet reports.
Automatic reversal of accruals and prepayments (plus reverse and post to specified periods).
Standard reports to incorporate monthly and year-to-date figures current/ previous year comparisons and budget comparison.
Multicurrency (desirable).
Multicompany and consolidation volumes.
User-friendly ad hoc report generation.
Enquiries against specified fields of the account code.
Ability to automatically output data to cost accounting/management reporting systems (if not updated in NL reporting).
Ability to have standard and tailored journals.
Archiving.
Fixed asset register with fixed asset listing variable depreciation rates and automatic postings of capital
Links to common spreadsheets, MS-DOS.

Most of the information generated by the system will be required for on-line enquiry. There will also be requirements for comprehensive reporting on the above facilities using an ad hoc report generator.

Key data for the nominal ledger includes

Company details, title, identification
Account
Cost centre codes
Budgets
Account details: number, type, sign, balance
Attribute field(s)

The expected data volumes are

	Average	Maximum
Ledger transactions per month	300	800
Characters in NL code	8	10
Elements in NL code	2	5
Number of NL accounts	200	500
Characters in Attribute field per account code	3	3
Accounting periods per year	12	12
Number of journal postings per period	20	50
Number of journal entries per period	300	800
Number of months of NL history	24	24
Number of weeks backward postings at year end	104	104

The system should provide audit transaction trails through every account showing all items, date, amount and source. Access to the master file should be restricted and a print of all changes produced.

9.2. Sales ledger

The sales ledger must be integrated with the sales order processing system. Invoices from this system may be in various currencies. The ledger should be on an open-item basis. It should provide statements and credit control letters. There should be a facility for blocking accounts and preventing order processing. The system should provide onscreen facilities and detailed ad hoc reporting on customer sales and profitability. The system should automatically update the nominal and costing ledger in real-time. On-line input and enquiry is essential.

Input will be automatically from the order processing system and by manual input. Cash and individual adjustments will be manually updated daily and weekly. This module should provide the following facilities:

Customer file detail
Automatic posting from invoicing
Posting for
 invoices
 credit notes
 adjustments
 receipts
Automatic production of invoices (if not carried out by sales order processing)
Part-payments and retention, ability to distinguish between sales and contract
Traceability of cash received and allocation to specific invoices
User-definable screens and reports
Memo fields available
Automatic selection and generation of a range of 'chaser' letters

Month-end reporting on aged open items
Comprehensive on-line enquiries
Automatic currency gain/loss facility

Data held in the sales ledger to include

General
> Discount policy
> Default customer terms
> Sterling conversion rates
> Job number and sales prices (if not provided in SOP) and cost of sales

Customer file
> Account number (alphanumeric)
> Area and sub-area codes
> Customer names
> Office addresses
> Multiple delivery addresses
> Local currency
> Sterling equivalent
> Default invoicing terms
> Blocked account

Invoicing
> Auto invoice numbering
> Contract number
> Invoice tax point date
> Product and sales details including cost of sales

Standard reports from the sales ledger should include

Customer lists
Ledger details and balance
Aged debt report
Overdue report
Sales analyses: monthly and cumulative
Home and export, based on area codes
VAT report
Credit control
Overbalance report

The current data volumes are expected to increase by a significant margin in the future. The volumes are the same as those for the sales orders.

9.3. Purchase ledger

The purchase ledger is to be maintained on the computer by manual input and by integration with the nominal ledger and the purchase order processing systems. The system proposed should maintain open balances and also provide an aged listing and payment prompt. The system will be used for both raw materials and all consumable goods and services. Separate subcontractor ledgers are essential.

The system should accept invoices, credit notes, cash to account, part-invoice payments and manual adjustments. It is desirable that the system should accept multicurrency invoices and maintain both currency and sterling equivalent balances. On-line, real-time updating facilities are required. The daily processing requirements are mainly day-to-day update of transactions and invoices. Cheque runs will normally be done weekly.

This module should also provide the following facilities:

User friendly screen and report generator
Supplier details and codes
Full enquiry facilities
Invoice matching to GRN – matched purchase orders
Traceability of cash paid and its allocation to suppliers' invoices
Stopping facility
Purchase ledger postings
    invoices
    credit notes
    adjustments
    payments
    goods received notes
Subcontract ledger facilities.
Payments
    suggested payments routine reflecting due dates for payment
Automated payments process incorporating cheque production and cash postings
Preparation of remittance advice
Memo fields to be available against invoice postings detail
Purchase invoice approval reference
BACS payment process incorporating BACS report and cost posting

At invoicing stage, the posting generated should have stock/cost/account code as appropriate. The default code is the code input at purchase order stage. The system should hold the normal supplier and invoice data, including

Supplier
    Name
    Address
    Credit terms
    Discount received
    Payment terms
    Turnover (recording facility)
Invoice
    Supplier invoice number
    Company invoice number

Company order number
Supplier order reference
Invoice date
Invoice amount
Invoice currency
Stopping facility

Required reports will include

Supplier lists
Ledger details and balances
Payment listings
Aged creditors report
Overdue payment report
Purchase analysis VAT report
Stopped invoices/account

Most reporting and printing will be done monthly with the ability to do small daily runs if required. The current volumes are

	Average	Maximum
POs in current financial year	5,000	10,000
Number of suppliers	600	1,000
Purchase orders outstanding	2,000	4,000
Comment lines per purchase order	10	2 × A4 page
Item lines per purchase order	5	50
Goods received per week	100	200
Number of invoices per week	500	1,000
Item lines per invoice	5	50
Cheques per month	400	1,000
Transactions per month	700	1,000

The ledger should be maintained through a control account and the facilities for cheque printing should have restricted access.

9.4. Cost accounting and management reporting

The system proposed should have comprehensive facilities for cost accounting and be linked to the nominal ledger and payroll. Data will be input automatically to this function from the production monitoring system (SFDC) ledger. The system will be required to produce

Standard production costing information and variances by
contract, subcontract and type
work centres
material

labour
overheads
Actual cost calculation
post to any period and report by any period
time period summaries
product summaries
build up of invoice costs on a sale basis (variances to show profit and loss)

Comprehensive reporting facilities must be provided using a user-friendly ad hoc report generator:

Details of time and attendance from the SFDC facilities to be linked to payroll.

■ Requirements comparison tables used for comparing MRP2 facilities

These tables appear on pages 506–9.

■ Requirements comparison tables used for comparing MRP2 facilities

MRP system requirement	N/A	Importance 1 = unimportant 10 = very important									Required by					Comments
												now	3mths	6mths	1yr	2yrs	
Inventory Control																	
Issue of general stock items	1	2	3	4	5	6	7	8	9	10		n	3	6	1	2	
Cycle counting	1	2	3	4	5	6	7	8	9	10		n	3	6	1	2	
Backflushing inventory usage	1	2	3	4	5	6	7	8	9	10		n	3	6	1	2	
Disposition codes in warehouse	1	2	3	4	5	6	7	8	9	10		n	3	6	1	2	
Three UOM conversions	1	2	3	4	5	6	7	8	9	10		n	3	6	1	2	
Calculation of safety stock	1	2	3	4	5	6	7	8	9	10		n	3	6	1	2	
Physical inventory	1	2	3	4	5	6	7	8	9	10		n	3	6	1	2	
Time-phased reorder point	1	2	3	4	5	6	7	8	9	10		n	3	6	1	2	
Reports on unplanned usage	1	2	3	4	5	6	7	8	9	10		n	3	6	1	2	
Analysis of EOQ/ABC codes	1	2	3	4	5	6	7	8	9	10		n	3	6	1	2	
Stock moves on a FIFO basis	1	2	3	4	5	6	7	8	9	10		n	3	6	1	2	
Stock adjustments for returns	1	2	3	4	5	6	7	8	9	10		n	3	6	1	2	
Inter-warehouse transfer	1	2	3	4	5	6	7	8	9	10		n	3	6	1	2	
Recognition of saleable parts	1	2	3	4	5	6	7	8	9	10		n	3	6	1	2	
Non-moving stock reports	1	2	3	4	5	6	7	8	9	10		n	3	6	1	2	
Printing of bar-code labels	1	2	3	4	5	6	7	8	9	10		n	3	6	1	2	
BOM routings																	
BOM effectivity by date	1	2	3	4	5	6	7	8	9	10		n	3	6	1	2	
Integration to OID	1	2	3	4	5	6	7	8	9	10		n	3	6	1	2	
Identify eng. revisions to w/o	1	2	3	4	5	6	7	8	9	10		n	3	6	1	2	
Engineering change control	1	2	3	4	5	6	7	8	9	10		n	3	6	1	2	
Part search facilities	1	2	3	4	5	6	7	8	9	10		n	3	6	1	2	
Customised screen formats	1	2	3	4	5	6	7	8	9	10		n	3	6	1	2	
Order specific BOMs	1	2	3	4	5	6	7	8	9	10		n	3	6	1	2	
Include material within BOM	1	2	3	4	5	6	7	8	9	10		n	3	6	1	2	
Flag BOM status i.e. incomplete	1	2	3	4	5	6	7	8	9	10		n	3	6	1	2	
50 character part description	1	2	3	4	5	6	7	8	9	10		n	3	6	1	2	

Customer order entry

Serial numbers on F. goods	2	1	6	3	7	10	9	8	7	6	5	4	3	2	1
Preshipment invoicing	2	1	6	3	7	10	9	8	7	6	5	4	3	2	1
Review stock status at OE	2	1	6	3	7	10	9	8	7	6	5	4	3	2	1
Dispatch note creation	2	1	6	3	7	10	9	8	7	6	5	4	3	2	1
Register invoices on despatch	2	1	6	3	7	10	9	8	7	6	5	4	3	2	1
Product class enhancement	2	1	6	3	7	10	9	8	7	6	5	4	3	2	1
Multicurrency	2	1	6	3	7	10	9	8	7	6	5	4	3	2	1
Allow comments on schedules	2	1	6	3	7	10	9	8	7	6	5	4	3	2	1
Multi-order despatch lines	2	1	6	3	7	10	9	8	7	6	5	4	3	2	1
Generate manual invoices	2	1	6	3	7	10	9	8	7	6	5	4	3	2	1
Pegging of orders to contract	2	1	6	3	7	10	9	8	7	6	5	4	3	2	1
Available to promise enquiry	2	1	6	3	7	10	9	8	7	5	5	4	3	2	1
Quotation creation	2	1	6	3	7	10	9	8	7	5	5	4	3	2	1
Sales acknowledgement	2	1	6	3	7	10	9	8	7	5	5	4	3	2	1
Internal sales order	2	1	6	3	7	10	9	8	7	5	5	4	3	2	1
Add carriage and packing details	2	1	6	3	7	10	9	8	7	5	5	4	3	2	1
Report shipment to end-users	2	1	6	3	7	10	9	8	7	5	5	4	3	2	1
Address labels	2	1	6	3	7	10	9	8	7	5	5	4	3	2	1
Order splitting for despatch	2	1	6	3	7	10	9	8	7	5	5	4	3	2	1
Sales forecast from previous	2	1	6	3	7	10	9	8	7	5	5	4	3	2	1

Purchasing control

Blanket purchase orders	2	1	6	3	7	10	9	8	7	5	5	4	3	2	1
Automatic GRN matching	2	1	6	3	7	10	9	8	7	5	5	4	3	2	1
Purchase requisitions from MRP	2	1	6	3	7	10	9	8	7	5	5	4	3	2	1
Flag for goods inward inspection	2	1	6	3	7	10	9	8	7	5	5	4	3	2	1
Vendor performance rating	2	1	6	3	7	10	9	8	7	5	5	4	3	2	1
Multiple vendor per part	2	1	6	3	7	10	9	8	7	5	5	4	3	2	1
Full PO detail on GRN	2	1	6	3	7	10	9	8	7	5	5	4	3	2	1
Identify preferred vendor	2	1	6	3	7	10	9	8	7	5	5	4	3	2	1
Vendor quotation management	2	1	6	3	7	10	9	8	7	5	5	4	3	2	1
Effective dates for price increases	2	1	6	3	7	10	9	8	7	5	5	4	3	2	1

	1	2	3	4	5	6	7	8	9	10	n	3	6	1	2
Monitor values of orders placed	1	2	3	4	5	6	7	8	9	10	n	3	6	1	2
Acknowledgement dates	1	2	3	4	5	6	7	8	9	10	n	3	6	1	2
Verification of nominal ledger code	1	2	3	4	5	6	7	8	9	10	n	3	6	1	2
Obsolete stock parameters	1	2	3	4	5	6	7	8	9	10	n	3	6	1	2
Return to supplier transaction	1	2	3	4	5	6	7	8	9	10	n	3	6	1	2
Prompts on PO cancellation	1	2	3	4	5	6	7	8	9	10	n	3	6	1	2
Multicurrency	1	2	3	4	5	6	7	8	9	10	n	3	6	1	2

Shop floor control

	1	2	3	4	5	6	7	8	9	10	n	3	6	1	2
Integration with SFDC system	1	2	3	4	5	6	7	8	9	10	n	3	6	1	2
Auto order closing with exception	1	2	3	4	5	6	7	8	9	10	n	3	6	1	2
Multiple resources staff/mc	1	2	3	4	5	6	7	8	9	10	n	3	6	1	2
Define ops requiring two staff	1	2	3	4	5	6	7	8	9	10	n	3	6	1	2
Group w/c load	1	2	3	4	5	6	7	8	9	10	n	3	6	1	2
Print actual location on pickslip	1	2	3	4	5	6	7	8	9	10	n	3	6	1	2
Back-issue enquiries by part	1	2	3	4	5	6	7	8	9	10	n	3	6	1	2
Operation defined as not monitored	1	2	3	4	5	6	7	8	9	10	n	3	6	1	2
Variation of future capacity	1	2	3	4	5	6	7	8	9	10	n	3	6	1	2
Workcentres defined as key	1	2	3	4	5	6	7	8	9	10	n	3	6	1	2
Operation tickets	1	2	3	4	5	6	7	8	9	10	n	3	6	1	2
Load limits by w/c +/- %	1	2	3	4	5	6	7	8	9	10	n	3	6	1	2
Bulk issue	1	2	3	4	5	6	7	8	9	10	n	3	6	1	2
No routing warnings on release	1	2	3	4	5	6	7	8	9	10	n	3	6	1	2
SFDC material control	1	2	3	4	5	6	7	8	9	10	n	3	6	1	2
SFDC job data control	1	2	3	4	5	6	7	8	9	10	n	3	6	1	2
SFDC bar-coding for material	1	2	3	4	5	6	7	8	9	10	n	3	6	1	2

Requirements planning

	1	2	3	4	5	6	7	8	9	10	n	3	6	1	2
Generate a file of orders	1	2	3	4	5	6	7	8	9	10	n	3	6	1	2
Forecasting at part level	1	2	3	4	5	6	7	8	9	10	n	3	6	1	2
Hard allocation of items	1	2	3	4	5	6	7	8	9	10	n	3	6	1	2
Net change MRP	1	2	3	4	5	6	7	8	9	10	n	3	6	1	2

	1	2	3	4	5	5	7	8	9	10	n	3	6	1	2
User-definable calendar	1	2	3	4	5	5	7	8	9	10	n	3	6	1	2
Trial kitting	1	2	3	4	5	5	7	8	9	10	n	3	6	1	2
Synchronised time buckets	1	2	3	4	5	5	7	8	9	10	n	3	6	1	2
Explosion stop code	1	2	3	4	5	5	7	8	9	10	n	3	6	1	2
Spread sales f/cast in weeks	1	2	3	4	5	5	7	8	9	10	n	3	6	1	2
Define w/c performance	1	2	3	4	5	5	7	8	9	10	n	3	6	1	2
User-definable time buckets	1	2	3	4	5	5	7	8	9	10	n	3	6	1	2
Gross requirements from jobsheets	1	2	3	4	5	5	7	8	9	10	n	3	6	1	2
Rough-cut capacity from MS	1	2	3	4	5	5	7	8	9	10	n	3	6	1	2
User-definable time fences	1	2	3	4	5	5	7	8	9	10	n	3	6	1	2
'What if' master schedule create	1	2	3	4	5	5	7	8	9	10	n	3	6	1	2

Appendix D
Procedures

■ **Engineering change**

Engineering change request forms follow on pages 511 and 512.

Engineering Change Request No [   ]

Change requested by _____ Date _____

Department Head _____ Date _____

Approval

Description of Change	Part No's	Drg No's

Attach additional information if required   Enter details of new purchased parts below

Models or Assemblies Affected

Reason for change	Type of Change	Effectivity	Material Disposition
☐ Error	☐ Emergency	☐ Immediate	☐ None
☐ Additional Information	☐ Mandatory	☐ Next time ordered	☐ Use up stock
☐ Design Improvement	☐ Elective	☐ See Remarks	☐ Scrap
☐ Vendor Required	☐ Routine		☐ Rework
☐ Customer Required			☐ Use elsewhere
☐ Latest Process			☐ Retain for Spares
☐ Clarification			☐ See Remarks
☐ Cost Reduction			
☐ Design rectification			

Remarks

Estimated saving

£

Saving ☐   Increase ☐

Details of New Purchased Parts Requested

Proposed Technical Description (max 50 characters)	Suggested Source	Stock Item Y/N	New Part No	Quantity

Engineering Approval		List No	Date	Issued by	Final Approval for Issue
	DO Issue List				
	Refrigeration Issue List				By _____
	Electrical Issue List				Date

Document No C01

Figure D1   Engineering change request form.

						Engineering Change Request No		

Date reviewed by committee		Accepted	Yes	No	Actual Cost	Saving		Increase	

If rejected – reason for rejection

Document received log	CCA	Design	Prod Eng	Prod Control	QA	Costing	

Records affected

Drawing No	Part No	Drawing /Spec Revised	Part Ledger Revised	Std Time Revised	Routing Revised	N/C Tapes Revised	Bom/Assy	Revised	Std Time Revised	Routing Revised

Manufacturing Jigs/Fixtures affected

Item	Drawing	Equip Mods	New Fixture

Inspection/Test Specs

Test Specification	New	Revised

Figure D2   Engineering change request form.

■ Procedure for cycle counting and inventory accuracy within the stores

1. Objective

To define the systems and procedures used by the company for inventory control within the stores to ensure stores records accuracy of at least 95%.

2. Scope

This procedure covers the inventory control system for determining the inventory accuracy of all parts within the Company.

3. Responsibility

Raising cycle count/recount card	Inventory control
Raising grief count card	Stores supervisor
Conducting stores count	Stores counter
Recording stores count	Stores supervisor
Adjust quantities in inventory system	Storer
Calculating/reporting inventory accuracy figures	Inventory control
Maintaining log of count cards and inventory accuracy records	Inventory control

4. Definitions and references

4.1　Cycle count: the regular, periodic, physical counting or recounting of all specific parts within the company stores.

4.2　Grief count: a count made when a discrepancy is found between the physical stock and stores records, outside a cycle count.

4.3　Inventory variance: the difference between the physical count and the computer registered count of a particular item.

4.4　Inventory control tolerance: all parts are categorised based on usage and cost:

A parts – High value (cost × usage)–	Counted quarterly
B parts – Medium value	– Counted bi-annually
C parts – Low value	– Counted over two-year period

Each category has an acceptable inventory variance as such: 0% for A, 2% for B, and 5% for C. Parts are within the inventory control tolerance if the count falls within the acceptable inventory variance level.

4.5   Inventory accuracy: the number of cycle counts within the inventory control tolerance divided by the number of counts undertaken expressed as a percentage.

5.   Procedures

5.1   Inventory control will decide which parts are to be counted daily by reviewing the three part categories which indicates when cycle count is due, i.e. quarterly, bi-annually or over a two-year period.

5.2   Inventory control will raise a cycle count card for each part to be counted or recounted on the following day. Parts for selection will include all parts for scheduled cycle count, specially requested parts and parts for grief count. The cycle count card will contain the following information:

■   Unique identification card number
■   Part number to be counted and description
■   Date actioned
■   Stores location(s) of part

5.3   Inventory control will issue each cycle count card to the stores supervisor on the day before counting.

5.4   The stores supervisor will distribute the cycle count card to the stores cycle counter, who will take a physical count at each location for each part designated on the cycle count card.

5.5   The cycle counter will record the actual quantity counted in each designated location in the appropriate section of the card and will indicate unrecorded issue or receipts to the stores.

5.6   The cycle counter will consult the stores supervisor with queries about any cycle count and relevant comments will be noted on the card if necessary.

5.7   Parts counted other than a physical hand count must have the method of counting entered in the comments section of the card (e.g. scale weight).

5.8   When the cycle count is complete, the cycle counter will return the cycle count card to the stores supervisor, who will check the following information has been recorded:

■   Counter's name
■   Date counted
■   Quantity of count
■   Acceptable reason for difference

5.9   The stores supervisor will sign the cycle count card when satisfied the information on the card is complete and forward the card to inventory control. Where a variation has been recorded, inventory control will record the total value of the variance on the card.

5.10 Details from the cycle count card will be used by inventory control to complete the cycle count report form. Information on the form will include

- Part number and location
- Value category (A, B or C)
- Cycle count quantity and computer balance quantity
- Variance: quantity and percent
- Date
- Value of count and +/− adjustment value

5.11 From these figures, inventory control will state whether the count is within tolerance and the inventory accuracy will be calculated from the total number of parts/locations counted. This will be recorded on the report form. The percentage value of adjustment will also be recorded.

5.12 Inventory control will maintain records of the daily inventory accuracy figure and prepare statistics for inclusion in the stores record accuracy report to indicate accuracy by week, month and year to date.

5.13 Inventory control may decide on a recount for significant records adjustment.

5.14 Where the acceptable tolerance is still not achieved by a recount then inventory control will notify the stores supervisor of the discrepancy for further investigations or action. If necessary, another recount should be conducted. If this produces no logical explanation for the discrepancy then the stores record should be adjusted.

5.15 Each cycle count card will be signed by inventory control after all details have been completed. Inventory control will retain the cards on file for at least three years.

5.16 The stores supervisor may initiate further stock counts within the stores if discrepancies with stock are discovered outside the normal cycle counts. These are known as grief counts and a card is completed by the storer in the same way as a cycle count. Similarly, the details forwarded to inventory control.

5.17 The storer is responsible for readjusting the on-line stock quantities of the stores inventory system following the results of a cycle/recount/grief count.

5.18 Management are responsible for approving the amendments to records as follows:

Approval	Value
Inventory controller	Up to £50
Production planning & BOM manager	Up to £100
Director of purchasing & distribution	Up to £250
Planning director	Above £250

6. Documentation and records

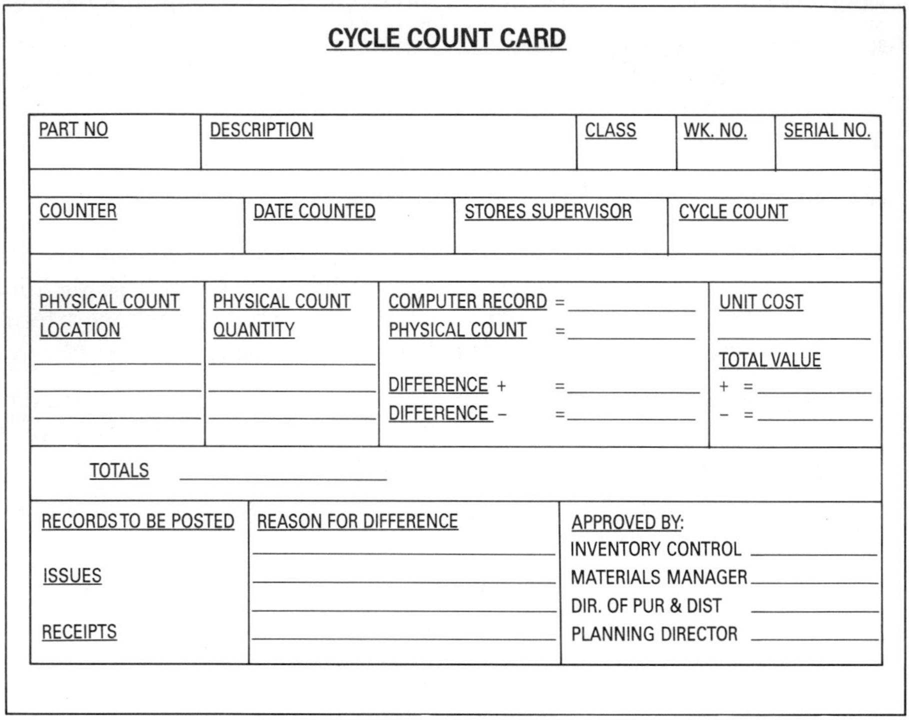

CYCLE COUNT CARD

PART NO	DESCRIPTION		CLASS	WK. NO.	SERIAL NO.

COUNTER	DATE COUNTED	STORES SUPERVISOR	CYCLE COUNT

PHYSICAL COUNT LOCATION	PHYSICAL COUNT QUANTITY	COMPUTER RECORD = _____	UNIT COST
		PHYSICAL COUNT = _____	_____
_____	_____		TOTAL VALUE
_____	_____	DIFFERENCE + = _____	+ = _____
		DIFFERENCE − = _____	− = _____

TOTALS _____

RECORDS TO BE POSTED	REASON FOR DIFFERENCE	APPROVED BY:
	_____	INVENTORY CONTROL _____
ISSUES	_____	MATERIALS MANAGER _____
	_____	DIR. OF PUR & DIST _____
RECEIPTS	_____	PLANNING DIRECTOR _____

Figure D3   Cycle count card.

PART NO.	LOCATION	A B C CLASS	PHYSICAL QUANTITY	COMPUTER QUANTITY	VARIANCE PERCENT	WITHIN TOLERANCE	VALUE OF COUNT £	NET VALUE OF ADJUSTMENT ± £

Total Parts within Tolerance (A) : _____

Total Parts Counted (B) : _____

Percentage Inventory Accuracy (A/B) × 100 : _____

Total Net Value of Adjustments (C) : _____

Total Value of Counts (D) : _____

Percentage Value Adjustment (C/D) × 100 : _____

Figure D4 Adjustment card summary.

517

■ Bill of material and job sheet accuracy measurement

1. Objective

To measure the bills of material (BOMs) and jobsheet accuracy as defined in the company with a view to achieving and maintaining accuracy of 98% or more and steps taken to correct any errors.

2. Scope

This procedure applies to the auditing of all selected BOMs and jobsheets produced by the company for the determination of accuracy.

3. Responsibility

Select BOMs/jobsheets for audit	OID supervisor
Raise BOM/jobsheet audit worksheet	OID supervisor
Correction of errors	BOM audit committee department member
Maintain/circulate details of accuracy	OID supervisor
Control of system and facilities	Planning director

4. Definitions and references

4.1   BOM audit committee: a committee of representatives from engineering, order interpretation, production, production engineering, production planning and control to review the accuracy of BOMs and jobsheets.

4.2   Engineering change control: see company standard procedure SP04.01 in the company procedures manual.

5. Procedures

5.1   The OID manager will select BOMs and jobsheets for auditing, generally at random, but sometimes at request for material or production requirements. A total of 20 selections should be made for each BOM audit committee meeting.

5.2   The OID supervior will raise a BOM audit worksheet for each BOM and jobsheet selected for audit. The following details will be entered at this stage:

Part No./Job No./Description

5.3   A BOM audit committee meeting will be convened once per month to audit the selection of BOMs and jobsheets presented by the OID supervisor. Each member will be supplied with a copy of each BOM audit worksheet.

5.4   The committee review each BOM/jobsheet to determine if the following data is correct:

- Parent unit of measure
- Component part numbers
- Component unit of measures
- Component quantities per parent
- Component costs
- Structure of document

5.5   The committee will record any information found to be incorrect on the worksheet with a description of the error and include the correct information.

5.6   The committee will record 'No errors detected' on the worksheet if no faults are found on an audit.

5.7   The OID supervisor will complete the additional information on the master BOM audit summary sheet:

- Members involved in audit, initialled by each
- Date of audit
- Part nos. audited and description
- Error/No error recorded
- Number of BOMs/jobsheets sampled
- Number of BOMs/jobsheets in error
- Accuracy percent: total number of bills correct divided by total number of bills audited × 100

5.8   The OID supervisor will present the accuracy figures to the monthly management meeting on a BOM accuracy performance graph updated monthly.

Copies of bills of material audit worksheets follow on pages 520 and 521.

BILLS OF MATERIAL AUDIT WORKSHEET

Sheet	of	Date / /1991	No Error Detected :-	Error detected :-	
Part description :-				Part no :-	
Component Part No.	Description of part in error			Information on B.O.M. is / should be	Actioned by Y/N

Figure D5 Bills of material audit worksheet.

BILLS OF MATERIAL AUDIT SUMMARY SHEET

Audit Date / /1991		BOM Accuracy	%
Members Present (Signature)			
Part No	Part Description	No Error	Error
No of BOMs Audited	Totals :–		

Figure D6 Bills of material audit summary sheet.

Appendix E

Costs of Quality

■ Costs of quality

The cost of quality consists of the price of nonconformance and the price of conformance. Nonconformance is associated with failure; all costs are incurred when the customer requirements are not satisfied. This includes failures inside and outside the company. Conformance costs are associated with ensuring conformance, e.g. inspection, appraisal, prevention.

Consider a domestic appliance company with sales of £27.18 million, expenditure of £26.77 million, and a profit of only £0.41 million. With 2,718 employees, a first pass at the cost of quality gave

	£m
Quality department: 187 staff	1.87
Field service: 281 staff	2.81
Manufacturing resources on rework: 62 staff	0.62
Warranty costs	1.27
Scrap costs	0.60
Total costs	7.17

Their cost of quality was over 26% of sales turnover. Reduction of any or all of these costs would directly benefit the profitability; it required full appraisal.

Bringing down the cost of quality often means spending more on prevention: training, procedures, planning, improvement teams, preventative maintenance and audits. It may also require changes to designs, processes and practices. However, these costs will always be much less than the costs of failure or nonconformance, which frequently surprise and shock management when fully defined. Costs of nonconformance vary by business sector but can equate to over 30% of turnover in many businesses.

Appraising your quality costs involves checking, auditing and measuring

522

all company activities and products for failure, complaints, rework or break-down of procedures, and including surveys of customers and suppliers.

From the quality development graph, the increase in prevention and appraisal costs leads to substantial reductions in internal and external failure costs, giving substantial overall improvement in total quality costs.

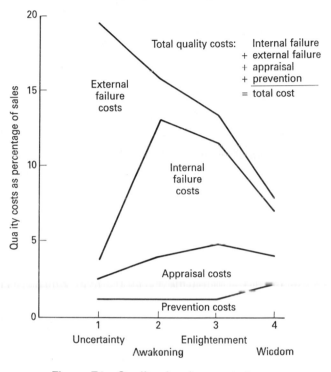

Figure E1 Quality development stages.

Appendix F

Costs/Benefit Analysis

One Year After MRP2

Implementation

■ Calculated efficiency measures

	1991	1992	1993
Percentage planned improvement over 1990	3.5	7	7
Percentage actual improvement	7.2	9	—
Actual saving £'000	374	468	—
60% attributed to MRP2	224	281	—

These improvements have been achieved by the management team working with the MRP2 procedures and systems to assist in each area of the business. MRP1 and MRP2 have provided the necessary information and tools that allow managers and staff to perform their jobs more effectively. However, efficiency has also been improved due to changes by production and personnel and other activities. Maintaining output with reduced staffing may not have been achieved without MRP2, but to attribute all the efficiency improvements directly to the capital investment seems inappropriate. We have apportioned 60% of the above efficiency improvements directly to the project and the rest to improved management practices.

Also, comparing percentage labour to sales, we have the following:

Year	Percentage	Comment
1990	14.1	
1991	13.2	6.4% improvement over 1990
1992	13.0	For the year to October, 7.8% improvement over 1990
		Forecast to year-end shows further improvement

■ Purchased parts shortages (back-orders)

Through the effective application and use of materials requirements planning, we have reduced our purchases shortages (part nos.) by over 80%. In early 1990, this was up to 50 parts per day, and is now down to 8 per day or less. This compares with St. Louis back-orders (proportionally) of 44 per day in 1992. This is also being achieved with a very low order book and backlog, which is known to adversely effect material planning.

■ Orders shipped complete and on time

Due to the improved systems and procedures being used, our on-time shipments are now averaging 99.3%, with 98.8% complete. This compares with the 1990 values shown.

	1990	1991	1992
Percentage on time	94.7	99.1	99.3
Variance on 1990	—	+4.6	+4.8
Percentage complete	92.1	98.9	98.8
Variance on 1990	—	+7.4	+7.3

■ Rapid response to the customer

The improved tools and operating procedures under MRP2 have allowed us to react much faster to customer enquiries and orders, which has allowed us to compete successfully in the 1992 recession. Examples of rapid response and shortening of internal lead time are

- Reduction by 80%, from 10 days to 2 days, to process MRP purchase orders, allowing us to order and obtain purchased materials much more rapidly.
- Reduction of the time taken to trial kit a job or case (to determine an effective delivery date) by 85%, dropping from an average of 8 hours to 1 hour.
- Reduction of lead time from order receipt to delivery (due to improved information processing of BOMs, schedules, etc.) by 3 working days, from 25 days to 22 days for most cases.
- Ability to take on close-in order changes.

These improvements are tangible but are difficult to measure in financial terms.

■ Inventory turns

The improvement of inventory turns was a prime objective of the project. Inventory is broken into three categories: raw material, work in process, and finished goods.

		1990	1991	1992*
Raw material (£m)		1.64	1.70	1.70
Work in process (£m)		1.72	1.77	1.63
Finished goods (£m)		2.13	1.57	1.84
	Total (£m)	5.49	5.04	5.17
Actual (+/−) (£m)		—	−0.450	−0.320
Target reduction (£m)		—	−0.085	−0.596
Change on 1990 (%)		—	−8.2	−5.8
Target reduction on 1990 (%)		—	−1.5	−9
Actual carry cost saved at 5% (£'000)		—	22	16

* Estimate at end of November.

This shows our total inventory reduction to be just slightly less than the target in 1992, with the potential to improve further in 1993 and beyond.

Raw material remains higher than desirable at present, but is planned to cope with short lead time customer orders, with the actual type of material held having been thoroughly reviewed and altered to give improved service. It also contains £120k of materials bought specifically for M&S orders that have been delayed. **Work in process** is staying low as part of our JIT/make-as-required operating policy.

Our decision in early 1992 to build additional stock cases in both remote and integral factories during a quiet period resulted in abnormally high finished goods, not sold by the end of 1992. Together with the recession and the resulting low level of sales, these have been the main reason for not achieving our planned inventory turns.

■ Inventory accuracy

As part of the project, we established systems and procedures to achieve more accurate inventory records and levels. Inventory accuracy rose to 77% in 1991, and to almost 90% by the end of 1992. We aim for over 95% accuracy in 1993.

As a direct result of this perpetual inventory (cycle counting) control accuracy, we were able to reduce the time taken for the 1992 annual stock-taking by several hours, saving £5,000, and have agreed with the auditors that the raw material stores will not need an annual audit in 1993, saving £4,000 and each year thereafter. This is being extended to finished goods, which will save a further £2,000 per annum from 1993.

		1991	1992	1993
Accuracy savings (£'000)		nil	5	6 (ongoing)
Planned project savings				
Staffing (£'000)		0	30	45
Efficiency (%)		182	364	364
		3.5	7	7
Inventory carry cost (£'000)		4	30	43
	Total (£'000)	186	424	452
Actual project savings				
Staffing (£'000)		—	20	35 (known)
Efficiency (%)		224	281	—
		7.2%	9%	—
Inventory carry cost (£'000)		22	16	—
Stock taking (£'000)		–	5	6 (known)
Total (£'000)		246	322	—
			(planned for 452 in 1993)	

	1991	1992
Variance against plan (£'000, +ve good)	+60	−102
Variance on plan (%)	+32	−24

■ Project payback and measurement

Effectively, the benefits for 1991 have accrued above plan by 32%, but have been behind plan for 1992 by 24%. The benefits are fully expected to increase to at least those of the projected £452 p.a. in 1993 and beyond, as we consolidate and improve the use of MRP2 within the company. The financial payback measurements are as follows:

	Plan	Latest actual
NPV ($)	1,195	1,106
IRR (%)	137.8	157.7
Payback (years)	1.2	1.1

■ Plans for 1993

Currently the remaining MRP2 system modules are being made live through November 1992. This will be followed by consolidation and the addition of some other system requirements that have been identified. This will give conclusion in the second quarter of 1993.

Index